Research Progress on Chitosan Applications

Research Progress on Chitosan Applications

Guest Editors

William Facchinatto
Sérgio Paulo Campana-Filho

Basel • Beijing • Wuhan • Barcelona • Belgrade • Novi Sad • Cluj • Manchester

Guest Editors

William Facchinatto
Department of Chemical
Engineering
University of Bath
Bath
United Kingdom

Sérgio Paulo Campana-Filho
Sao Carlos Institute of Chemistry
University of Sao Paulo
Sao Carlos
Brazil

Editorial Office
MDPI AG
Grosspeteranlage 5
4052 Basel, Switzerland

This is a reprint of the Special Issue, published open access by the journal *Polymers* (ISSN 2073-4360), freely accessible at: www.mdpi.com/journal/polymers/special_issues/B15HY77K3M.

For citation purposes, cite each article independently as indicated on the article page online and using the guide below:

Lastname, A.A.; Lastname, B.B. Article Title. *Journal Name* **Year**, *Volume Number*, Page Range.

ISBN 978-3-7258-3098-5 (Hbk)
ISBN 978-3-7258-3097-8 (PDF)
https://doi.org/10.3390/books978-3-7258-3097-8

© 2025 by the authors. Articles in this book are Open Access and distributed under the Creative Commons Attribution (CC BY) license. The book as a whole is distributed by MDPI under the terms and conditions of the Creative Commons Attribution-NonCommercial-NoDerivs (CC BY-NC-ND) license (https://creativecommons.org/licenses/by-nc-nd/4.0/).

Contents

Preface vii

William M. Facchinatto and Sérgio Paulo Campana-Filho
Closing Editorial: Research Progress on Chitosan Applications
Reprinted from: *Polymers* 2024, 16, 3527, https://doi.org/10.3390/polym16243527 1

Linh Doan, Tam T. T. Nguyen, Khoa Tran and Khanh G. Huynh
Surface Modifications of Superparamagnetic Iron Oxide Nanoparticles with Chitosan, Polyethylene Glycol, Polyvinyl Alcohol, and Polyvinylpyrrolidone as Methylene Blue Adsorbent Beads
Reprinted from: *Polymers* 2024, 16, 1839, https://doi.org/10.3390/polym16131839 4

Linh Doan, Quynh N. Le, Khoa Tran and An H. Huynh
Surface Modifications of Silver Nanoparticles with Chitosan, Polyethylene Glycol, Polyvinyl Alcohol, and Polyvinylpyrrolidone as Antibacterial Agents against *Staphylococcus aureus*, *Pseudomonas aeruginosa*, and *Salmonella enterica*
Reprinted from: *Polymers* 2024, 16, 1820, https://doi.org/10.3390/polym16131820 30

Brenda M. Ipinza-Concha, Luciano Dibona-Villanueva, Denis Fuentealba, Alexander Pinilla-Quispe, Daniel Schwantes and María A. Garzón-Nivia et al.
Effect of Chitosan–Riboflavin Bioconjugate on Green Mold Caused by *Penicillium digitatum* in Lemon Fruit
Reprinted from: *Polymers* 2024, 16, 884, https://doi.org/10.3390/polym16070884 44

Omar M. Khubiev, Anton R. Egorov, Daria I. Semenkova, Darina S. Salokho, Roman A. Golubev and Nkumbu D. Sikaona et al.
Rhodamine B-Containing Chitosan-Based Films: Preparation, Luminescent, Antibacterial, and Antioxidant Properties
Reprinted from: *Polymers* 2024, 16, 755, https://doi.org/10.3390/polym16060755 57

Igor D. Zlotnikov, Ivan V. Savchenko and Elena V. Kudryashova
Specific FRET Probes Sensitive to Chitosan-Based Polymeric Micelles Formation, Drug-Loading, and Fine Structural Features
Reprinted from: *Polymers* 2024, 16, 739, https://doi.org/10.3390/polym16060739 72

Katia Celina Santos Correa, William Marcondes Facchinatto, Filipe Biagioni Habitzreuter, Gabriel Henrique Ribeiro, Lucas Gomes Rodrigues and Kelli Cristina Micocci et al.
Activity of a Recombinant Chitinase of the *Atta sexdens* Ant on Different Forms of Chitin and Its Fungicidal Effect against *Lasiodiplodia theobromae*
Reprinted from: *Polymers* 2024, 16, 529, https://doi.org/10.3390/polym16040529 93

Wen-Nee Tan, Benedict Anak Samling, Woei-Yenn Tong, Nelson Jeng-Yeou Chear, Siti R. Yusof and Jun-Wei Lim et al.
Chitosan-Based Nanoencapsulated Essential Oils: Potential Leads against Breast Cancer Cells in Preclinical Studies
Reprinted from: *Polymers* 2024, 16, 478, https://doi.org/10.3390/polym16040478 110

Carmina Ortega-Sánchez, Yaaziel Melgarejo-Ramírez, Rogelio Rodríguez-Rodríguez, Jorge Armando Jiménez-Ávalos, David M. Giraldo-Gomez and Claudia Gutiérrez-Gómez et al.
Hydrogel Based on Chitosan/Gelatin/Poly(Vinyl Alcohol) for In Vitro Human Auricular Chondrocyte Culture
Reprinted from: *Polymers* 2024, 16, 479, https://doi.org/10.3390/polym16040479 127

Emerson Durán, Andrónico Neira-Carrillo, Felipe Oyarzun-Ampuero and Carolina Valenzuela
Thermosensitive Chitosan Hydrogels: A Potential Strategy for Prolonged Iron Dextran Parenteral Supplementation
Reprinted from: *Polymers* 2023, 16, 139, https://doi.org/10.3390/polym16010139 146

Naira Geovana Camilo, Alex da Rocha Gonçalves, Larissa Pinzan Flauzino, Cristiane Martins Rodrigues Bernardes, Andreza Maria Fábio Aranha and Priscilla Cardoso Lazari-Carvalho et al.
Influence of Chitosan 0.2% in Various Final Cleaning Methods on the Bond Strength of Fiberglass Post to Intrarradicular Dentin
Reprinted from: *Polymers* 2023, 15, 4409, https://doi.org/10.3390/polym15224409 161

Lixia Huang, Yifei Jiang, Xinying Chen, Wenqi Zhang, Qiuchen Luo and Siyan Chen et al.
Supramolecular Responsive Chitosan Microcarriers for Cell Detachment Triggered by Adamantane
Reprinted from: *Polymers* 2023, 15, 4024, https://doi.org/10.3390/polym15194024 172

Dulce J. García-García, G. F. Pérez-Sánchez, H. Hernández-Cocoletzi, M. G. Sánchez-Arzubide, M. L. Luna-Guevara and E. Rubio-Rosas et al.
Chitosan Coatings Modified with Nanostructured ZnO for the Preservation of Strawberries
Reprinted from: *Polymers* 2023, 15, 3772, https://doi.org/10.3390/polym15183772 184

Chanothai Hengtrakool, Supreya Wanichpakorn and Ureporn Kedjarune-Leggat
Chitosan Resin-Modified Glass Ionomer Cement Containing Epidermal Growth Factor Promotes Pulp Cell Proliferation with a Minimum Effect on Fluoride and Aluminum Release
Reprinted from: *Polymers* 2023, 15, 3511, https://doi.org/10.3390/polym15173511 203

Fiona Milano, Anik Chevrier, Gregory De Crescenzo and Marc Lavertu
Injectable Lyophilized Chitosan-Thrombin-Platelet-Rich Plasma (CS-FIIa-PRP) Implant to Promote Tissue Regeneration: In Vitro and Ex Vivo Solidification Properties
Reprinted from: *Polymers* 2023, 15, 2919, https://doi.org/10.3390/polym15132919 215

Preface

This reprint, inspired by the Special Issue "Research Progress on Chitosan Applications" that was published in *Polymers*, celebrates the remarkable advancements in the field of chitosan-based materials. Chitosan, a natural polymer with versatile properties, has emerged as a cornerstone for developing innovative solutions across the biomedical, environmental, and industrial domains.

The contributions compiled here represent a diverse array of research endeavors that collectively illustrate the transformative potential of chitosan. From targeted drug delivery and tissue engineering to sustainable food preservation and agricultural biocontrol, the applications of chitosan that are showcased in this book underscore the material's adaptability and impact. Highlights include breakthroughs in encapsulation technologies for enhanced therapeutic efficacy, the development of antimicrobial coatings for food safety, and the creation of bioactive hydrogels for regenerative medicine.

Particularly noteworthy are the studies that harness chitosan's unique properties for environmental remediation, including the design of adsorbent materials for water purification and eco-friendly fungicides for agricultural use. These innovations not only address pressing global challenges but also reflect a growing commitment to sustainable and ethical practices in materials science.

This reprint would not have been possible without the tireless efforts of the contributing authors and the rigorous feedback provided by the expert reviewers. Their collective dedication ensured that the contents within these pages are both cutting-edge and scientifically robust.

We hope that this reprint serves as a valuable resource for researchers, practitioners, and students, inspiring further exploration into the vast potential of chitosan and related biomaterials. By disseminating these insights, we aim to contribute to the ongoing dialogue in polymer science and to foster the development of innovative solutions that benefit society at large.

William Facchinatto and Sérgio Paulo Campana-Filho
Guest Editors

Editorial

Closing Editorial: Research Progress on Chitosan Applications

William M. Facchinatto [1,*] and Sérgio Paulo Campana-Filho [2,*]

1 Aveiro Institute of Materials, CICECO, Department of Chemistry, University of Aveiro, St. Santiago, 3810-193 Aveiro, Portugal
2 São Carlos Institute of Chemistry, University of São Paulo, IQSC/USP, Ave. Trabalhador São-Carlense, 400, São Carlos 13560-970, SP, Brazil
* Correspondence: williamfacchinatto@alumni.usp.br (W.F.); scampana@iqsc.usp.br (S.P.C.-F.)

1. Introduction

Chitosan has attracted significant attention due to its versatile properties, which make it an ideal candidate for varied biomedical and industrial applications. This Special Issue of *Polymers* will present recent innovations and challenges in developing chitosan-based biomaterials, addressing applications in drug delivery, tissue engineering, antimicrobial treatments, food preservation, and environmental remediation. The 14 papers included in this Special Issue explore novel strategies to optimize the structural, physical–chemical, mechanical, and bioactive properties of chitosan composites, highlighting the material's prospectives.

2. An Overview of the Published Articles

The articles in this Special Issue highlight chitosan's potential use in targeted therapies, regenerative medicine, and sustainable practices. Tan et al. discuss the encapsulation of essential oils in chitosan nanoparticles used for breast cancer treatment, where nanoencapsulation significantly enhances bioavailability and therapeutic efficacy against breast cancer cells [1]. Complementing this therapeutic approach, Milano et al. report on freeze-dried chitosan-based implants containing thrombin and platelet-rich plasma, optimized for tissue regeneration applications, showing efficient solidification properties that are very favorable in orthopedic surgeries [2].

Hengtrakool et al. present findings on modified chitosan resin–glass ionomer cement for odontological applications, which facilitates the prolonged release of bioactive molecules, promoting cellular proliferation while ensuring minimal fluoride release [3]. García-García et al. explore the use of chitosan with nanostructured ZnO for strawberry preservation, where the composite coating maintains fruit quality while extending shelf life due to its antimicrobial properties [4].

Addressing non-enzymatic cell detachment methods, Huang et al. introduce responsive chitosan microcarriers utilizing host–guest interactions, which allow for controlled cell detachment [5]. This innovation is beneficial in cell expansion and bioreactor systems, particularly where non-invasive cell retrieval is critical. Camilo et al. study the role of chitosan in enhancing bond strength in odontological applications, focusing on the material's role in reinforcing adhesion between dentin and fiberglass pillars [6].

Durán et al. highlight thermosensitive chitosan hydrogels for prolonged iron supplementation, providing a solution for controlled nutrient release, particularly beneficial in veterinary applications [7]. Ortega-Sánchez et al. contribute to the field of tissue engineering with chitosan-based hydrogels designed for chondrocyte culture, showing structural compatibility and bioactivity conducive to auricular cartilage repair [8].

Correa et al. examine the antifungal properties of chitinase and chitin platforms, which exhibit strong fungicidal effects against *Lasiodiplodia theobromae*, a phytopathogen [9]. Their findings have potential for agricultural biocontrol. Zlotnikov et al. investigate Förster resonance energy transfer (FRET) probes within chitosan micelles, optimizing drug-loading efficiency and structural integrity for drug delivery applications [10].

Khubiev et al. introduce Rhodamine B-infused chitosan films with luminescent and antibacterial properties, suitable for packaging and clinical environments where antimicrobial efficacy and visual detection are beneficial [11]. Ipinza-Concha et al. present a chitosan–riboflavin bioconjugate effective against green mold in citrus fruits, offering a natural alternative to synthetic fungicides [12].

Doan et al. further expand on chitosan's versatility by detailing two studies [13,14]. The first one explores silver nanoparticle composites with chitosan, polyethylene glycol, polyvinyl alcohol, and polyvinylpyrrolidone, optimized as potent antibacterial agents against pathogens such as *Staphylococcus aureus*, *Pseudomonas aeruginosa*, and *Salmonella enterica*. This study underscores the material's potential uses in antimicrobial applications [13]. The second study presents superparamagnetic iron oxide nanoparticles modified with chitosan and similar polymers, demonstrating their efficacy as methylene blue adsorbents in water treatment applications. This innovative approach highlights chitosan's role in addressing environmental issues [14].

These 14 articles collectively illustrate the chitosan-based material's adaptability, discussing potential modifications to enhance its mechanical and bioactive properties for diverse applications, ranging from environmental sustainability to medical therapies.

3. Conclusions

This Special Issue underscores the breadth of chitosan-based innovations and their potential to address real-world challenges. From advanced drug delivery systems to eco-friendly preservation solutions, the studies collectively highlight chitosan's adaptability, biodegradability, and efficacy. As research advances, the further optimization of chitosan's properties could lead to more targeted and efficient applications in both industrial and clinical settings.

Funding: This research received no external funding.

Acknowledgments: We extend our gratitude to the authors for their contributions and the reviewers for their invaluable feedback and efforts in ensuring the high quality of this Special Issue. Additionally, we thank the editorial team at *Polymers* for their support throughout the publication process.

Conflicts of Interest: The authors declare no conflicts of interest regarding the publication of this Special Issue.

References

1. Tan, W.-N.; Samling, B.A.; Tong, W.-Y.; Chear, N.J.-Y.; Yusof, S.R.; Lim, J.-W.; Tchamgoue, J.; Leong, C.-R.; Ramanathan, S. Chitosan-Based Nanoencapsulated Essential Oils: Potential Leads against Breast Cancer Cells in Preclinical Studies. *Polymers* **2024**, *16*, 478. [CrossRef]
2. Milano, F.; Chevrier, A.; De Crescenzo, G.; Lavertu, M. Injectable Lyophilized Chitosan-Thrombin-Platelet-Rich Plasma (CS-FIIa-PRP) Implant to Promote Tissue Regeneration: In Vitro and Ex Vivo Solidification Properties. *Polymers* **2023**, *15*, 2919. [CrossRef] [PubMed]
3. Hengtrakool, C.; Wanichpakorn, S.; Kedjarune-Leggat, U. Chitosan Resin-Modified Glass Ionomer Cement Containing Epidermal Growth Factor Promotes Pulp Cell Proliferation with a Minimum Effect on Fluoride and Aluminum Release. *Polymers* **2023**, *15*, 3511. [CrossRef] [PubMed]
4. García-García, D.J.; Pérez-Sánchez, G.F.; Hernández-Cocoletzi, H.; Sánchez-Arzubide, M.G.; Luna-Guevara, M.L.; Rubio-Rosas, E.; Krishnamoorthy, R.; Morán-Raya, C. Chitosan Coatings Modified with Nanostructured ZnO for the Preservation of Strawberries. *Polymers* **2023**, *15*, 3772. [CrossRef] [PubMed]
5. Huang, L.; Jiang, Y.; Chen, X.; Zhang, W.; Luo, Q.; Chen, S.; Wang, S.; Weng, F.; Xiao, L. Supramolecular Responsive Chitosan Microcarriers for Cell Detachment Triggered by Adamantane. *Polymers* **2023**, *15*, 4024. [CrossRef]
6. Camilo, N.G.; Gonçalves, A.d.R.; Flauzino, L.P.; Bernardes, C.M.R.; Aranha, A.M.F.; Lazari-Carvalho, P.C.; Carvalho, M.A.d.; Oliveira, H.F.d. Influence of Chitosan 0.2% in Various Final Cleaning Methods on the Bond Strength of Fiberglass Post to Intrarradicular Dentin. *Polymers* **2023**, *15*, 4409. [CrossRef] [PubMed]
7. Durán, E.; Neira-Carrillo, A.; Oyarzun-Ampuero, F.; Valenzuela, C. Thermosensitive Chitosan Hydrogels: A Potential Strategy for Prolonged Iron Dextran Parenteral Supplementation. *Polymers* **2024**, *16*, 139. [CrossRef] [PubMed]

8. Ortega-Sánchez, C.; Melgarejo-Ramírez, Y.; Rodríguez-Rodríguez, R.; Jiménez-Ávalos, J.A.; Giraldo-Gomez, D.M.; Gutiérrez-Gómez, C.; Rodriguez-Campos, J.; Luna-Bárcenas, G.; Velasquillo, C.; Martínez-López, V.; et al. Hydrogel Based on Chitosan/Gelatin/Poly(Vinyl Alcohol) for In Vitro Human Auricular Chondrocyte Culture. *Polymers* **2024**, *16*, 479. [CrossRef] [PubMed]
9. Correa, K.C.S.; Facchinatto, W.M.; Habitzreuter, F.B.; Ribeiro, G.H.; Rodrigues, L.G.; Micocci, K.C.; Campana-Filho, S.P.; Colnago, L.A.; Souza, D.H.F. Activity of a Recombinant Chitinase of the Atta sexdens Ant on Different Forms of Chitin and Its Fungicidal Effect against Lasiodiplodia theobromae. *Polymers* **2024**, *16*, 529. [CrossRef] [PubMed]
10. Zlotnikov, I.D.; Savchenko, I.V.; Kudryashova, E.V. Specific FRET Probes Sensitive to Chitosan-Based Polymeric Micelles Formation, Drug-Loading, and Fine Structural Features. *Polymers* **2024**, *16*, 739. [CrossRef] [PubMed]
11. Khubiev, O.M.; Egorov, A.R.; Semenkova, D.I.; Salokho, D.S.; Golubev, R.A.; Sikaona, N.D.; Lobanov, N.N.; Kritchenkov, I.S.; Tskhovrebov, A.G.; Kirichuk, A.A.; et al. Rhodamine B-Containing Chitosan-Based Films: Preparation, Luminescent, Antibacterial, and Antioxidant Properties. *Polymers* **2024**, *16*, 755. [CrossRef] [PubMed]
12. Ipinza-Concha, B.M.; Dibona-Villanueva, L.; Fuentealba, D.; Pinilla-Quispe, A.; Schwantes, D.; Garzón-Nivia, M.A.; Herrera-Défaz, M.A.; Valdés-Gómez, H.A. Effect of Chitosan–Riboflavin Bioconjugate on Green Mold Caused by Penicillium digitatum in Lemon Fruit. *Polymers* **2024**, *16*, 884. [CrossRef] [PubMed]
13. Doan, L.; Le, Q.N.; Tran, K.; Huynh, A.H. Surface Modifications of Silver Nanoparticles with Chitosan, Polyethylene Glycol, Polyvinyl Alcohol, and Polyvinylpyrrolidone as Antibacterial Agents against Staphylococcus aureus, Pseudomonas aeruginosa, and Salmonella enterica. *Polymers* **2024**, *16*, 1820. [CrossRef] [PubMed]
14. Doan, L.; Nguyen, T.T.T.; Tran, K.; Huynh, K.G. Surface Modifications of Superparamagnetic Iron Oxide Nanoparticles with Chitosan, Polyethylene Glycol, Polyvinyl Alcohol, and Polyvinylpyrrolidone as Methylene Blue Adsorbent Beads. *Polymers* **2024**, *16*, 1839. [CrossRef]

Disclaimer/Publisher's Note: The statements, opinions and data contained in all publications are solely those of the individual author(s) and contributor(s) and not of MDPI and/or the editor(s). MDPI and/or the editor(s) disclaim responsibility for any injury to people or property resulting from any ideas, methods, instructions or products referred to in the content.

Surface Modifications of Superparamagnetic Iron Oxide Nanoparticles with Chitosan, Polyethylene Glycol, Polyvinyl Alcohol, and Polyvinylpyrrolidone as Methylene Blue Adsorbent Beads

Linh Doan [1,2,3,*], Tam T. T. Nguyen [1,2,3], Khoa Tran [2,3] and Khanh G. Huynh [2,4]

1. Department of Chemical Engineering, International University—Vietnam National University, Ho Chi Minh City 70000, Vietnam
2. Nanomaterials Engineering Research & Development (NERD) Laboratory, International University—Vietnam National University, Ho Chi Minh City 70000, Vietnam
3. School of Chemical and Environmental Engineering, International University—Vietnam National University, Ho Chi Minh City 70000, Vietnam
4. School of Biomedical Engineering, International University—Vietnam National University, Ho Chi Minh City 70000, Vietnam
* Correspondence: dhlinh@hcmiu.edu.vn

Abstract: Due to the negative impacts the dye may have on aquatic habitats and human health, it is often found in industrial effluent and poses a threat to public health. Hence, to solve this problem, this study developed magnetic adsorbents that can remove synthetic dyes like methylene blue. The adsorbent, in the form of beads, consists of a polymer blend of chitosan, polyethylene glycol, polyvinyl alcohol, polyvinylpyrrolidone, and superparamagnetic iron oxide nanoparticles (average size of 19.03 ± 4.25 nm). The adsorption and desorption of MB from beads were carried out at pH values of 7 and 3.85, respectively. At a concentration of 9 mg/L, the loading capacity and the loading amount of MB after 5 days peaked at $29.75 \pm 1.53\%$ and 297.48 ± 15.34 mg/g, respectively. Meanwhile, the entrapment efficiency of MB reached $29.42 \pm 2.19\%$ at a concentration of 8 mg/L. The cumulative desorption capacity of the adsorbent after 13 days was at its maximum at $7.72 \pm 0.5\%$. The adsorption and desorption kinetics were evaluated.

Keywords: adsorption; desorption; kinetics; methylene blue; synthetic dye; wastewater treatment

1. Introduction

Water is an important natural resource in the world, and ensuring its efficient management is crucial for safeguarding its future sustainability for both the environment and human survival [1]. Over the years, significant contamination has been noted, and certain technologies employed for remediation may generate secondary contaminants or byproducts, which can worsen environmental pollution [2]. Contaminants in wastewater originate from two main sources. The initial reason is natural processes like volcanic activity, soil erosion, and rock weathering, and the second source is human activities such as waste disposal, urban runoff, mining, printed circuit board manufacturing, agriculture, metal surface treatment, fuel combustion, textile dyeing, semiconductor production, and others [3,4].

To protect the diverse range of plant and animal life it supports, it is essential to implement measures that prevent contamination from both organic and inorganic pollutants [5]. Over time, there has been a consistent observation of severe contamination, with dyes often being cited among the persistent organic and mineral pollutants reported [6]. Manufactured on a global scale in significant quantities and diverse forms [7], dyes are categorized according to the source of their materials (natural or synthetic) and the nature of their chromophore or autochrome groups [6]. This classification contributes to their potential to enter water bodies, where they can hinder light penetration, resulting in a substantial detrimental effect on ecosystems by reducing photosynthetic activity.

Additionally, dyes pose severe risks as they are highly harmful and carcinogenic. Their accumulation in certain aquatic organisms presents a notable environmental threat, alongside the potential for adverse effects on human health, such as skin irritation, allergic dermatitis, cancer, and genetic mutations [8]. In industries, a significant volume of vibrant wastewater is generated, often containing toxins, resistant to biodegradation, and posing environmental sustainability challenges [9]. Several examples of artificial coloring agents encompass aniline blue, alcian blue, basic fuchsin, methylene blue (MB), crystal violet, toluidine blue, and congo red [10]. MB is among the harmful dyes utilized across different industries.

Methylene blue (MB) is among the harmful dyes utilized across different industries. MB, with the chemical formula $C_{16}H_{18}N_3SCl$ [11], is an aromatic heterocyclic basic dye. It is alternatively known as a cationic or primary thiazine dye. The presence of negative polar sites on water molecules results in an electrostatic attraction towards the cationic dye, causing the positive ions to separate and form a stable solution with water at ambient temperature [11].

MB has diverse applications across various industries, notably in textiles, where it serves as a dye for cotton, wool, and silk [12]. Additionally, MB is primarily used in the textile industry to impart a vibrant blue color to garments. Its chemical properties allow it to adhere effectively to the interstitial gaps of cotton fibers, ensuring that the dye remains stable and durable on the fabric [11]. This strong adherence and stability make MB one of the most popular and frequently used dyes in apparel manufacturing. Its widespread application is due to its ability to produce a consistent and long-lasting color, making it a staple choice for dyeing various types of clothing.

Since 7×10^7 tons of synthetic dyes are dumped into the environment annually, numerous methods have been attempted across various technologies in wastewater treatment to ensure the safety of water supplies for consumption [13]. These methods include coagulation [14], electrochemical processes [15], biological treatment [16], adsorption [17], and photocatalytic activity [18]. Nonetheless, water treatment technologies demand extended treatment durations and incur substantial operational expenses. Adsorption seems to stand out as one of the most prevalent methods for eliminating MB [11]. The process of adsorption is recognized for its efficacy in treatment owing to its operational simplicity, cost-efficiency, versatility, and responsiveness towards harmful contaminants. Numerous adsorbents have been utilized for eliminating various categories of dyes. Among the frequently employed adsorbents are activated carbons, plant or lignocellulosic residues, clays, and biopolymers [19,20].

Moreover, nano-adsorbents prove to be a remarkably efficient technology for removing organic dyes from both water and wastewater. Their reduced size and expanded adsorptive surface area significantly enhance their effectiveness in this application. Some researchers have enhanced the adsorption capacity of materials by modifying them through combinations with other chemicals and substances. Examples of nano-adsorbents such as nanofibers, graphene, metal oxides, and carbon nanotubes have the potential to enhance water and wastewater treatment processes [21]. In the field of water treatment, the inclusion of additional nanoparticles (NPs) may facilitate the process of separation during the elimination of artificial dyes from water owing to their magnetic properties [22]. As a result, superparamagnetic iron oxide nanoparticles (SPIONs), also known as Fe_3O_4, are among the most frequently utilized materials.

Superparamagnetic NPs can find applications in diverse fields through surface modifications with various materials. In order to prevent aggregation and maintain the stability of the nanoparticles, SPIONs can be coated with organic compounds such as acids, polysaccharides, or polymers. However, various studies have shown that certain undesirable aggregates of SPIONs may exhibit reduced stability, biocompatibility, and effectiveness. Consequently, there has been research into combining SPIONs with other stabilizing agents, such as carbon compounds [23–25]. Therefore, modifying SPIONs with polymers is essential to improve their adsorption capabilities for MB, a synthetic textile dye that is both toxic and carcinogenic. The incorporation of polymers onto the surface of SPIONs can significantly enhance their interaction with MB molecules, thereby increasing the efficiency of the adsorption process. This modification is vital for the development of advanced materials aimed at effectively removing MB from wastewater, thereby reducing its harmful environmental and health impacts.

Although SPION nanoparticles possess a significant surface area, incorporating carbon-based materials like activated charcoal (AC), graphene oxide, and carbon nanotubes can substantially enhance the adsorption capacity of synthetic dyes [21–23]. This improvement is due to the larger surface area, superior adsorption properties, and favorable chemical structure of these carbon-based materials. Additionally, hydrophobic SPIONs can be produced through various methods, including co-precipitation, hydrothermal, and the sol-gel approach. Among these, co-precipitation stands out as the quickest method because of its controlled size and magnetic properties, yielding the highest amount of magnetite and offering an easy synthesis process by minimizing maghemite formation, despite some drawbacks such as particle agglomeration [26]. SPIONs enhance the separation process due to their magnetic properties and offer a considerable surface area.

To avoid aggregation, SPIONs can be modified using polymers. Recently, chitosan was used to modify SPION. Chitosan (CS) is a cationic polyelectrolyte prepared by N-deacetylation of chitin, also named poly (β-1-4)-2-amino-2-deoxy-D-glucopyranose [27]. CS exhibits non-toxic, hydrophilic, biocompatible, biodegradable, and antibacterial properties. These characteristics have spurred its versatile application across various sectors, including biomedicine, cosmetics, food, and textiles [28]. CS has been identified as an appropriate natural polymer for the adsorption of metal ions [29], as the amino (–NH_2) and hydroxyl (–OH) groups present on the chitosan chain can function as sites for chelating metal ions. Nevertheless, the behavior of CS is greatly influenced by the pH level, leading to its ability to transition between a gel state and a dissolved state based on the pH values [19]. Hence, CS has shown promise as an adsorbent in wastewater treatment.

Multiple research studies have focused on altering the surface of chitosan (CS) by chemical means, either through uniform or varied crosslinking processes involving di- or polyfunctional substances. This has been aimed at enhancing mechanical properties, adsorption capacity, or preventing the dissolution of chitosan in acidic environments. Li et al. developed a material composed of CS-coated magnetic mesoporous silica NPs and applied it effectively for the removal of MB from water [30]. Furthermore, Hoa et al. published a paper focusing on the adsorption characteristics of porous beads made of hydroxyapatite/graphene oxide/chitosan toward MB [31]. In addition, the adsorption capacity of SPION-based MB adsorbents can also be influenced by the presence of CS. According to some studies, the adsorption capabilities of SPION/CS/graphene oxide and SPION/PVA/CS/graphene oxide may be 30.01 and 36.4 mg/g, respectively [25,32]. Thus, the significant effectiveness of combining CS with SPIONs holds promise for wastewater treatment.

Furthermore, polyethylene glycol (PEG), also referred to as $H(OCH_2CH_2)_nOH$, is a synthetic polymer consisting of repetitive ethylene glycol units. When dissolved, each ethylene glycol unit binds to approximately two water molecules, resulting in a molecular size that is 5–10 times greater than that of proteins or other macromolecules of comparable molecular weights [33,34].

PEG finds widespread application in tissue engineering, drug delivery, electronics, and fluorescence detection owing to its antimicrobial characteristics and non-toxic nature [35]. PEG is a widely used agent for enhancing the biocompatibility of material surfaces that interact with cells. The modification of polyvinyl chloride (PVC) resin with PEG enhanced its blood compatibility [36]. Furthermore, PEG has the capability to serve as active sites by virtue of the polyethylene oxide chains' capacity to create durable complexes with metal cations resembling crown ethers. PEG is appealing for chemical synthesis due to its ease of functionalization with various groups such as azides, thiols, carboxylic acids, hydroxyls, and epoxides [37]. In order to uphold electroneutrality, it is necessary for these complexes of PEG-metal cations to incorporate a corresponding anion into the organic phase. This ensures the anion is present for potential interactions with the organic reactants [38]. Due to its high solubility in water and non-absorption by the human body, PEG is frequently used in various medical applications. It is utilized in anticancer drugs, organ preservation, and tablet formulation, where it functions as a lubricant and binder [35]. PEG's higher viscosity compared to other meltable binders makes it suitable for achieving melt agglomeration through the immersion mechanism. Some PEGs are available as solid beads. For example, melt agglomeration studies can utilize PEG beads of various sizes to investigate the mechanisms behind the formation and growth of agglomerates under these processing conditions [39].

Moreover, PEG, an affordable synthetic non-ionic polymer, enhances the pore volume and dispersion of the composite material. When incorporated with CS, this organic porogen selectively dissolves within the polymeric matrix, generating macroporous networks [40].

On the other hand, polyvinyl alcohol (PVA) is a semi-crystalline or linear synthetic polymer that appears creamy or whitish with a tasteless, odorless, non-toxic, and thermostable nature, typically found in granular or powdered form [41]. It is a derivative of a vinyl polymer linked solely by C–C bonds [42].

PVA possesses excellent formability and is extensively utilized as a carrier to produce PVA composites with specific mechanical strength [43]. PVA is commercially available in various grades, distinguished by viscosity and degree of hydrolysis [42]. Moreover, PVA exhibits excellent solubility in water because of the abundance of hydroxyl groups in its molecular structure and can undergo biodegradation within a relatively brief period. PVA offers additional benefits such as high biocompatibility, hydrophilicity, and the capacity to form fibers capable of retaining significant quantities of water and/or biological solutions, all while preserving their structural integrity under deformation [44]. Polyvinyl alcohol (PVA) is a synthetic polymer characterized by its semi-crystalline or linear structure, which presents a creamy or whitish appearance. It possesses the attributes of being tasteless, odorless, non-toxic, and exhibiting thermostable properties. This polymer is commonly encountered in either granular or powdered form [42]. For both the strength and formation of beads, a pH range of 4–6 is necessary. However, a higher pH level is preferable for enhancing the strength and stability of PVA beads.

Several publications have demonstrated the use of PVA in wastewater treatment for the removal of toxic dyes in different forms. PVA beads and CS/PVA hydrogel beads have been extensively studied as efficient adsorbents for heavy metal ions and dyes [45]. Jeong et al. conducted a study assessing the effectiveness of a water purification system composed of PVA gel beads incorporating photosynthetic bacteria [46]. In pursuit of the study's objectives, PVA surfaces were employed for the adsorption of MB.

Moreover, polyvinylpyrrolidone (PVP, also known as povidone or povidone) is denoted by the molecular formula $(C_6H_9N_O)_n$, derived from its monomer N-vinylpyrrolidone. It is a bulky, linear homopolymer, a non-toxic, non-ionic polymer containing functional groups such as C=O, C-N, and CH_2 [47].

It is accessible in different molecular weights, and it exhibits high solubility in both water and various organic solvents, commonly employed in NP synthesis. These attributes, coupled with PVP's capacity to form complexes with polar molecules and its biodegradability, have resulted in its widespread industrial usage, notably in pharmaceuticals and processed foods [37]. PVP is a preferred alternative for medication delivery systems due to its biocompatibility.

In addition to being a hydrophilic polymer, PVP exhibits great solubility in solvents with varied polarity, good binding capabilities, and a stabilizing impact for suspensions and emulsions. PVP is considered to be physiologically suitable for both animal and human use. Its utilization as a blood plasma extender, carrier for drugs, suspending agent (specifically as a protective colloid), and aid for tableting has been observed in both the United States and Europe [48]. The solubility of PVP in water and various non-aqueous solvents is attributed to the presence of the highly polar amide group located within the pyrrolidone ring, along with the apolar methylene and methine groups found in the ring and along its backbone [49]. PVP may create hydrogen bonds with CS, functioning as a cross-linker, which improves its mechanical characteristics. The presence of PVP improves the thermal stability of CS.

Hence, in response to this challenge, the aim of the study on surface modifications of SPIONs with chitosan (CS), polyethylene glycol (PEG), polyvinyl alcohol (PVA), and polyvinylpyrrolidone (PVP) as MB adsorbent beads is to develop novel materials for removing this hazardous dye from aqueous solutions. Additionally, this study can be used as the foundation to further develop dual-function materials with applications in dye removal and antimicrobial agents since the polymer blend that was used in this research exhibited antimicrobial activity [50].

2. Materials and Methods

2.1. Materials

Sodium hydroxide (NaOH), polyethylene glycol—1000 (PEG), iron (II) chloride tetrahydrate ($FeCl_2 \cdot 4H_2O$), iron (III) chloride hexahydrate ($FeCl_3 \cdot 6H_2O$), hydrochloric acid (HCl), and ammonia solution (NH_4OH) were purchased from Xilong Scientific Co., Ltd. (Shantou, China). Chitosan (CS), polyvinylpyrrolidone (PVP K30) from Shanghai Zhanyun Chemical Co., Ltd. (Shanghai, China). Glacial acetic acid (AA) was purchased from RCI Labscan (Bangkok, Thailand). Polyvinyl alcohol (PVA) was purchased from Wuxi Yatai United Chemical Co., Ltd. (Shantou, China) All chemicals were used as received.

2.2. Methods

2.2.1. Synthesis of SPION and M8C

SPION was synthesized individually, similar to previous publications, without any modifications [51]. The synthesis of the polymer blend M8C followed a procedure outlined in a previous study by Linh et al., with specific modifications as detailed [50]. Specifically, the M8C variant was prepared by preparing PEG, PVA, and PVP individually by adding the polymers with DI under constant stirring and heating at 80 °C. These polymers were synthesized by adding 2 g of each polymer to 50 mL of DI. Additionally, the CS mixture was synthesized by mixing 1 g of CS in 70 mL of a 3% AA solution under constant stirring and heating at 80 °C. Then, these polymers were blended together under constant stirring and heating at 80 °C.

2.2.2. Synthesis of SPION/M8C Composite

The SPION/M8C composite was initially synthesized by mixing 30 mL of the M8C blend with 0.5 g of SPION. Then, the mixture was sonicated for 30 min. Then, the mixture was carefully added dropwise to a 9 M NaOH solution, allowing for controlled incorporation. Subsequently, the beads were extensively washed with distilled water (DW) until the pH of the filtrate reached neutrality. After washing, the beads were transferred to a glass dish and left to air dry overnight in an oven set at 80 °C.

2.2.3. MB Adsorption

Multiple experiments were conducted using MB solutions at various concentrations to establish the calibration curve for adsorption. Following this, the concentration of nanoparticles was measured at room temperature over a specific duration. Specifically, 25 mL of MB solution at concentrations of 8 mg/L, 9 mg/L, and 10 mg/L, all adjusted to a pH of 7.0, were combined with 0.2 g of the composite beads in a 50 mL falcon tube. Subsequently, the tube was left to stand at room temperature for a period of 5 days.

Samples—aliquots—were collected and transferred into the cuvette using plastic pipettes and a neodymium magnet every 120 h to ensure enough observation intervals and analyzed using UV–Vis spectroscopy (Hach DR6000, Loveland, CO, USA). After measuring the absorbance, the aliquots were transferred back into the falcon tubes. To maintain data integrity and improve result accuracy, this adsorption experiment was carefully performed three times, ensuring consistency and dependability in the collected data.

2.2.4. MB Desorption

Following the end of the adsorption experiment, neodymium magnets were used to remove an aliquot from each falcon tube, ensuring effective separation of the loaded adsorbents. Subsequently, a solution with a pH of 3.85 was prepared by combining DW with 2 M HCl. Each falcon tube was then filled with 25 mL of this acidic solution to initiate the desorption experiment at room temperature.

During the desorption process, the loaded adsorbents were allowed to release the MB for 13 days while maintaining a pH of 3.85 throughout the experiment. Following the required desorption interval, the samples were analyzed using UV–Vis spectroscopy to determine the amount of MB released. After analysis, the aliquots were reintroduced into their respective falcon tubes for further testing. To confirm the accuracy and uniformity of the data, this desorption operation was rigorously performed three times.

2.2.5. Calculation

As shown in Equations (1)–(19), the loading amount (Q_t), loading capacity (%LC), entrapment efficiency (%EE), adsorption kinetics models (pseudo 1st order—PFO, pseudo 2nd order—PSO), intraparticle diffusion, Elovich kinetic model, and desorption models (0th order, Higuchi, and Korsmeyer–Peppas) were used to fit the experimental data similar to previous publications [25,51].

$$\text{Loading amount } Q_t = \frac{(C_0 - C_t)V}{m} \quad (1)$$

$$\text{Loading capacity } \%LC = \frac{\text{Weight of MB adsorbed on to the particles (mg)}}{\text{Weight of particles (mg)}} \times 100 \quad (2)$$

$$\text{Entrapment Efficiency } \%EE = 100 \times \frac{\text{Weight of MB adsorbed on to the particles (mg)}}{\text{Weight of MB initially fed (mg)}} \quad (3)$$

$$\text{PFO nonlinear } Q_t = Q_e\left(1 - e^{-kt}\right) \quad (4)$$

$$\text{PFO linear } \log(Q_e - Q_t) = \log Q_e - \left(\frac{k_1}{2.303}\right)t \quad (5)$$

$$\text{PSO nonlinear } Q_t = Q_e\left(1 - e^{-kt}\right) \quad (6)$$

$$\text{PSO linear type I } Q_t = \frac{k_2 Q_e^2 t}{1 + k_2 Q_e t} \quad (7)$$

$$\text{PSO linear type II } \frac{1}{Q_t} = \frac{1}{k_2 Q_e^2} + t/Q_e \quad (8)$$

$$\text{PSO linear type III } \frac{1}{Q_t} = \left(\frac{1}{k_2 Q_e^2}\right)\frac{1}{t} + \frac{1}{Q_e} \quad (9)$$

$$\text{PSO linear type IV } Q_t = Q_e - \left(\frac{1}{k_2 Q_e}\right)\frac{Q_t}{t} \quad (10)$$

$$\text{Zeroth-order linear } M_t = k_0 t \quad (11)$$

$$\text{Zeroth-order nonlinear } M_t = k_0 t \quad (12)$$

$$\text{Korsmeyer–Peppas linear } \log\left(\frac{M_t}{M_\infty}\right) = \log(k_{KP}) + n_{KP}\log(t) \quad (13)$$

$$\text{Korsmeyer–Peppas nonlinear } \frac{M_t}{M_\infty} = (k_{KP})(t^{n_{KP}}) \quad (14)$$

$$\text{Higuchi linear } \log(M_t) = \log(k_H) + 0.5\log(t) \quad (15)$$

$$\text{Higuchi nonlinear } M_t = k_H t^{1/2} \quad (16)$$

$$\text{Intraparticle diffusion } Q_t = I + k_i t^{1/2} \quad (17)$$

$$\text{Simplified Elovich model } Q_t = \beta \ln(\alpha\beta) + \beta \ln t \quad (18)$$

$$\text{Chi-square test } \chi^2 = \sum_{i=1}^{m} \frac{\left(Q_{t,exp} - Q_{t,calc}\right)^2}{Q_{t,calc}} \quad (19)$$

2.2.6. Characterization

The use of field emission scanning electron microscopy (FE-SEM, Hitachi, SU8000, Tokyo, Japan), Fourier transform infrared spectroscopy (FTIR, Tensor 27, Bruker, MA, USA) was important to confirm the successful construction of the SPION/M8C beads material. The FE-SEM provided detailed images revealing the material's morphology, while FTIR identified the iron oxide nanoparticle formation. XRD analysis confirmed the characteristics of the iron oxide nanoparticles, demonstrating their superparamagnetic nature. A vibrating sample magnetometer (VSM, Tensor 27, Bruker, Germany) may be used to confirm the existence of superparamagnetism in iron nanoparticles. Moreover, techniques such as Brunauer–Emmett–Teller (BET) or Barrett–Joyner–Halenda (BJH) can assess the surface area, pore size, and volume. In addition, the swelling ratio of the beads can be determined and calculated using the following equation [52]:

$$\text{Swelling ratio} = \frac{w_{wet} - w_{dry}}{w_{dry}} \quad (20)$$

where w_{wet} is the mass of wet beads and w_{dry} is the mass of dry beads.

3. Results and Discussion

3.1. FE-SEM Analysis

Scanning electron microscopes (SEM) are extensively utilized for the study and analysis of microparticle and nanoparticle imaging and the characterization of solid materials. A specific type of SEM, known as a field emission scanning electron microscope (FE-SEM), uses negatively charged electrons emitted from a field emission source instead of light for imaging. Figure 1 represents M8C's (polymer blend of PVA, PVP, PEG, and CS) morphology.

It is clearly seen that the combination of several polymer components with varying characteristics would likely give M8C's surface an uneven or rough appearance, based on Figure 1. Pores or other visible characteristics may be present, depending on the circumstances and manner of production.

M8C had some porosity, depending on composition and processing technique. The creation of porous structures during the drying or crosslinking processes, the existence of empty spaces between polymer particles, or trapped gas bubbles during synthesis might all contribute to this porosity.

On the other hand, the morphology of the superparamagnetic iron oxide nanoparticles (SPIONs) is illustrated in Figure 2.

Based on the FE-SEM images, the morphology, as shown in Figure 2a, the size distribution, as shown in Figure 2b, and the average size of SPIONs were estimated. The results were 19.03 ± 4.23 nm, which is in line with previous publications [53]. When zooming into higher magnifications, individual SPIONs came into focus, revealing their spherical or quasi-spherical morphology. Additionally, the size distribution is measured in terms of nanometers (nm), with a range of 10 to 30 nm. It can be seen from this distribution that the majority of particles have sizes between 18 and 20 nm.

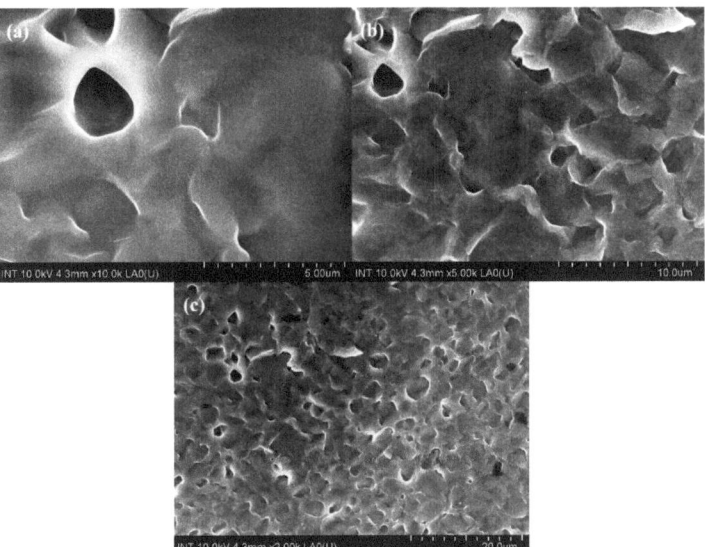

Figure 1. FE-SEM images of M8C at (**a**) 5 µm, (**b**) 10 µm, and (**c**) 20 µm.

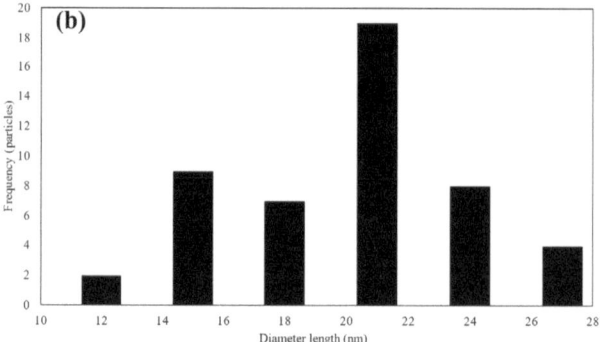

Figure 2. (**a**) The morphology of SPIONs at 500 nm and (**b**) the normal size distribution of SPIONs.

In addition, Figure 3 depicts the morphology of SPION/M8C beads.

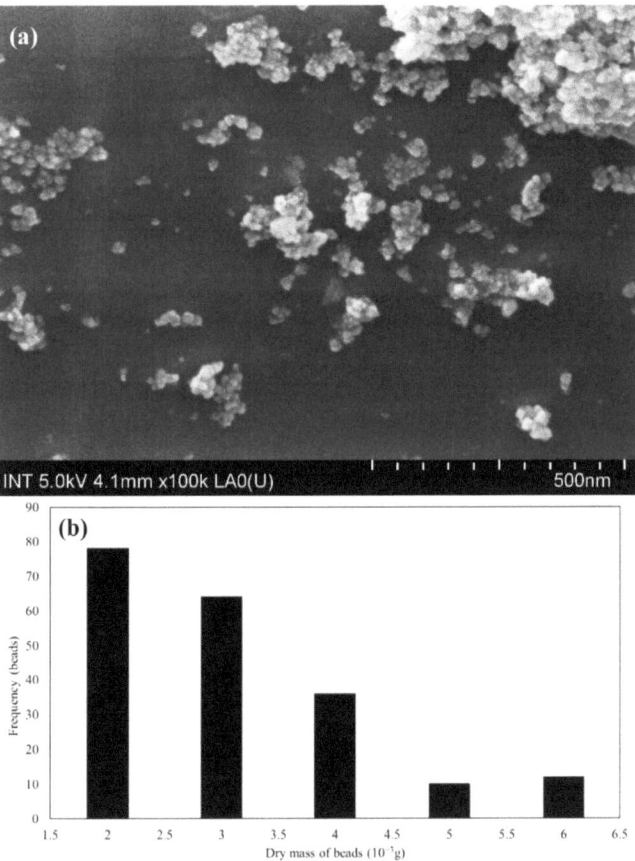

Figure 3. (a) FESEM of SPION/M8C beads at 500 nm and (b) the normal mass distribution of SPIONs.

Using ImageJ software (v.1.53t) to analyze the FE-SEM images, the average size was found to be 22.32 ± 4.44 nm, which is displayed in Figure 3a. The illustrations showed that the composite structure was heterogeneous, with various polymer components and SPIONs dispersed throughout the beads. It demonstrated an uneven distribution of SPIONs in the polymer. Furthermore, observable surface protrusions or abnormalities might mean that SPIONs or polymer aggregates were present on the bead surface.

This distribution analysis in Figure 3b revealed that the majority of the synthesized beads had a mass ranging between 0.002 and 0.004 g after undergoing the drying process. These beads were presented in significant quantities, indicating a relatively uniform synthesis process. However, a smaller subset of the beads exhibited a mass between 0.005 and 0.006 g. The variation in bead mass can be attributed to the dissolution and distribution behavior of the SPIONs during the synthesis. During the sonication process involving the M8C polymer blend and SPIONs, the typically insoluble SPION particles became well-dispersed throughout the mixture. This effective dispersion ensured that each droplet of the mixture, when introduced into the NaOH solution, incorporated a consistent amount of SPIONs, thereby increasing its mass.

3.2. FTIR Analysis

FTIR proves to be a useful tool for quantifying and identifying the functional groups within nanoparticles, thus confirming the material's final structure. Following the synthesis of SPION/M8C beads, the sample underwent FTIR analysis, as shown in Figure 4.

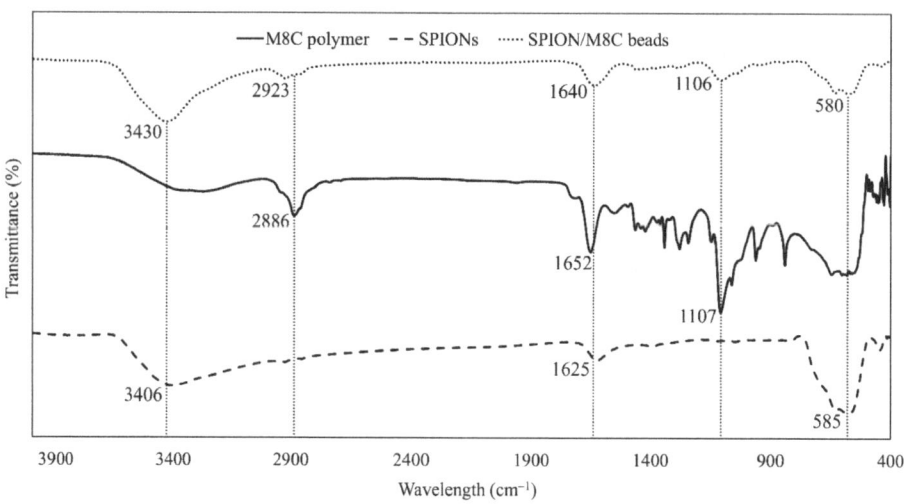

Figure 4. FTIR spectrum of M8C, SPIONs, and SPION/M8C beads.

As illustrated in Figure 4, the peaks of SPION/M8C beads underwent analysis. The intensity of O–H symmetric stretching stood out, likely due to the low concentration of pure polymers in water. However, the characteristic peaks of each component could be distinctly identified, including the vibrations of Fe–O in SPIONs and SPION/M8C beads, the Amide I band, CH_2 asymmetric stretching vibration, and the N–H groups of CS + AA. Based on Figure 4, the FTIR spectrum was further analyzed and shown in Table 1.

Table 1. FTIR spectrum of M8C, SPIONs, and SPION/M8C beads.

Materials	Wavelength (cm^{-1})	Functional Group	References
M8C	2886	Asymmetric C–H stretching vibration	[54]
	1652	C=O stretching vibration of PVP, O–H bending mode of the –OH groups (due to the high amount of water), C=O stretching vibration of PEG, C=O stretching (Amide I) of CS + AA	[50]
	1107	C–O–C symmetric stretching of PEG free amino group –NH$_2$ at the C2 position of glucosamine in CS + AA	[50]
SPIONs	3406	O–H stretching	[55]
	1625	Carbonyl C=O stretching band	[55]
	585	Vibration Fe–O	[56]

Table 1. *Cont.*

Materials	Wavelength (cm^{-1})	Functional Group	References
SPION/M8C beads	3430	–OH stretching of PVA, PVP, PEG –NH groups of CS + AA	[50]
	2923	Asymmetric C–H stretching vibration	[54]
	1640	C=O stretching vibration of PVP, O–H bending mode of the –OH groups (due to the high amount of water), C=O stretching vibration of PEG, C=O stretching (Amide I) of CS + AA	[50]
	1106	C–O–C symmetric stretching of PEG free amino group –NH$_2$ at the C2 position of glucosamine in CS + AA	[50]
	580	Stretching of Fe–O Formation of FeO	[56]

As shown in Table 1, upon analysis of the FTIR spectrum of M8C, it was evident that there were four distinct peaks observed at 2886 cm^{-1}, 1652 cm^{-1}, and 1107 cm^{-1}. These peaks were associated with specific molecular vibrations, specifically an asymmetric C–H stretching vibration [57], C=O stretching vibration, and C–O–C symmetric stretching.

The vibrational spectra of SPIONs exhibited the emergence of three peaks. The first peak, at 3406 cm^{-1}, was attributed to O–H stretching, while the carbonyl C=O stretching band and the vibration of Fe–O were observed at 1625 cm^{-1} and 585 cm^{-1}, respectively.

In the FTIR spectrum for the composite of SPION/M8C beads, the observed peaks closely align with those of the individual components, though some slight shifts are present. It is noteworthy that the peak observed at 3430 cm^{-1} likely corresponds to the O–H stretching present in all components, whereas the peak at 2923 cm^{-1} may be associated with the asymmetric C–H stretching vibration. Additionally, the peak at 1640 cm^{-1} corresponded to the carbonyl stretching vibration. Moreover, the absorption band observed at 1106 cm^{-1} likely corresponds to the free amino group located at the glucosamine C2 position of CS, or it may be indicative of C–O–C stretching.

Additionally, the characteristic vibration of Fe–O in SPIONs is also evident in the spectrum of the nanocomposites, with peaks appearing at both 583 cm^{-1} and 580 cm^{-1}. These features further highlight the presence and interaction of the various functional groups within the composite material. Due to the presence of surfactant on the particles' surface, the peaks may have shifted by a few cm^{-1} compared to the compositional peaks in the FTIR spectra of other samples. Consequently, this FTIR analysis confirmed that the IONPs were indeed SPIONs and that the compound was successfully synthesized.

3.3. XRD Analysis

XRD is a commonly used technique for studying material crystalline structure. It offers comprehensive details on a sample's structural characteristics, crystallinity, and phase composition of SPIONs and SPION/M8C beads, as shown in Figure 5.

Figure 5b displays X-ray diffractograms showcasing distinct peaks located at specific angles such as 30.47°, 35.76°, 53.85°, 57.41°, and 63.01° [11]. These peaks, characteristic of magnetite, served as significant evidence indicating the nanoparticles' composition as SPIONs. Similarly, in Figure 5c, the X-ray diffractograms of SPION/M8C beads exhibited notable peaks at angles of 30.05°, 35.53°, 53.60°, 57.18°, and 62.83° [11]. These pronounced peaks strongly suggest the presence of iron within the SPION/M8C bead composite, reinforcing its composition analysis.

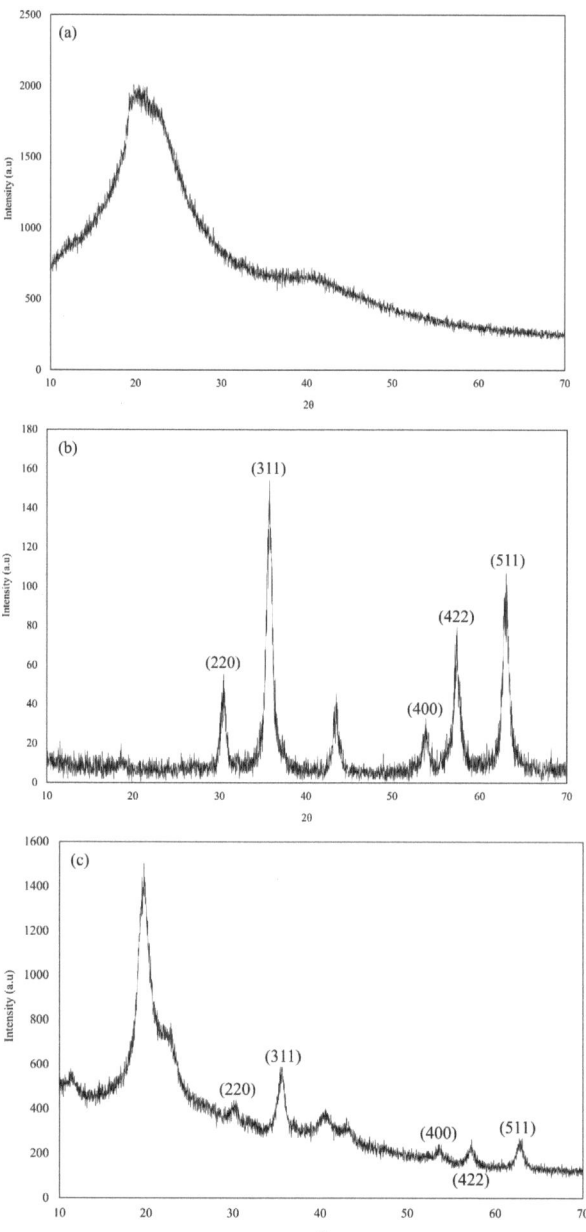

Figure 5. Overlay X-ray diffractogram of (**a**) M8C, (**b**) SPIONs, and (**c**) SPION/M8C beads.

The prominent peaks detected at angles of 30°, 35°, 54°, 57°, and 63° in the XRD pattern of SPIONs correspond precisely to the crystallographic planes (*hkl*) (220), (311), (400), (422), and (511) within the spinel cubic lattice structure characteristic of SPIONs [11]. This correlation strongly suggests that the magnetic core material (M8C) has been effectively adorned with SPIONs, demonstrating the successful decoration of the M8C with

SPIONs. Combining the results of FE-SEM, FTIR, and XRD, further investigations should be conducted to determine whether the SPION was modified on the surface permanently.

3.4. VSM Analysis

The VSM curves showed that the IONPs behaved in a superparamagnetic manner, as observed in Figure 6. In line with other assessments, this result confirms that the IONPs are superparamagnetic [25].

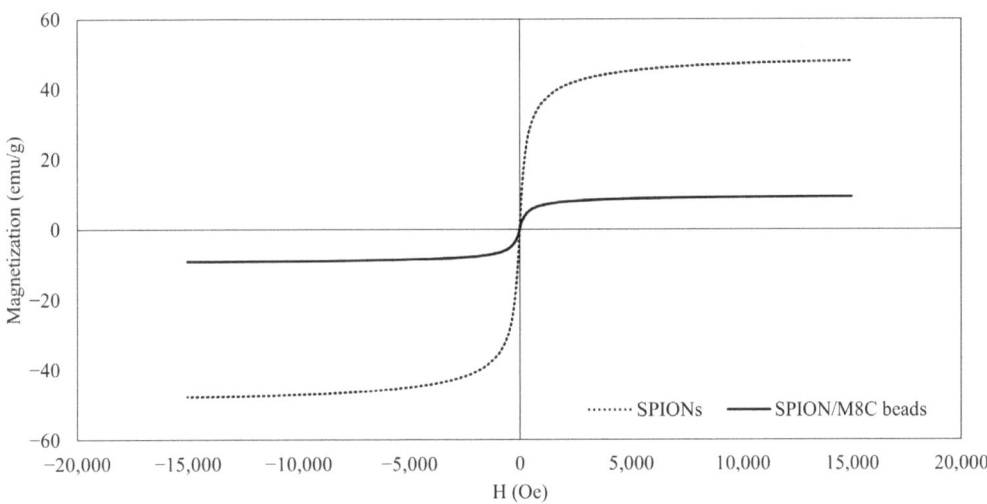

Figure 6. VSM results of SPIONs and SPION/M8C beads.

Similarly, even after the surface modification of SPIONs with CS, PEG, PVA, and PVP, the superparamagnetic properties remained intact. However, there were significant decreases in magnetization, likely due to the presence of the polymer shell coatings surrounding the nanoparticles.

The magnetic properties of both SPIONs and SPION/M8C beads were assessed using VSM at room temperature. Across all samples, the magnetization curves exhibited characteristic S-shaped profiles in response to the applied magnetic field, with no observable remanence or coercivity. These findings were indicative of superparamagnetic behavior across the board. Specifically, the magnetization curve depicted in Figure 6 for SPIONs showcases this superparamagnetic behavior. Furthermore, the saturation magnetization values, averaging around 51 emu/g, aligned closely with approximately 54% of the bulk magnetite content [58].

3.5. BET/BJH Analysis

The volume of adsorbate uptake initiates at a relative pressure and gradually rises with higher relative pressures, as shown by the adsorption–desorption isotherm, as described in Figure 7. This pattern shows how the adsorbate and the adsorbent interact dynamically over a variety of pressures, increasing the material's ability to retain the adsorbate as the relative pressure rises.

The adsorption/desorption isotherm analysis of SPION/M8C beads revealed a type IV isotherm, indicating adsorbents with wide pore size distributions [11,25]. This structural characterization suggests a complex pore architecture, which could contribute to enhanced adsorption properties and surface reactivity. It is characterized by pores ranging in size from 2 to 50 nm [25]. The average pore diameter of the synthesized SPION/M8C beads, determined using the BJH method, is summarized in Table 2.

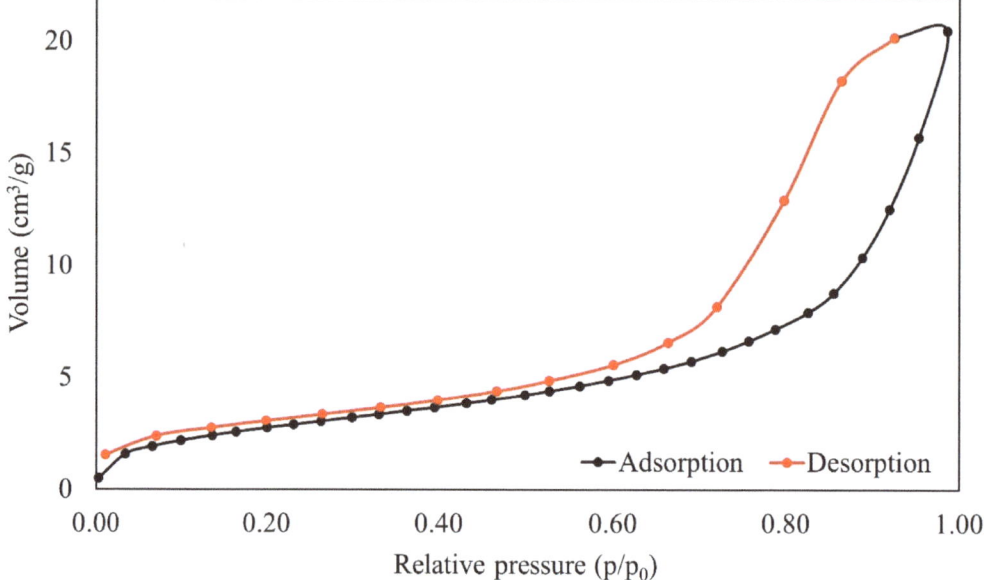

Figure 7. The BET results for relative pressure over the volume of SPION/M8C beads.

Table 2. The BJH and BET analyses of SPION/M8C beads.

	Adsorption	Desorption
Surface area (m^2/g)	25.89	16.71
Pore diameter (Å)	9.82	90.28
Pore volume (cm^3/g)	0.036	0.034

The results obtained from the BJH and BET tests yield valuable insights into the surface characteristics of the material, as outlined in Table 2. Notably, it was discovered that the surface areas for desorption and adsorption were measured at 25.89 m^2/g and 16.71 m^2/g, respectively.

Moreover, examining the SPION/M8C beads sample revealed peak values for both total pore volume and pore diameter during adsorption, reaching 0.036 cm^3/g and 9.82 Å, respectively. Similarly, the pore diameter and volume during desorption were determined to be 90.281 Å and 0.034 cm^3/g, respectively. These findings collectively contribute to a comprehensive understanding of the material's structural characteristics and their correlation with its adsorption capabilities.

3.6. Swelling Ratio Analysis

The swelling ratio (SR) of beads, including polymer and hydrogel beads, is determined similarly to other materials. It quantifies the extent to which the beads swell upon absorbing a liquid relative to their original dry state.

Cycle I beads had a wet weight of 0.0398 ± 0.005 g and a dry weight of 0.0032 ± 0.001 g. During cycle II, the wet weight reduced to 0.0243 ± 0.004 g, whereas the dry weight was 0.0029 ± 0.001 g. These findings revealed a significant drop in wet and dry weights between the two cycles.

Figure 8 presents the average SR for each cycle. In cycle I, the average SR was 12.78 ± 4.48, whereas in cycle II, it dropped to 8.24 ± 3.14. These findings meant that the beads were capable of absorbing and retaining nearly 13 times their weight in the swelling medium throughout cycle I, demonstrating their remarkable absorbency. Thus, the comparison of the two cycles highlighted the dynamic nature of the beads' swelling behavior as well as their capacity to efficiently absorb and hold moisture.

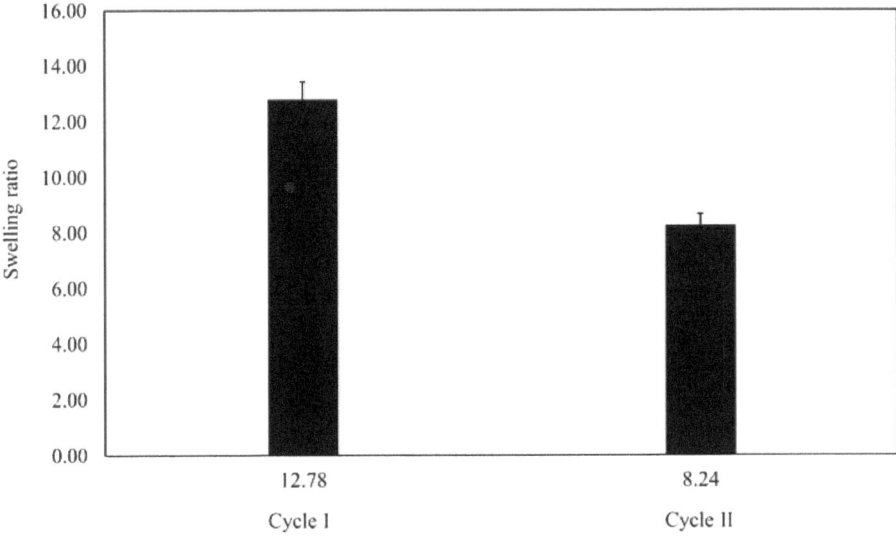

Figure 8. The average SR of cycle I and cycle II.

Based on Figure 8, the average swelling ratio after the second cycle is lower than the first cycle. Specifically, the swelling ratio distribution of each cycle can be seen in Figure 9.

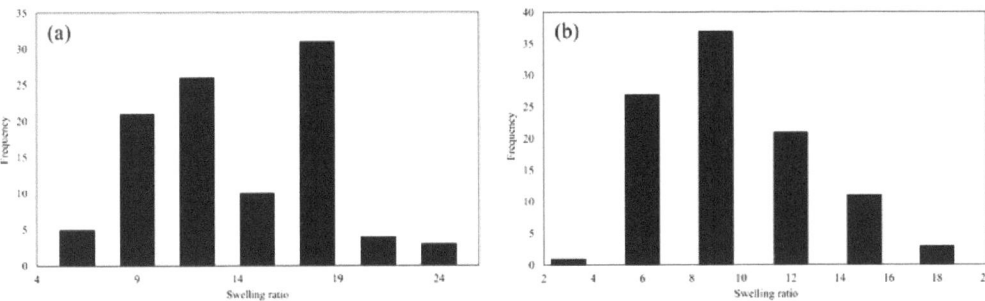

Figure 9. The swelling ratio of (a) cycle I and (b) cycle II.

According to Figure 9, it was observed that during cycle I, the majority of beads displayed a swelling ratio spanning from 10 to 20. Beads with a swelling ratio of around 10 were found to absorb liquid approximately 10 times their dry weight or volume, signifying a moderate level of absorbency. Conversely, beads with a swelling ratio around 20 exhibited the ability to absorb liquid up to 20 times their dry weight or volume, indicating a notably higher absorbent capacity.

Furthermore, the size and form of the beads had a considerable impact on their swelling behavior. Smaller beads had a higher surface area-to-volume ratio, which improved their absorption capabilities. This indicated that smaller beads may have a more noticeable swelling propensity than bigger ones due to the greater surface area available for interaction with the liquid medium.

3.7. Adsorption

At hourly intervals, the loading amount (Q_t), entrapment efficiency (%EE), and percent loading capacity (%LC) for the adsorption process were identified and presented in Table 3. These calculations were carried out using the UV–Vis spectrometry results for MB, which were obtained at a wavelength of 664 nm.

Table 3. The Q_t, %LC, and %EE of SPION/M8C beads adsorbed MB after 120 h.

Initial MB Concentration (mg/L)	Q_t (mg/g)	%LC	%EE
8	278.68 ± 18.65	27.87 ± 1.87	29.42 ± 2.19
9	297.48 ± 15.34	29.75 ± 1.53	25.98 ± 1.29
10	192.82 ± 23.18	19.28 ± 2.32	15.39 ± 1.83

According to Table 3, the highest Q_t and %LC of MB at 120 h were observed at an initial MB concentration of 9 mg/L. Specifically, the Q_t reached 297.48 ± 15.34 mg/g, and the %LC was 29.75 ± 1.53%. At this concentration, the adsorbent demonstrated optimal performance in terms of the amount of MB it could adsorb and the efficiency of its capacity utilization.

On the other hand, the highest %EE was recorded at an initial MB concentration of 8 mg/L, with a value of 29.42 ± 2.19%. This indicated that, at this concentration, a significant proportion of MB was successfully entrapped within the adsorbent relative to the initial amount.

Conversely, at an initial MB concentration of 10 mg/L, the values for Q_t, %LC, and %EE were the lowest among the concentrations tested. The Q_t was 192.82 ± 23.18 mg/g, the %LC was 19.28 ± 2.32%, and the %EE was 15.39 ± 1.83%. These lower values suggested that, at higher MB concentrations, the adsorption efficiency and capacity utilization of the adsorbent decreased, possibly due to saturation effects or limitations in the adsorbent's ability to handle higher concentrations effectively. As shown in Figure 10, the Q_t, %LC, and %EE were evaluated over time.

After one day at room temperature, the amount of MB increased significantly in three different concentrations, as depicted in Figure 10a, roughly 100 mg/g. These percentages correspond to concentration increases of 8, 9, and 10 mg/L. It is evident from this that the adsorption capacity is influenced by concentration.

The patterns of the %LC of MB in Figure 10b and the quantity of MB in Figure 10a were identical. When comparing the Q_t and %LC results after 24 and 72 h, they were nearly twice as high. Moreover, it could be clearly seen that the starting concentration of 9 mg/L, Q_t, and %LC both achieved their maximum value and kept growing until they reached equilibrium at 10 mg/L.

As shown in Figure 10c, the concentration of 8 mg/L produced the greatest percentage for the %EE of MB after 24 h. The %EE had reached about 15% at this stage. Over time, this proportion grew steadily, and before the system achieved equilibrium at 120 h, the %EE sharply rose to 30%.

Nonetheless, it is evident that the Q_t, %LC, and %EE of the 10 mg/L concentration consistently fall to the bottom. As can be seen, all three values (Q_t, %LC, and %EE) rose daily but either held steady after 72 h or even slightly declined. This is because an absorbance of a concentration greater than 10 is outside the range, and a concentration of 10 mg/L is the maximum concentration.

Based on the loading amount over time, kinetic models were evaluated, as shown in Table 4 and Figure 11.

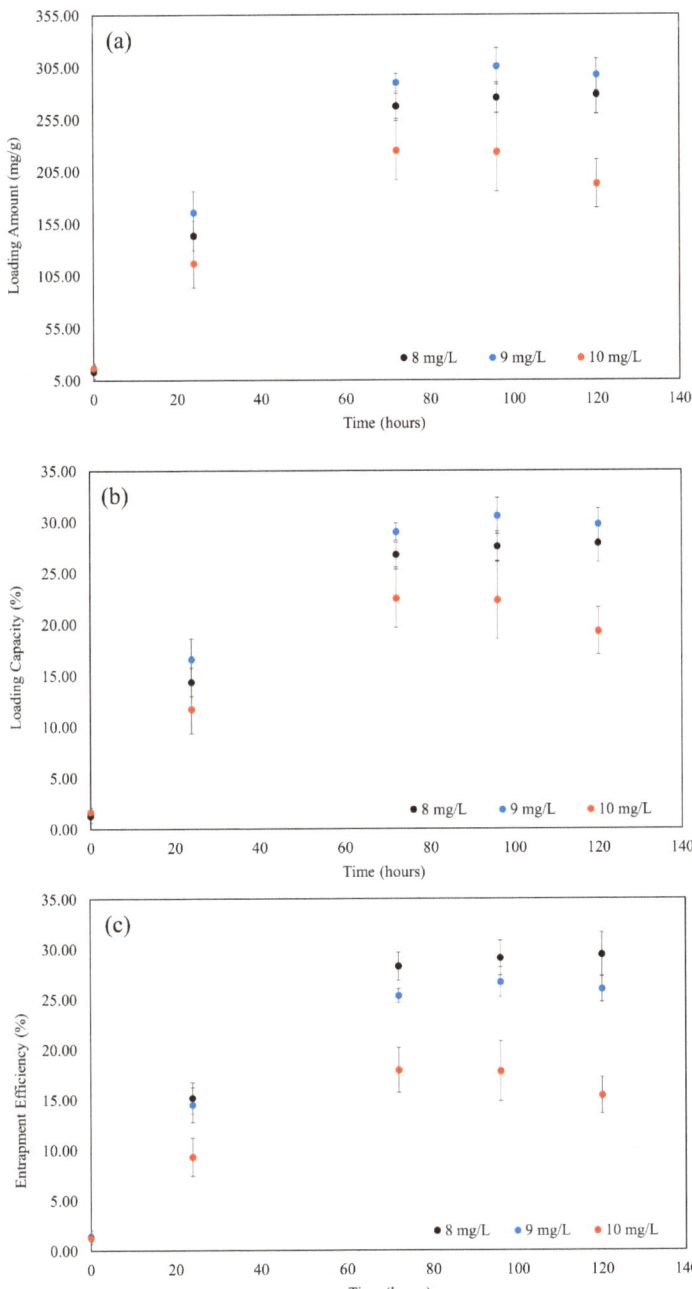

Figure 10. (**a**) The adsorption amount with time; (**b**) the %LC with time; and (**c**) the %EE with time.

Table 4. The pseudo-order kinetic model of SPION/M8C beads adsorbed MB after 120 h.

Model		Initial MB Concentration (mg/L)		
		8	9	10
Pseudo 1st order nonlinear	Q_e	247.92	270.96	194.56
	k_1	1.00	1.00	1.00
	χ^2	52.35	49.15	40.00
Pseudo 2nd order nonlinear	Q_e	247.92	270.96	194.56
	k_2	171.82	171.82	171.82
	χ^2	52.35	49.15	40.00
Pseudo 2nd order linear	Q_e	12.59	13.48	8.94
	k_2	0.17	0.18	0.39
	R^2	0.96	0.97	0.95

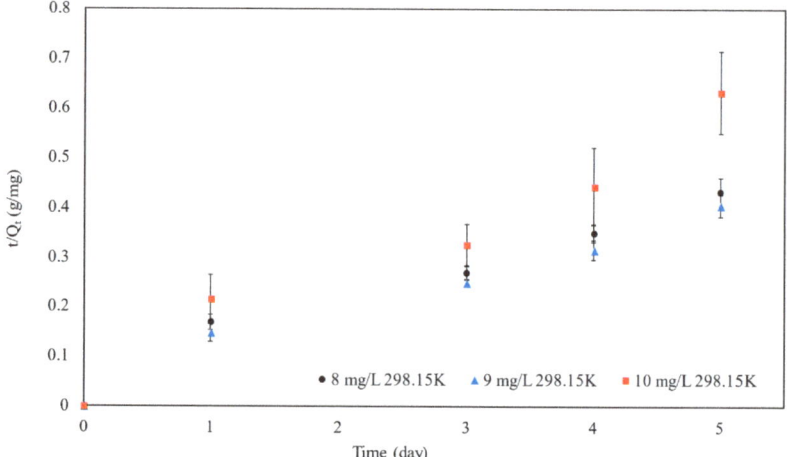

Figure 11. The pseudo-second-order kinetic model of SPION/M8C beads adsorbed MB.

The R^2 values obtained from the linear plot for pseudo-second-order kinetics, corresponding to MB concentrations of 8, 9, and 10 mg/L, were extracted from both Table 4 and Figure 11, revealing values of 0.96, 0.97, and 0.95, respectively. Additionally, it was noted that the Q_e and χ^2 values derived from both pseudo-first-order and pseudo-second-order nonlinear models remained consistent. However, the adsorption kinetic model is best fitted as a pseudo-second-order linear model.

According to the pseudo-second-order kinetic model, it was inferred that MB molecules undergo chemisorption onto the surface of particles through the exchange of electrons, implying the involvement of valence forces between MB and SPION/M8C beads. Notably, this model suggests that the adsorption rate is influenced by concentration, with a quicker adsorption rate observed at lower MB concentrations compared to higher ones. This trend is supported by the observed reduction in the values of the rate constant (k_2) as the concentration of MB increased, indicating a slower adsorption process at higher concentrations.

Additionally, the intraparticle diffusion kinetic model (IPD model) was also evaluated, as shown in Figure 12, to investigate the adsorption mechanism of MB onto SPION/M8C beads, particularly if the film diffusion process is one of the adsorption processes.

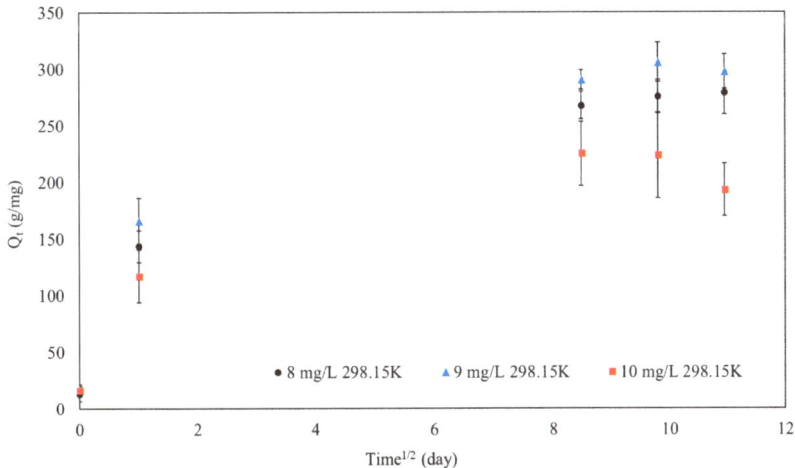

Figure 12. The IPD model of SPION/M8C beads adsorbed MB.

As demonstrated in Figure 12, the adsorption capabilities of MB onto SPION/M8C beads may be fitted via intraparticle diffusion. Weber's intraparticle diffusion model may be used to compute the intraparticle diffusion rate. In the case when I = 0, intraparticle diffusion is the adsorption process. Table 5 displays the computed values of k_{IPD} and I.

Table 5. The IPD kinetic model of SPION/M8C beads adsorbed MB after 120 h.

	Initial MB Concentration (mg/L)		
	8	9	10
I	67.65	79.91	61.41
k_{IPD}	21.18	22.37	15.47
R^2	0.87	0.85	0.80

Table 5 observes R^2 values of 0.87, 0.85, and 0.80 for MB concentrations of 8 mg/L, 9 mg/L, and 10 mg/L, based on a linear plot of pseudo-second-order kinetics. These results show that the pseudo-second-order model fitted the experimental data quite well, especially at lower MB concentrations.

However, the analysis of the IPD model indicated that the intercept (I) did not equal zero. The difference proved that IPD had an important role in the adsorption process. The non-zero intercept suggested that the adsorption mechanism includes film and intraparticle diffusion [51,59].

Additionally, the Elovich kinetic model was also evaluated, as shown in Figure 13 and Table 6.

Figure 13 and Table 6 exhibit substantial R^2 values from the Elovich model, indicating energy heterogeneity at the adsorbent surface. The R^2 values of 0.95, 0.93, and 0.71 for MB concentrations of 8 mg/L, 9 mg/L, and 10 mg/L, respectively, confirm a strong fit for the model, particularly at lower concentrations.

Understanding the efficiency of the adsorption process relies heavily on the initial sorption rates, represented by the α values. At MB concentrations of 8, 9, and 10 mg/L, the initial sorption rates were determined to be 479.8, 616.74, and 514.30 mg/g·min per day. These data illustrated that the adsorbent had a high initial capacity for MB absorption.

As the concentration of MB rose from 8 mg/L to 10 mg/L, the β values, which represented the activation energy for chemisorption, fluctuated between 0.01 mg/g and 0.02 mg/g. The range of β values provided the energy required for chemisorption and suggested that the adsorption process was regulated by the concentration of MB. The range in these values emphasized the adsorption mechanism's complexity and the importance of surface energy heterogeneity in the process.

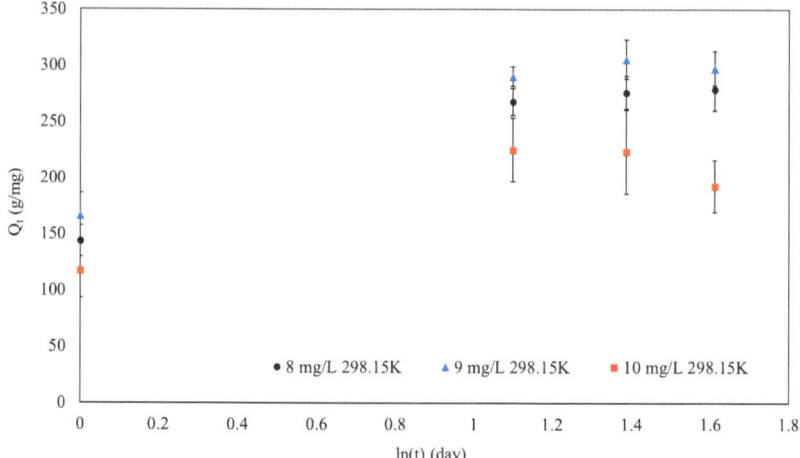

Figure 13. The Elovich model of SPION/M8C beads adsorbed MB.

Table 6. The Elovich kinetic model of SPION/M8C beads adsorbed MB after 120 h.

	Initial MB Concentration (mg/L)		
	8	9	10
α	479.86	616.74	514.30
β	0.01	0.01	0.02
R^2	0.95	0.93	0.71

3.8. Desorption

Even though the release of MB occurred at a constant temperature and a pH of 3.85, the release percentage of MB varied depending on initial loading conditions, such as temperature and starting MB concentration, as demonstrated in Table 7.

Table 7. The percentage release average of MB from SPION/M8C beads after 13 days.

	Initial MB Concentration (mg/L)		
	8	9	10
% Release average	3.00	4.32	7.72
% Release standard deviation	0.10	0.26	0.50

The release percentage of MB rose with increasing initial MB loading concentration. While the maximum release percentage was recorded at an initial concentration of 10 mg/L of MB, the trend indicated that increasing the initial MB concentration typically resulted in a greater release percentage. Additionally, the desorption kinetics were examined to investigate the desorption process, as shown in Figure 14.

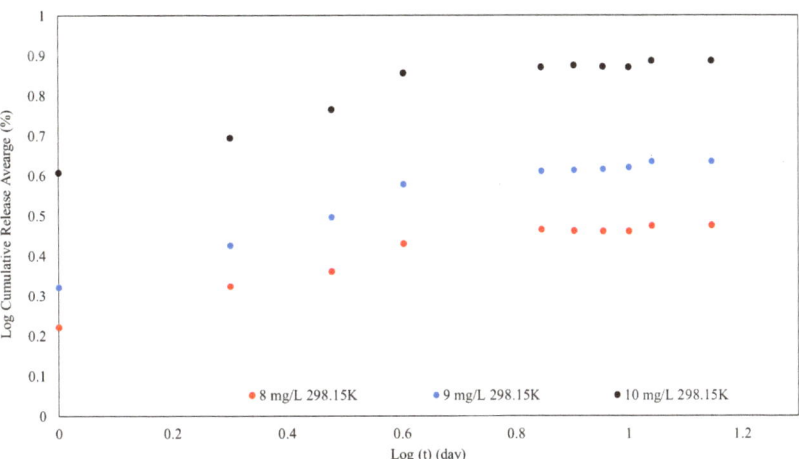

Figure 14. The releasing kinetics after 13 days of MB on SPION/M8C beads.

Figure 14 expresses the progression from an initial fast-release phase to a more continuous and regulated release of MB from SPION/M8C beads. This found that the desorption process was controlled by variables such as the initial MB concentration.

The presence of SPIONs and polymers such as CS, PEG, PVA, and PVP could be responsible for the release profile seen in the graph. SPIONs may increase release rates in the presence of external magnetic fields, whereas polymers create a network that regulates diffusion and release kinetics.

Moreover, the best-fitting kinetic models were determined by combining the cumulative release percentage, the desorption kinetics models, and the χ^2 values, according to Figure 15.

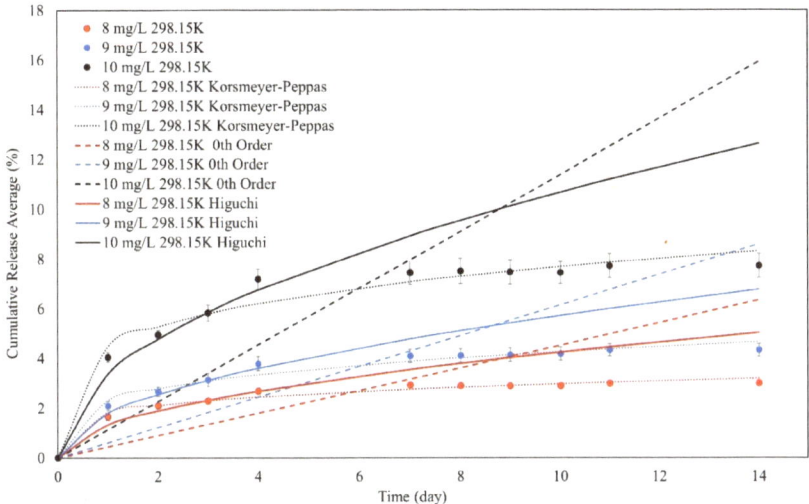

Figure 15. The Korsmeyer–Peppas, zeroth order, and Higuchi kinetic models for the desorption of SPION/M8C beads for MB.

As shown in Figure 15, some of the models can fit the experimental data, and some cannot. To determine the most-fitted model and determine the desorption process, several parameters must be calculated, as shown in Table 8.

Table 8. The Korsmeyer–Peppas, zeroth order, and Higuchi kinetic model calculations for the desorption of SPION/M8C beads for MB.

Initial MB Concentration (mg/L)	KP		0th Order	Higuchi	χ^2		
	k_{KP}	n_{KP}	k_0	k_H	KP	0th Order	Higuchi
8	1.83	0.21	0.45	1.34	0.07	9.60	0.21
9	2.33	0.26	0.61	1.81	0.13	11.06	0.16
10	4.47	0.24	1.14	3.38	0.30	22.35	0.42

Based on the data in Table 8, the Korsmeyer–Peppas models provide release exponents (n values) that are less than 0.45, demonstrating a quasi-Fickian release mechanism. The quasi-Fickian release refers to a diffusion-controlled process in which the material is released from the matrix by a mix of diffusion and other processes. These variables may include polymer matrix relaxation, bead swelling, or other non-Fickian transport processes [60]. Furthermore, the release kinetics exhibit time-dependent behavior, as indicated by the values of n_{KP}. This shows that the release rate changes over time, implying a dynamic interaction between the SPION/M8 beads and the released MB molecules.

All three of the models' χ^2 values were less than 1, but the zeroth order had the biggest, suggesting that the models did not fairly represent the experimental data. The KP's χ^2 value was the lowest. As a result, the KP model suited the desorption in this experiment the best.

4. Conclusions

In conclusion, the beads developed through the combination of polymers M8C, comprising CS, PEG, PVA, and PVP, encapsulating SPIONs represent a novel and cost-effective technique utilizing adsorption. This innovative approach offers antibacterial properties and ensures safety, presenting a promising solution to environmental challenges.

The synthesized SPIONs exhibited a diameter range of 19.03 ± 4.25 nm and a saturation magnetization value of approximately 51 emu/g. Moreover, the SPION/M8C beads displayed a total pore volume and pore diameter for adsorption peaking at 0.036 cm^3/g and 9.822 Å, respectively. This was accompanied by the observation of a type IV isotherm, indicating adsorbents with wide pore size distributions.

Furthermore, the average SR of the beads exhibited a reduction from 12.78 ± 4.48 in cycle I to 8.24 ± 3.14 in cycle II, indicating a shift in the swelling behavior of the beads over successive cycles. Depending on the initial concentrations of MB, which were 8, 9, and 10 mg/L, the Q_t reached values of 278.68 ± 18.65, 297.48 ± 15.34, and 192.82 ± 23.18 mg/g, respectively. The %LC was recorded as 27.87 ± 1.87%, 29.75 ± 1.53%, and 19.28 ± 2.32%, while the %EE values were 29.42 ± 2.19%, 25.98 ± 1.29%, and 15.39 ± 1.83%, respectively. The pseudo-second-order kinetic model suggested that MB would chemisorb to the surface of the particles through the transfer of electrons between SPION/M8C beads and MB.

The cumulative release percentage of MB ranged from 3.00 ± 0.10% to 7.72 ± 0.50% when released at pH 3.85 over a duration of 13 days. Notably, the KP model was found to accurately represent the desorption kinetics observed in this experiment for the release of MB from SPION/M8C beads.

Author Contributions: Performing experiment, data analysis, and writing—review and editing, K.T.; conceptualization, methodology, data analysis, writing—original draft preparation, and writing—review and editing, L.D.; performing experiments and data analysis: T.T.T.N. and K.G.H. All authors have read and agreed to the published version of the manuscript.

Funding: This research received no external funding.

Institutional Review Board Statement: Not applicable.

Data Availability Statement: All data that support the findings of this study are included within the article.

Conflicts of Interest: The authors declare no conflicts of interest.

References

1. Watkinson, A.J.; Murby, E.J.; Costanzo, S.D. Removal of Antibiotics in Conventional and Advanced Wastewater Treatment: Implications for Environmental Discharge and Wastewater Recycling. *Water Res.* **2007**, *41*, 4164–4176. [CrossRef]
2. Stackelberg, P.E.; Furlong, E.T.; Meyer, M.T.; Zaugg, S.D.; Henderson, A.K.; Reissman, D.B. Persistence of Pharmaceutical Compounds and Other Organic Wastewater Contaminants in a Conventional Drinking-Water-Treatment Plant. *Sci. Total Environ.* **2004**, *329*, 99–113. [CrossRef]
3. Chowdhary, P.; Raj, A.; Bharagava, R.N. Environmental Pollution and Health Hazards from Distillery Wastewater and Treatment Approaches to Combat the Environmental Threats: A Review. *Chemosphere* **2018**, *194*, 229–246. [CrossRef]
4. Launay, M.A.; Dittmer, U.; Steinmetz, H. Organic Micropollutants Discharged by Combined Sewer Overflows—Characterisation of Pollutant Sources and Stormwater-Related Processes. *Water Res.* **2016**, *104*, 82–92. [CrossRef]
5. Younas, F.; Mustafa, A.; Farooqi, Z.U.R.; Wang, X.; Younas, S.; Mohy-Ud-Din, W.; Ashir Hameed, M.; Mohsin Abrar, M.; Maitlo, A.A.; Noreen, S.; et al. Current and Emerging Adsorbent Technologies for Wastewater Treatment: Trends, Limitations, and Environmental Implications. *Water* **2021**, *13*, 215. [CrossRef]
6. Grigoraș, C.-G.; Simion, A.-I.; Drob, C. Hydrogels Based on Chitosan and Nanoparticles and Their Suitability for Dyes Adsorption from Aqueous Media: Assessment of the Last-Decade Progresses. *Gels* **2024**, *10*, 211. [CrossRef]
7. Altıntıg, E.; Ates, A.; Angın, D.; Topal, Z.; Aydemir, Z. Kinetic, Equilibrium, Adsorption Mechanisms of RBBR and MG Dyes on Chitosan-Coated Montmorillonite with an Ecofriendly Approach. *Chem. Eng. Res. Des.* **2022**, *188*, 287–300. [CrossRef]
8. Chatterjee, S.; Ohemeng-Boahen, G.; Sewu, D.D.; Osei, B.A.; Woo, S.H. Improved Adsorption of Congo Red from Aqueous Solution Using Alkali-Treated Goethite Impregnated Chitosan Hydrogel Capsule. *J. Environ. Chem. Eng.* **2022**, *10*, 108244. [CrossRef]
9. Ehrampoush, M.H.; Moussavi, G.R.; Ghaneian, M.T.; Rahimi, S.; Ahmadian, M. Removal of Methylene Blue (MB) Dye from Textile Synthetic Wastewater Using TiO2/UV-C Photocatalytic Process. *Austr. J. Basic Appl. Sci.* **2010**, *4*, 4279–4285.
10. Oladoye, P.O.; Ajiboye, T.O.; Omotola, E.O.; Oyewola, O.J. Methylene Blue Dye: Toxicity and Potential Elimination Technology from Wastewater. *Results Eng.* **2022**, *16*, 100678. [CrossRef]
11. Doan, L. Modifying Superparamagnetic Iron Oxide Nanoparticles as Methylene Blue Adsorbents: A Review. *ChemEngineering* **2023**, *7*, 77. [CrossRef]
12. Khan, I.; Saeed, K.; Zekker, I.; Zhang, B.; Hendi, A.H.; Ahmad, A.; Ahmad, S.; Zada, N.; Ahmad, H.; Shah, L.A.; et al. Review on Methylene Blue: Its Properties, Uses, Toxicity and Photodegradation. *Water* **2022**, *14*, 242. [CrossRef]
13. Al-Tohamy, R.; Ali, S.S.; Li, F.; Okasha, K.M.; Mahmoud, Y.A.-G.; Elsamahy, T.; Jiao, H.; Fu, Y.; Sun, J. A Critical Review on the Treatment of Dye-Containing Wastewater: Ecotoxicological and Health Concerns of Textile Dyes and Possible Remediation Approaches for Environmental Safety. *Ecotoxicol. Environ. Saf.* **2022**, *231*, 113160. [CrossRef]
14. Jabbar, K.Q.; Barzinjy, A.A.; Hamad, S.M. Iron Oxide Nanoparticles: Preparation Methods, Functions, Adsorption and Coagulation/Flocculation in Wastewater Treatment. *Environ. Nanotechnol. Monit. Manag.* **2022**, *17*, 100661. [CrossRef]
15. Song, S.; Fan, J.; He, Z.; Zhan, L.; Liu, Z.; Chen, J.; Xu, X. Electrochemical Degradation of Azo Dye C.I. Reactive Red 195 by Anodic Oxidation on Ti/SnO2–Sb/PbO2 Electrodes. *Electrochim. Acta* **2010**, *55*, 3606–3613. [CrossRef]
16. Selim, M.T.; Salem, S.S.; Mohamed, A.A.; El-Gamal, M.S.; Awad, M.F.; Fouda, A. Biological Treatment of Real Textile Effluent Using Aspergillus Flavus and Fusarium Oxysporium and Their Consortium along with the Evaluation of Their Phytotoxicity. *JoF* **2021**, *7*, 193. [CrossRef]
17. Essa, W.K.; Yasin, S.A.; Abdullah, A.H.; Thalji, M.R.; Saeed, I.A.; Assiri, M.A.; Chong, K.F.; Ali, G.A.M. Taguchi L25 (54) Approach for Methylene Blue Removal by Polyethylene Terephthalate Nanofiber-Multi-Walled Carbon Nanotube Composite. *Water* **2022**, *14*, 1242. [CrossRef]
18. Mungondori, H.H.; Tichagwa, L.; Katwire, D.M.; Aoyi, O. Preparation of Photo-Catalytic Copolymer Grafted Asymmetric Membranes (N-TiO2-PMAA-g-PVDF/PAN) and Their Application on the Degradation of Bentazon in Water. *Iran. Polym. J.* **2016**, *25*, 135–144. [CrossRef]
19. Wan Ngah, W.S.; Teong, L.C.; Hanafiah, M.A.K.M. Adsorption of Dyes and Heavy Metal Ions by Chitosan Composites: A Review. *Carbohydr. Polym.* **2011**, *83*, 1446–1456. [CrossRef]
20. Tiris, M.; Topkaya, B.; Bahadir, M. Editorial: Clean Soil Air Water 2/2009. *CLEAN Soil Air Water* **2009**, *37*, 101–102. [CrossRef]
21. Mustapha, S.; Ndamitso, M.M.; Abdulkareem, A.S.; Tijani, J.O.; Shuaib, D.T.; Ajala, A.O.; Mohammed, A.K. Application of TiO2 and ZnO Nanoparticles Immobilized on Clay in Wastewater Treatment: A Review. *Appl. Water Sci.* **2020**, *10*, 49. [CrossRef]
22. Altıntıg, E.; Altundag, H.; Tuzen, M.; Sarı, A. Effective Removal of Methylene Blue from Aqueous Solutions Using Magnetic Loaded Activated Carbon as Novel Adsorbent. *Chem. Eng. Res. Des.* **2017**, *122*, 151–163. [CrossRef]

23. Sharma, V.; Singh, H.; Guleria, S.; Bhardwaj, N.; Puri, S.; Arya, S.K.; Khatri, M. Application of Superparamagnetic Iron Oxide Nanoparticles (SPIONs) for Heavy Metal Adsorption: A 10-Year Meta-Analysis. *Environ. Nanotechnol. Monit. Manag.* **2022**, *18*, 100716. [CrossRef]
24. Ahamad, T.; Naushad, M.; Eldesoky, G.E.; Al-Saeedi, S.I.; Nafady, A.; Al-Kadhi, N.S.; Al-Muhtaseb, A.H.; Khan, A.A.; Khan, A. Effective and Fast Adsorptive Removal of Toxic Cationic Dye (MB) from Aqueous Medium Using Amino-Functionalized Magnetic Multiwall Carbon Nanotubes. *J. Mol. Liq.* **2019**, *282*, 154–161. [CrossRef]
25. Quach, T.P.T.; Doan, L. Surface Modifications of Superparamagnetic Iron Oxide Nanoparticles with Polyvinyl Alcohol, Chitosan, and Graphene Oxide as Methylene Blue Adsorbents. *Coatings* **2023**, *13*, 1333. [CrossRef]
26. Kim, D.K.; Zhang, Y.; Voit, W.; Rao, K.V.; Muhammed, M. Synthesis and Characterization of Surfactant-Coated Superparamagnetic Monodispersed Iron Oxide Nanoparticles. *J. Magn. Magn. Mater.* **2001**, *225*, 30–36. [CrossRef]
27. Teixeira-Costa, B.E.; Andrade, C.T. Chitosan as a Valuable Biomolecule from Seafood Industry Waste in the Design of Green Food Packaging. *Biomolecules* **2021**, *11*, 1599. [CrossRef]
28. Kumar, M.N.V.R.; Muzzarelli, R.A.A.; Muzzarelli, C.; Sashiwa, H.; Domb, A.J. Chitosan Chemistry and Pharmaceutical Perspectives. *Chem. Rev.* **2004**, *104*, 6017–6084. [CrossRef]
29. Eiden, C.A.; Jewell, C.A.; Wightman, J.P. Interaction of Lead and Chromium with Chitin and Chitosan. *J. Appl. Polym. Sci.* **1980**, *25*, 1587–1599. [CrossRef]
30. Li, Y.; Zhou, Y.; Nie, W.; Song, L.; Chen, P. Highly Efficient Methylene Blue Dyes Removal from Aqueous Systems by Chitosan Coated Magnetic Mesoporous Silica Nanoparticles. *J. Porous Mater.* **2015**, *22*, 1383–1392. [CrossRef]
31. Hoa, N.V.; Minh, N.C.; Cuong, H.N.; Dat, P.A.; Nam, P.V.; Viet, P.H.T.; Phuong, P.T.D.; Trung, T.S. Highly Porous Hydroxyapatite/Graphene Oxide/Chitosan Beads as an Efficient Adsorbent for Dyes and Heavy Metal Ions Removal. *Molecules* **2021**, *26*, 6127. [CrossRef]
32. Tran, H.V.; Bui, L.T.; Dinh, T.T.; Le, D.H.; Huynh, C.D.; Trinh, A.X. Graphene Oxide/Fe3O4/Chitosan Nanocomposite: A Recoverable and Recyclable Adsorbent for Organic Dyes Removal. Application to Methylene Blue. *Mater. Res. Express* **2017**, *4*, 035701. [CrossRef]
33. Banerjee, S.S.; Aher, N.; Patil, R.; Khandare, J. Poly(Ethylene Glycol)-Prodrug Conjugates: Concept, Design, and Applications. *J. Drug Deliv.* **2012**, *2012*, 103973. [CrossRef]
34. Harris, J.M.; Chess, R.B. Effect of Pegylation on Pharmaceuticals. *Nat. Rev. Drug Discov.* **2003**, *2*, 214–221. [CrossRef]
35. Schiller, L.R.; Emmett, M.; Santa Ana, C.A.; Fordtran, J.S. Osmotic Effects of Polyethylene Glycol. *Gastroenterology* **1988**, *94*, 933–941. [CrossRef]
36. Balakrishnan, B.; Kumar, D.S.; Yoshida, Y.; Jayakrishnan, A. Chemical Modification of Poly(Vinyl Chloride) Resin Using Poly(Ethylene Glycol) to Improve Blood Compatibility. *Biomaterials* **2005**, *26*, 3495–3502. [CrossRef]
37. *Polyvinylpyrrolidone Excipients for Pharmaceuticals*; Springer: Berlin/Heidelberg, Germany, 2005; ISBN 978-3-540-23412-8.
38. Alem, M.; Teimouri, A.; Salavati, H.; Kazemi, S. Central Composite Design Optimization of Methylene Blue Scavenger Using Modified Graphene Oxide Based Polymer. *Chem. Methodol.* **2017**, *1*, 55–73. [CrossRef]
39. Seo, A.; Schæfer, T. Melt Agglomeration with Polyethylene Glycol Beads at a Low Impeller Speed in a High Shear Mixer. *Eur. J. Pharm. Biopharm.* **2001**, *52*, 315–325. [CrossRef]
40. Trikkaliotis, D.G.; Christoforidis, A.K.; Mitropoulos, A.C.; Kyzas, G.Z. Adsorption of Copper Ions onto Chitosan/Poly(Vinyl Alcohol) Beads Functionalized with Poly(Ethylene Glycol). *Carbohydr. Polym.* **2020**, *234*, 115890. [CrossRef]
41. Abdullah, Z.W.; Dong, Y.; Davies, I.J.; Barbhuiya, S. PVA, PVA Blends, and Their Nanocomposites for Biodegradable Packaging Application. *Polym. Plast. Technol. Eng.* **2017**, *56*, 1307–1344. [CrossRef]
42. Aslam, M.; Kalyar, M.A.; Raza, Z.A. Polyvinyl Alcohol: A Review of Research Status and Use of Polyvinyl Alcohol Based Nanocomposites. *Polym. Eng. Sci.* **2018**, *58*, 2119–2132. [CrossRef]
43. Zhao, H.; Qiu, S.; Wu, L.; Zhang, L.; Chen, H.; Gao, C. Improving the Performance of Polyamide Reverse Osmosis Membrane by Incorporation of Modified Multi-Walled Carbon Nanotubes. *J. Membr. Sci.* **2014**, *450*, 249–256. [CrossRef]
44. Gupta, V.K.; Tyagi, I.; Agarwal, S.; Sadegh, H.; Shahryari-ghoshekandi, R.; Yari, M.; Yousefi-nejat, O. Experimental Study of Surfaces of Hydrogel Polymers HEMA, HEMA–EEMA–MA, and PVA as Adsorbent for Removal of Azo Dyes from Liquid Phase. *J. Mol. Liq.* **2015**, *206*, 129–136. [CrossRef]
45. Habiba, U.; Afifi, A.M.; Salleh, A.; Ang, B.C. Chitosan/(Polyvinyl Alcohol)/Zeolite Electrospun Composite Nanofibrous Membrane for Adsorption of Cr6+, Fe3+ and Ni2+. *J. Hazard. Mater.* **2017**, *322*, 182–194. [CrossRef]
46. Jeong, S.K.; Cho, J.-S.; Kong, I.-S.; Jeong, H.D.; Kim, J.K. Purification of Aquarium Water by PVA Gel-Immobilized Photosynthetic Bacteria during Goldfish Rearing. *Biotechnol. Bioproc. E* **2009**, *14*, 238–247. [CrossRef]
47. Engström, J.U.A.; Helgee, B. Functional Beads of Polyvinylpyrrolidone: Promising Support Materials for Solid-Phase Synthesis. *Macro Chem. Phys.* **2006**, *207*, 605–614. [CrossRef]
48. Siggia, S. The Chemistry of Polyvinylpyrrolidone-Iodine**General Aniline and Film Corporation, Easton, Pa. *J. Am. Pharm. Assoc. (Sci. Ed.)* **1957**, *46*, 201–204. [CrossRef]
49. Koczkur, K.M.; Mourdikoudis, S.; Polavarapu, L.; Skrabalak, S.E. Polyvinylpyrrolidone (PVP) in Nanoparticle Synthesis. *Dalton Trans.* **2015**, *44*, 17883–17905. [CrossRef]
50. Doan, L.; Tran, K. Relationship between the Polymer Blend Using Chitosan, Polyethylene Glycol, Polyvinyl Alcohol, Polyvinylpyrrolidone, and Antimicrobial Activities against Staphylococcus Aureus. *Pharmaceutics* **2023**, *15*, 2453. [CrossRef]

51. Doan, L. Surface Modifications of Superparamagnetic Iron Oxide Nanoparticles with Polyvinyl Alcohol and Graphite as Methylene Blue Adsorbents. *Coatings* **2023**, *13*, 1558. [CrossRef]
52. Mac Kenna, N.; Calvert, P.; Morrin, A. Impedimetric Transduction of Swelling in pH-Responsive Hydrogels. *Analyst* **2015**, *140*, 3003–3011. [CrossRef]
53. Farjadian, F.; Moradi, S.; Hosseini, M. Thin Chitosan Films Containing Super-Paramagnetic Nanoparticles with Contrasting Capability in Magnetic Resonance Imaging. *J. Mater. Sci. Mater. Med.* **2017**, *28*, 47. [CrossRef]
54. Kharazmi, A.; Faraji, N.; Mat Hussin, R.; Saion, E.; Yunus, W.M.M.; Behzad, K. Structural, Optical, Opto-Thermal and Thermal Properties of ZnS–PVA Nanofluids Synthesized through a Radiolytic Approach. *Beilstein J. Nanotechnol.* **2015**, *6*, 529–536. [CrossRef]
55. Sodipo, B.K.; Aziz, A.A. A Sonochemical Approach to the Direct Surface Functionalization of Superparamagnetic Iron Oxide Nanoparticles with (3-Aminopropyl)Triethoxysilane. *Beilstein J. Nanotechnol.* **2014**, *5*, 1472–1476. [CrossRef]
56. Jalilian, A.R.; Panahifar, A.; Mahmoudi, M.; Akhlaghi, M.; Simchi, A. Preparation and Biological Evaluation of [67Ga]-Labeled-Superparamagnetic Nanoparticles in Normal Rats. *Radiochim. Acta* **2009**, *97*, 51–56. [CrossRef]
57. Fernandes Queiroz, M.; Melo, K.; Sabry, D.; Sassaki, G.; Rocha, H. Does the Use of Chitosan Contribute to Oxalate Kidney Stone Formation? *Mar. Drugs* **2014**, *13*, 141–158. [CrossRef]
58. Book Reviews. Announcements. *Corros. Rev.* **1997**, *15*, 533–559. [CrossRef]
59. Yao, Y.; Xu, F.; Chen, M.; Xu, Z.; Zhu, Z. Adsorption Behavior of Methylene Blue on Carbon Nanotubes. *Bioresour. Technol.* **2010**, *101*, 3040–3046. [CrossRef]
60. Kaur, M.; Datta, M. Diclofenac Sodium Adsorption onto Montmorillonite: Adsorption Equilibrium Studies and Drug Release Kinetics. *Adsorpt. Sci. Technol.* **2014**, *32*, 365–387. [CrossRef]

Disclaimer/Publisher's Note: The statements, opinions and data contained in all publications are solely those of the individual author(s) and contributor(s) and not of MDPI and/or the editor(s). MDPI and/or the editor(s) disclaim responsibility for any injury to people or property resulting from any ideas, methods, instructions or products referred to in the content.

Article

Surface Modifications of Silver Nanoparticles with Chitosan, Polyethylene Glycol, Polyvinyl Alcohol, and Polyvinylpyrrolidone as Antibacterial Agents against *Staphylococcus aureus*, *Pseudomonas aeruginosa*, and *Salmonella enterica*

Linh Doan [1,2,3,*], Quynh N. Le [1,2,3], Khoa Tran [2,3] and An H. Huynh [1,2,3]

1. Department of Chemical Engineering, International University—Vietnam National University Ho Chi Minh City, Ho Chi Minh City 70000, Vietnam
2. Nanomaterials Engineering Research & Development (NERD) Laboratory, International University—Vietnam National University Ho Chi Minh City, Ho Chi Minh City 70000, Vietnam; tldkhoa@hcmiu.edu.vn
3. School of Chemical and Environmental Engineering, International University—Vietnam National University Ho Chi Minh City, Ho Chi Minh City 70000, Vietnam
* Correspondence: dhlinh@hcmiu.edu.vn

Citation: Doan, L.; Le, Q.N.; Tran, K.; Huynh, A.H. Surface Modifications of Silver Nanoparticles with Chitosan, Polyethylene Glycol, Polyvinyl Alcohol, and Polyvinylpyrrolidone as Antibacterial Agents against *Staphylococcus aureus*, *Pseudomonas aeruginosa*, and *Salmonella enterica*. *Polymers* **2024**, *16*, 1820. https://doi.org/10.3390/polym16131820

Academic Editors: Sérgio Paulo Campana-Filho and William Facchinatto

Received: 22 May 2024
Revised: 18 June 2024
Accepted: 25 June 2024
Published: 27 June 2024

Copyright: © 2024 by the authors. Licensee MDPI, Basel, Switzerland. This article is an open access article distributed under the terms and conditions of the Creative Commons Attribution (CC BY) license (https://creativecommons.org/licenses/by/4.0/).

Abstract: In medicine, the occurrence of antibiotic resistance was becoming a critical concern. At the same time, traditional synthesis methods of antibacterial agents often lead to environmental pollution due to the use of toxic chemicals. To address these problems, this study applies the green synthesis method to create a novel composite using a polymer blend (M8) consisting of chitosan (CS), polyethylene glycol (PEG), polyvinyl alcohol (PVA), polyvinylpyrrolidone (PVP), and silver nanoparticles. The results show that the highest ratio of AgNO$_3$:M8 was 0.15 g/60 mL, which resulted in a 100% conversion of Ag$^+$ to Ag^0 after 10 h of reaction at 80 °C. Hence, using M8, Ag nanoparticles (AgNPs) were synthesized at the average size of 42.48 ± 10.77 nm. The AgNPs' composite (M8Ag) was used to inhibit the growth of *Staphylococcus aureus* (SA), *Pseudomonas aeruginosa* (PA), and *Salmonella enterica* (SAL). At 6.25% dilution of M8Ag, the growth of these mentioned bacteria was inhibited. At the same dilution percentage of M8Ag, PA was killed.

Keywords: *S. aureus*; *P. aeruginosa*; *S. enterica*; antimicrobial; composite

1. Introduction

Microorganisms have a significant impact on human health, particularly bacteria prevalent in numerous surroundings such as food and drinking water, among the most common of which are *Staphylococcus aureus* (SA), *Pseudomonas aeruginosa* (PA), and *Salmonella enterica* (SAL). All of them can cause infections, ranging in intensity and variety, from skin infections to gastrointestinal and respiratory infections [1–3].

Staphylococcus aureus (SA) is a Gram-positive cocci in the *Micrococcaceae* family which develops gleaming, smooth, complete, elevated, and transparent colonies with a golden color and 1 to 4 mm in diameter [4]. SA generates tissue-degrading enzymes such as protease, lipase, and hyaluronidase [5]. These bacterial metabolites may help in the spread of infection to nearby tissues. *S. aureus* has a predisposition for spreading to certain organs, including the bones, joints, kidneys, and lungs [6]. In addition, SA can cause skin infections (such as boils and cellulitis), respiratory infections, and food poisoning [7].

Pseudomonas aeruginosa (PA) is a Gram-negative, heterotrophic, motile bacteria of the *Pseudomonadaceae* family [8]. PA is a rod-shaped bacteria of around 1–5 μm in length and 0.5–1.0 μm in width [8] that forms large, opaque, flat colonies with uneven borders. It is known to cause opportunistic infections, especially in people with weaker immune systems or cystic fibrosis [8]. *P. aeruginosa* has the potential to cause infections in a variety of organs, including the lungs, urinary tract, and skin. On the clinical level, patients with impaired immune systems, including patients with cystic fibrosis, HIV/AIDS, cancer, burn and eye injuries, and non-healing diabetic wounds, are at greatest risk [8].

Salmonella enterica (*SAL*), as a group, is a Gram-negative, non-spore-forming prokaryote belonging to the Enterobacteriaceae family [9]. Salmonella range in size from 0.7–1.5 mm to 2.2–5.0 mm, and colonies commonly measure 2–4 mm in diameter [10]. *SAL* infection can cause a systemic illness known as enteric fever, an intestinal infection known as gastroenteritis, or a blood infection in humans known as bacteremia [11].

SA, *PA*, and *SAL* growth can be inhibited, or these bacteria can even be killed, using antibiotics [12,13]. However, using antibiotics can cause antibiotic resistance (ABR) [14]. Hence, a material that could inhibit the growth of these bacteria, or even kill these bacteria, without causing ABR phenomena would be ideal. To achieve this, researchers used various antibacterial agents that were not conventional and/or traditional antibiotics such as polymers, metal oxides, or other types of composites. One of the most common antibacterial agents are metal oxides and other metal-type particles such as silver particles [2,3,15–18], iron oxide nanoparticles [19–21], calcium oxide nanoparticles [22], zinc oxide and copper oxide nanoparticles [13,23,24]. However, out of all metal oxides or metal composites, silver particles are the most common antibacterial material.

Metals like silver are widely employed as active agents to create antibacterial surfaces due to their potent antibacterial properties, even at extremely low doses [25]. Metals, especially silver (Ag), have long been known as efficient antibacterial compounds capable of destroying bacteria [26]. Hence, silver has been utilized for its antibacterial properties even before the pharmaceutical antibiotic revolution. Polymer–metal composites have evolved as a highly effective technique for various surface applications. These silver polymeric materials can also be used in the form of gels or patches for topical applications such as cosmetics or medicine administration and require a balance of physical strength and antibacterial activity [27]. The biological activity of silver particles is influenced by factors such as surface chemistry, size, shape, particle morphology and composition, coating/capping, agglomeration, and dissolution rate. Additionally, particle reactivity in solution, cell type, and the type of reducing agents used for silver particle synthesis play crucial roles in determining their antibacterial efficacy [28].

The methods for synthesizing silver particles, including biological, chemical, and physical approaches, each have their advantages and disadvantages [29]. For instance, the chemical method uses hazardous materials, generates toxic byproducts, requires a lot of energy, and has the potential to harm the environment [28,29]. The physical synthesis process requires sophisticated equipment and procedures, consumes a lot of energy, lacks stabilizing or capping chemicals to avoid agglomeration, and has limits in managing the size and form of silver particles [28,29]. The most significant advantage of silver particles is their environmental friendliness. Green synthesis approaches use reducing biological agents, such as plant extracts or microbial components, to aid in the reduction of silver ions into silver particles. These approaches frequently remove the requirement for external capping and stabilizing agents, minimizing the usage of potentially dangerous compounds throughout the synthesis process [29]. Furthermore, green synthesis methods are seen as more sustainable and ecofriendly than standard chemical synthesis processes, harmonizing with the worldwide effort to achieve sustainable development goals while lowering environmental impacts [29].

Green synthesis and the antimicrobial activities of silver particles involved in it were investigated extensively in materials such as CS/PVA/Ag [30,31], CS/Ag [32–34], PVP/CS/Ag [35], PEG/Ag [36,37], PVP/Ag [38], CS/PEG/Ag [39–41], and PVA/Ag [12,33,42]. According to Linh et al. [43], a novel polymeric material, M8, which can inhibit the growth of *SA*, consists of PVA, PVP, PEG, and CS. Combining the ability to synthesize silver particles using CS, PVA, PVP, and PEG individually and the anti-*SA* polymer blend M8, in this study, a novel antibacterial agent using silver particles and M8 were synthesized.

2. Materials and Methods

2.1. Materials

All the chemicals were from the same source as previous study [43]. Additionally, silver nitrate ($AgNO_3$, 99.8%) was bought from Shanghai Zhanyun Chemical Co., Ltd. (Shanghai, China). All chemicals were used as received. The School of Biotechnology, International University—Vietnam National University (Ho Chi Minh City, Vietnam) provided *Staphylococcus aureus* strain ATCC 29523 and *Salmonella enterica* ATCC 14028. The Research center for Infectious Disease, International University—Vietnam National University (Ho Chi Minh City, Vietnam) provided *Pseudomonas aeruginosa* strain ATCC 9027.

2.2. Methods

2.2.1. Synthesis of M8/Ag Composite

First, the M8 synthesis was based on the publication [43]. Then, 60 mL of M8 was heated to 80 °C. Specifically, the PVA, PVP, and PEG were separately dissolved in distilled water (DI) at a concentration of 0.02 g/mL. CS was mixed with a 3% acetic acid solution at an amount of 0.01 g/mL. To dissolve polymers completely, stirring and heating for a duration of 40 min was carried out. Next, the polymers using a PVA:PVP:PEG:CS:DI ratio of 1:1:1:1:6 (v/v) were stirred regularly for 1 h to achieve an ideal mixture, referred to as M8. Then, different masses of $AgNO_3$ were added into the polymeric mixture as shown in Table 1.

Table 1. Mass of $AgNO_3$ in M8.

Samples	M8 (mL)	$AgNO_3$ (g)
S1	60	0.05
S2	60	0.075
S3	60	0.1
S4	60	0.15
S5	60	0.2
S6	60	0.25
S7	60	0.3

After adding the silver nitrate, the mixtures were heated and stirred at 80 °C for 6 h. To determine the conversion percentage of Ag^+ to Ag^0, every hour, the aliquots were measured using UV–Vis spectrophotometer (JASCO V-730, Tokyo, Japan) at a scanning speed of 40 nm/min and the wavelength range of 400 nm to 500. Then, the sample with the highest conversion was allowed to react for 10 h to determine the reaction time required to reach equilibrium.

2.2.2. Characterization

The M8Ag and M8 composite were characterized to investigate their chemical and physical properties. Fourier-transform infrared spectroscopy (FTIR, LUMOS, Bruker, Billerica, MA, USA) was utilized to identify the functional groups present in compounds. Subsequently, field emission scanning electron microscopy (FE-SEM, Hitachi SU8000, Tokyo, Japan) was employed to investigate the surface morphology (particle sizes and shapes). Following this, X-ray diffraction spectroscopy (XRD, Bruker D-76187, Karlsruhe, Germany) was employed to analyze the phase composition and crystal structure of the material. Additionally, energy dispersive X-ray spectroscopy (EDX, JEOL JED-2300, Tokyo, Japan) was used to analyze the elemental composition of the composite.

2.2.3. Antibacterial Activity

First, the culture broth and the bacteria were prepared in a manner similar to that described in our previous publication [43]. From each isolate agar plate, a single morphologically comparable colony was chosen and transferred into a glass tube containing 10 mL of Mueller Hinton broth (Himedia, Maharashtra, India) (MHB). The tubes were incubated at 35 °C for 24 h. Subsequently, optical density at a wavelength of 600 nm (OD600) was measured using Hach DR6000 spectrophotometer (Hach, Loveland, CO, USA), and the bacterial suspension was diluted to achieve an OD600 of 0.01 [44]. This process was repeated for three types of bacteria: SA, PA, and SAL. Then, to determine the minimum inhibition concentration (MIC) and the bactericidal effects, the MIC experiment was performed in a manner similar to that described in the previous publication, without any modifications [43]. After obtaining the inhibition percentage, the three best-performing antibiotic concentrations (M8Ag and M8) were chosen and spread evenly onto separate Mueller Hinton agar plates. Subsequently, the plates were incubated at 35 °C for 24 h, and then we identified the presence of bacteria colonies on each dish.

3. Results and Discussion

3.1. Determining the Conversion Percentage of Ag^+ to Ag^0 from $AgNO_3$ and M8

Measuring the absorbance at 455 nm, samples with $AgNO_3$ masses ranging from 0.05 to 0.15 g continued to exhibit ongoing reactions after 6 h, achieving a 100% conversion rate, as shown in Figure 1.

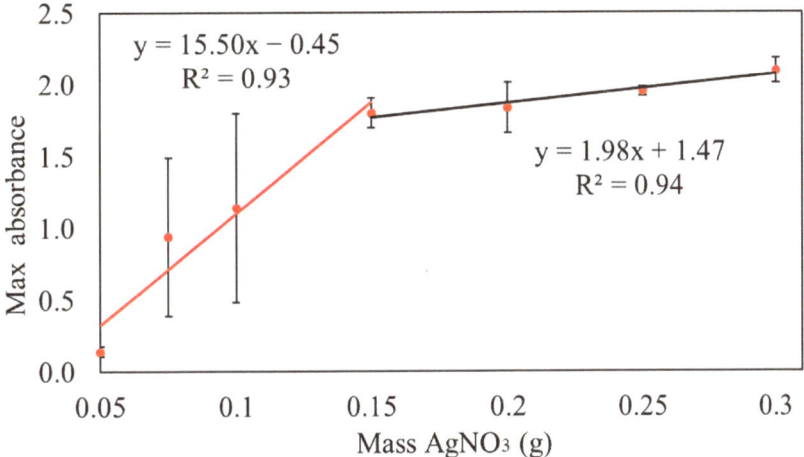

Figure 1. Absorbance of samples after 6 h at wavelength of 455 nm.

The mechanism of converting Ag^+ to Ag^0 in the M8 polymer blend can be explained as the electron was exchanged. M8 availability was limited, and the increases in the concentrations of $AgNO_3$ caused an increase in Ag^+ concentration. Insufficient M8 led to the reduction of Ag^+ to Ag^0 conversion, which accounted for the observed slope variations and the appearance of two different trends in the graph [45]. The conversion percentage of Ag^+ to Ag^0 in sample S1 was excessively low, while samples S2 and S3 exhibited large error bars, rendering the data unreliable. Conversely, samples S5, S6, and S7 were deemed unsuitable for further analysis as they were nearing equilibrium, which produces less Ag^+ than S4. Hence, sample S4 was determined to be the maximum mass of silver nitrate to obtain 100% conversion of Ag^+ to Ag^0. However, to determine the equilibrium time to produce Ag^0, sample S4 reacted for 10 h.

Figure 2 shows the absorbance of the M8/$AgNO_3$ (S4 sample) solution for 10 h, with absorbance measurements taken at hourly intervals.

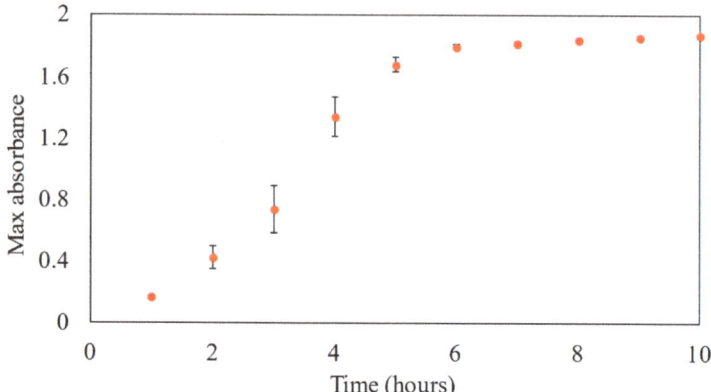

Figure 2. Absorbance of samples for 10 h at wavelength of 455 nm.

As shown in Figure 2, initially, within the first 6 h of the reaction, a rapid increase in absorbance values was observed. This indicated that the reaction between M8 and AgNO$_3$ was proceeding vigorously and with high efficiency during this initial phase. However, after the first 6 h, the rate of growth in absorbance values began to slow down, gradually stabilizing in the subsequent hours. This might suggest the approach of the reaction to the equilibrium state at 10 h. Based on the previous literature, all the components of M8 can reduce Ag$^+$ to Ag^0 [12,30–42]. However, to determine which component can reduce the silver ions the most, the reduction mechanisms, and the synergistic effects between these polymers in the reduction process, further investigation should be conducted in the future.

Afterward, the reflectance percentage of sample S4 from 200 to 700 nm was investigated, as shown in Figure 3.

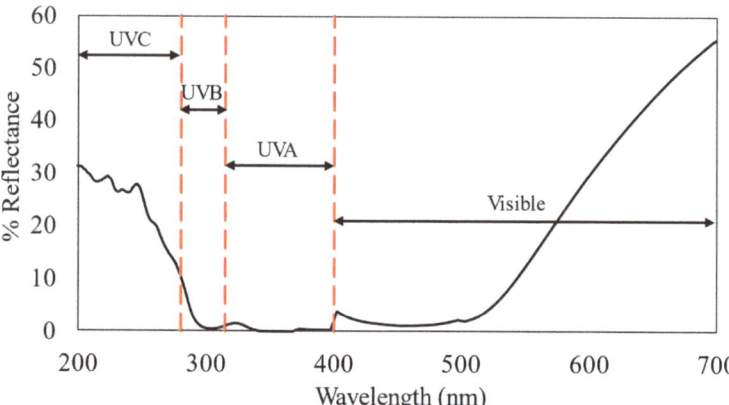

Figure 3. Reflectance percentage (200–700 nm) of M8Ag after reacting for 10 h.

As illustrated in Figure 3, the S3 composite displayed reflectance in the UV-C range (below 280 nm) ranging from 11% to 31%, below the 11% reflectance in the UV-B range (280–315 nm), and less than the 2% reflectance in the UV-A range (315–400 nm). Reflectance in the visible light spectrum ranged from 2% to 65%. These findings suggested that the material demonstrates moderate reflectivity in visible light.

3.2. Characterization

3.2.1. FE-SEM

Using FE-SEM, the morphology of M8 and M8Ag were determined, as shown in Figure 4.

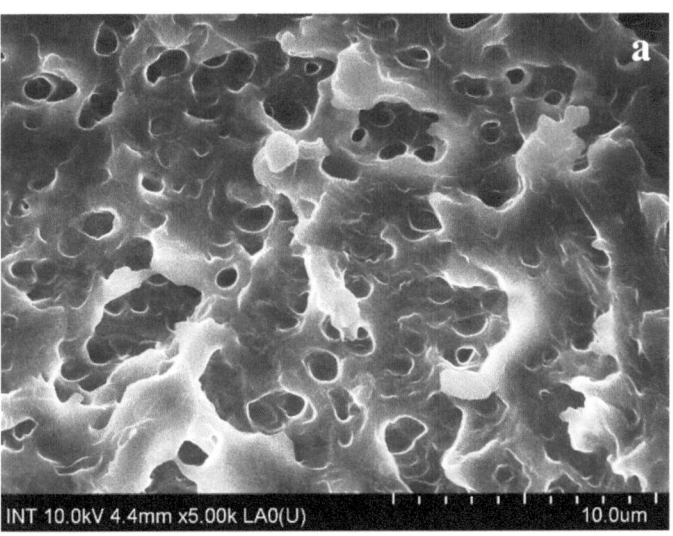

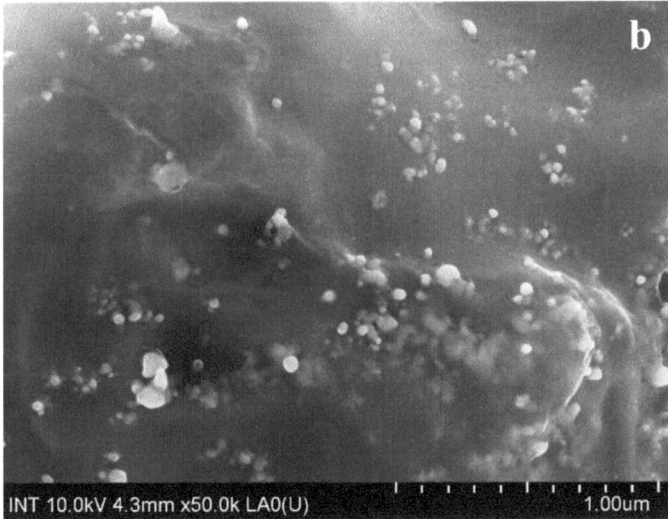

Figure 4. FE-SEM image of (**a**) M8 and (**b**) M8Ag.

As shown in Figure 4a, M8 exhibited pores without any discernible small dots. In contrast, M8Ag (Figure 4b) displayed small round dots, indicating the presence of these particles. Using ImageJ software (version 1.53e) and the FE-SEM images shown in Figure 4b, the size and size distribution of the silver particles were determined, as shown in Figure 5.

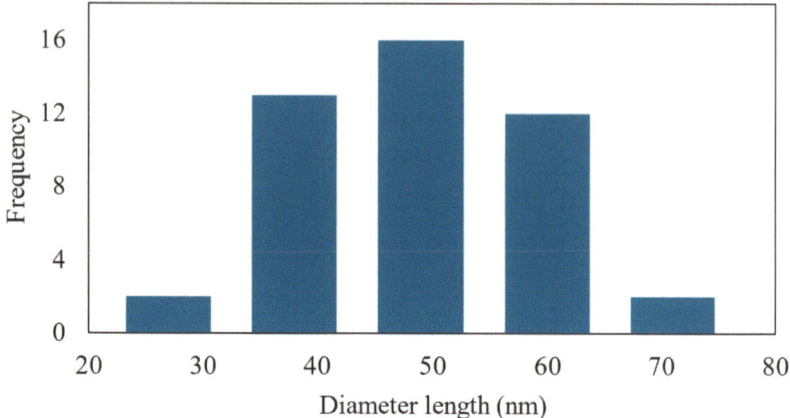

Figure 5. Size distribution of AgNPs in M8Ag composite.

As shown in Figure 5, the size distribution of particles in M8Ag sample was determined to range from 35 nm to 50 nm, with the highest frequency observed in the 40–45 nm size range. Additionally, there was a noticeable trend towards larger sizes, indicating the presence of a small number of particles exceeding 50 nm. Hence, the average size of the AgNPs was 42.48 ± 10.77 nm. However, to confirm whether these nanoparticles were silver necessitated the use of the XRD method.

3.2.2. XRD

XRD was used to analyze the phase composition and crystal structure of M8 and M8Ag. The XRD analysis is shown in Figure 6.

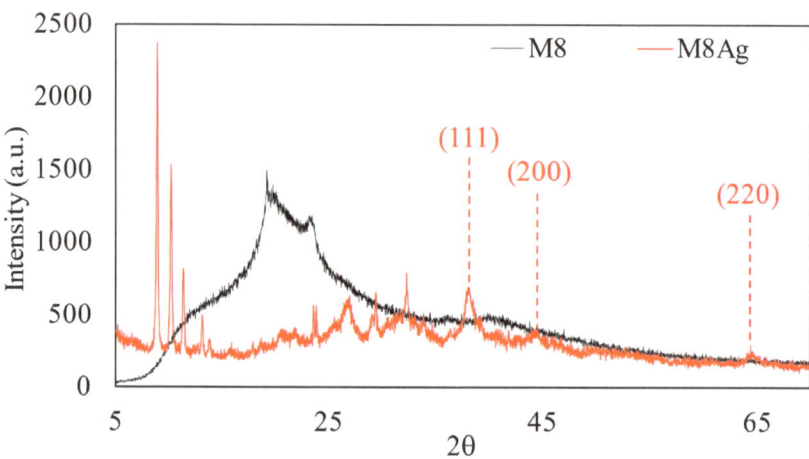

Figure 6. XRD analysis of M8 and M8Ag composite.

As shown in Figure 6, the peaks at 2θ values of 38°, 44°, and 65° correspond to the *hkl* indices of (111), (200), and (220), indicating spherical and crystalline Ag particles [32,46]. Using the Scherrer equation (the shape factor and X-ray wavelength are 0.94 and 1.5406 Å, respectively) [47], the determined average crystallite size of Ag was 44.53 nm, consistent with the results obtained from FE-SEM measurements (42.48 ± 10.77 nm). Hence, using XRD and FE-SEM, silver nanoparticles (AgNPs) were presented in M8–polymer blends consisting of CS, PVA, PEG, and PVP. To be certain, EDX was used once again to identify whether the particles were constituted of Ag and to calculate their percentage inside M8.

3.2.3. EDX

EDX was used to analyze the elemental composition of M8 and M8Ag, as shown in Table 2.

Table 2. Mass percentage in M8 and M8Ag.

Element	M8 (%)	M8Ag (%)
C	55.51 ± 0.22	18.12 ± 0.11
N	4.36 ± 0.27	9.10 ± 0.25
O	39.05 ± 0.44	30.20 ± 0.38
Na	0.23 ± 0.02	-
Si	0.53 ± 0.03	-
Ca	0.10 ± 0.02	-
Fe	0.23 ± 0.05	-
Cl	-	0.15 ± 0.02
K	-	3.20 ± 0.07
Ag	-	39.22 ± 0.34
Total	100	100

In Table 2, it can be seen that Ag provided 39.22 ± 0.34% of the mass in M8Ag. This suggests that the remainder of the matrix was composed of polymers (PEG, PVP, PVA, and CS). Additionally, M8 exhibits a minimal presence of sodium (Na), silicon (Si), calcium (Ca), and iron (Fe), all below 1%, implying the impurities of the raw materials, as well as the synthesis process, which contributed negligibly to the material's composition. Conversely, M8Ag reveals the presence of chlorine (Cl) at 0.15 ± 0.02% and potassium (K) at 3.20 ± 0.07%, elements absent in M8. However, during the AgNPs' synthesis, no chemicals that contain Cl and K were used. Hence, these elements can be considered as contamination during the synthesis. Additionally, the atom fraction of Ag in M8Ag was determined, as shown in Table 3.

Table 3. Atom percentage in M8 and M8Ag.

Element	M8 (%)	M8Ag (%)
C	62.38 ± 0.24	33.57 ± 0.21
N	4.2 ± 0.26	14.44 ± 0.4
O	32.94 ± 0.37	42 ± 0.53
Na	0.13 ± 0.01	-
Si	0.26 ± 0.02	-
Ca	0.03 ± 0.01	-
Fe	0.06 ± 0.01	-
Cl	-	0.1 ± 0.01
K	-	1.79 ± 0.04
Ag	-	8.1 ± 0.07
Total	100	100

As shown in Tables 2 and 3, and combined with the data from FE-SEM and XRD, 39.22 ± 0.34% (mass fraction) and 8.1 ± 0.07% (atomic fraction) of silver nanoparticles (AgNPs) at the size of 42.48 ± 10.77 nm (FE-SEM) or 44.53 nm (XRD) were presented in M8–polymer blends consisting of CS, PVA, PEG, and PVP.

3.2.4. FTIR

FTIR was utilized to identify the functional groups present in compounds (Figure 7).

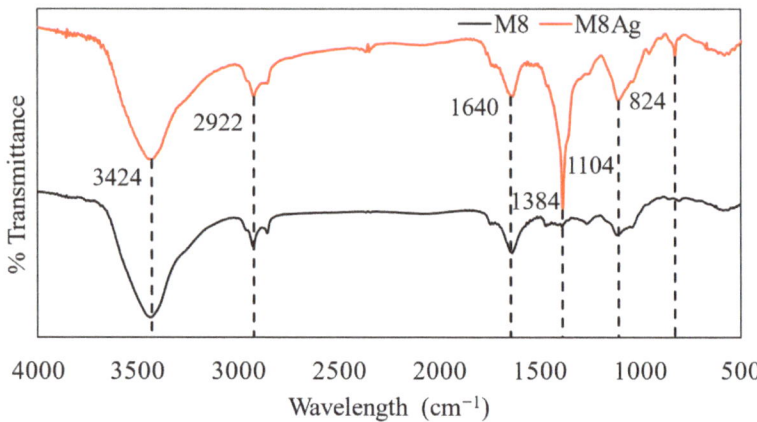

Figure 7. FTIR spectra of M8 and M8Ag.

For M8, the peak at 3424 cm^{-1} may correspond to the stretching vibration of -OH groups with primary -NH groups of CS [43,48]. The peak observed at 1640 cm^{-1} could be attributed to various functional groups, including the C=O stretching vibration of PVP, the bending mode of O-H groups (due to water presence), the C=O stretching vibration of PEG, or the C=O stretching (Amide I) of CS + AA [43,48]. Additionally, the peak at 1104 cm^{-1} might correspond to the stretching vibrations of C-O bonds, the symmetric stretching of C-O-C in PEG, or the shift of the free amino group (-NH$_2$) at the C2 position of glucosamine in CS + AA [43,48].

Since M8Ag was synthesized based on M8, it shares similarities in the functional groups present in compounds. In contrast, the FTIR spectra of the silver nanoparticles in the M8Ag sample revealed prominent peaks at 2922 cm^{-1}, 1640 cm^{-1}, and 1384 cm^{-1}. A sharp and intense absorption band at 1640 cm^{-1} was attributed to the stretching vibration of the (NH) C=O group. As mentioned, another much sharper and more intense peak at 1384 cm^{-1} in the M8Ag compared to M8 indicated the C-C and C-N stretching. Furthermore, the presence of a sharp peak at 2922 cm^{-1} was assigned to the stretching vibration of C-H and C-H (methoxy compounds). Additionally, in the M8Ag sample, a new distinct peak was found at 824 cm^{-1}. Due to the similarities, it is safe to conclude that in the M8Ag sample, M8 still exists, and some changes in the bonds between the polymers in M8 and the silver particles led to some shifts, increases, and changes in the peaks. This indicated the M8Ag was successfully synthesized.

3.3. Antibacterial Activity

Figure 8 presents the minimum inhibitory concentration (MIC) results for three types of bacteria: *SA*, *SAL*, and *PA* from 5% to 50% dilution.

For *SA* bacteria, M8Ag had an MIC$_{50}$ value of 6.25%. Comparing to M8 from previous publication, M8Ag has a smaller MIC$_{50}$ value [43]. However, M8Ag did not display any inhibitory activity at MIC$_{90}$, regardless of the dilution percentage. Comparing to the M8 from previous publication, M8Ag has inferior inhibition ability [43]. In the case of *SAL* bacteria, M8Ag exhibited an MIC$_{50}$ value of 6.25% and an MIC$_{90}$ value of 12.5%, indicating its efficacy against *SAL* bacteria. However, against *SAL*, M8 has the MIC$_{90}$ at 25% dilution. This indicates that M8Ag has lower MIC$_{90}$ than M8. For *PA* bacteria, M8Ag demonstrated MIC$_{50}$ and MIC$_{90}$ values of 6.25%. On the other hand, against *PA*, M8 does not have MIC90, indicating that M8Ag is superior to M8. Hence, overall, M8Ag is a better antibacterial agent than M8. Notably, from Figure 8, Table 4 focuses on MIC$_{50}$ and MIC$_{90}$ results, referring to the minimum concentrations at which 50% and 90% inhibition of bacterial growth were achieved for M8Ag, respectively.

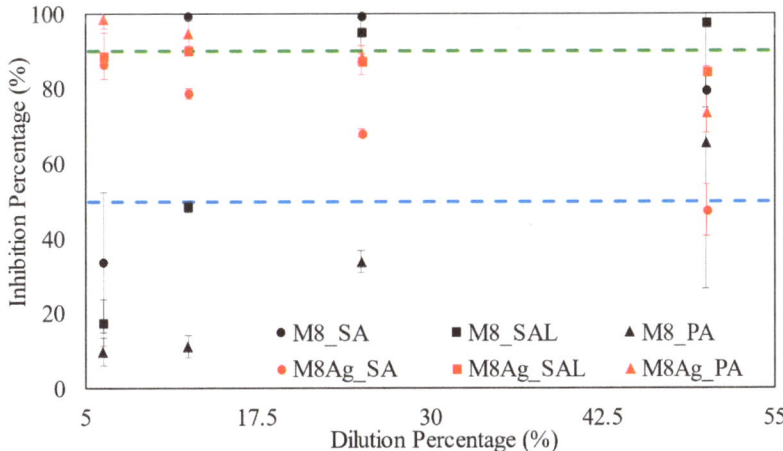

Figure 8. MIC results: the blue line represents MIC$_{50}$, green line represents MIC$_{90}$. Triangle, square, and round dots represent *PA*, *SAL*, and *SA*, respectively. Red color and black color represent M8Ag and M8, respectively.

Table 4. MIC results of M8Ag.

Bacteria	MIC$_{50}$	MIC$_{90}$
SA	6.25%	-
SAL	6.25%	12.5%
PA	-	6.25%

Table 4 clearly indicates that M8Ag exhibited superior antimicrobial activity against *SAL* and *PA* compared to *SA*, due to its lower MIC$_{90}$ values for inhibiting bacterial growth.

Using MIC results, the aliquots at concentrations of 12.5% and 6.25% dilution were spread on agar plates to assess the bactericidal abilities of M8Ag. The findings revealed that neither dilution concentration could effectively kill *SA* and *SE* bacteria, as shown in Table 5.

Table 5. Bactericidal effects of M8Ag (−) and (+) means negative and positive bacteria growth.

Bacteria	12.50%	6.25%
SA	−++	+++
SAL	+++	+++
PA	+++	−−−

As shown in Table 5, 6.25% M8Ag demonstrated the ability to eradicate *PA* bacteria. Based on both MIC and the bactericidal effect results, 6.25% M8Ag exhibited a strong capability to eliminate *PA*, indicating that *PA* was the most susceptible bacteria to M8Ag. For *SA* and *SAL*, M8Ag only inhibited their growth and did not completely eradicate them.

M8 included PVA, PVP, PEG, and CS. CS and PVA were crucial in their ability to fight against a variety of bacteria [33,43]. However, PVA alone had poor stability in water properties; to overcome this limitation, PVA should be mixed with PVP, which covers the surface particles to produce a stable colloid [43]. The combination of PEG and PVA produced strong interactions via hydrogen bond formation, which improved the mixture's overall stability and properties [43]. Meanwhile, the PVA/PEG blend decreased adhesion to bacterial surfaces, which added to the particles' antibacterial abilities [49,50]. Specifically, the hydrophilicity of PVA worked as a regulating element, lowering protein adsorption and cell adhesion [49]. Hence, PVA appeared to operate as a barrier against bacterial adherence on the catheter surface [49]. In addition, PEG worked as a reducing agent, speeding up the reduction of silver ions to silver particles and allowing for the efficient production of silver particles with improved antibacterial activity [40].

On the other hand, silver particles released silver ions continuously, acting as a mechanism of antibacterial activity [51]. These ions adhered to cell walls and membranes due to electrostatic attraction and sulfur protein affinity [52]. This adhesion enhanced membrane permeability, disrupting the bacterial envelope. Inside cells, silver ions deactivated respiratory enzymes, generating reactive oxygen species, which halted adenosine triphosphate (ATP) production and caused deoxyribonucleic acid (DNA) modification [51]. The interaction between silver ions and the sulfur and phosphorus components of DNA can disrupt DNA replication and cell reproduction and potentially lead to microbial termination, given the significance of sulfur and phosphorus in DNA's structure [51].

M8Ag combined the advantages of M8 with Ag particles, considerably increasing its antibacterial efficiency. More specifically, the presence of M8 promoted the conversion of Ag^+ to Ag^0. In contrast, in an acidic environment (3% acetic acid is used to produce chitosan), bacteria may hydrolyze nearby ester segments and create Ag particles, resulting in enhanced Ag production [52]. As a result, the antibacterial efficiency increases in proportion to the number of Ag particles [51].

Gram-negative bacteria have thinner cellular walls than Gram-positive ones, and the strong cellular wall may limit nanoparticle penetration into cells [53]. Hence, M8Ag was more effective against Gram-negative bacteria. However, while both *SAL* and *PA* are Gram-negative bacteria, the antibacterial efficacy of M8Ag against *SAL* is not as strong as against *PA*, which was shown in MIC_{90} and bactericidal effect results. Hence, the antibacterial mechanisms, interaction between the antimicrobial agents and the bacteria, and further investigation for other possible explanations for the aforementioned phenomenon should be the subject of future studies and research. Additionally, researchers should consider studying the cell cytotoxicity and toxicity on mice to provide much deeper understanding of M8Ag.

4. Conclusions

The combination of M8 (including PVA, PVP, PEG, and CS) and silver nitrate created a novel composite (M8Ag) that was environmentally friendly, possessed antibacterial properties, and addressed the issue of antibiotic resistance. M8Ag, with an $AgNO_3$:M8 ratio of 0.15 g/60 mL, achieved 100% conversion of Ag^+ to Ag^0 after reacting at 80 °C for 10 h. However, the reduction mechanism, as well as determining the main reducing agent(s) in M8, should be investigated in the future.

Silver nanoparticles were successfully synthesized using M8 polymer blends, with a mass fraction of $39.22 \pm 0.34\%$ and an atomic fraction of 8.1 ± 0.07. The nanoparticles had a size of 42.48 ± 10.77 nm according to FE-SEM analysis and 44.53 nm based on XRD measurements.

The antibacterial efficacy of M8Ag at diluted concentrations varied across bacterial strains. Specifically, for *SA*, a concentration of 6.25% was effective at MIC_{50}. *SAL* exhibited susceptibility at both MIC_{50} and MIC_{90}, with concentrations of 6.25% and 12.5%, respectively. *PA*, on the other hand, demonstrated sensitivity at MIC_{90}, with a concentration of 6.25%.

In conclusion, at 6.25% M8Ag, this dilute concentration served as the MIC_{90} for *PA*, while for the other two strains, it approached the MIC_{50} threshold. This suggested that M8Ag exhibited the strongest antibacterial activity against *PA*. To elucidate further, when spread on agar dishes, 6.25% M8Ag can eliminate *PA* but not *SA* and *SAL*. To elucidate the principle and assess the antibacterial potential of M8Ag against both Gram-negative and Gram-positive bacteria, further investigation into other potential explanations for this observed phenomenon could be pursued as part of future studies. Additionally, the cells' cytotoxicity and toxicity on mice should be investigated in the future.

Author Contributions: Performing experiment, data analysis, writing—review and editing, K.T.; conceptualization, methodology, data analysis, writing—original draft preparation, writing—review and editing, L.D.; performing experiments, data analysis: Q.N.L. and A.H.H. All authors have read and agreed to the published version of the manuscript.

Funding: No internal and external funding received.

Institutional Review Board Statement: Not applicable.

Data Availability Statement: All data that support the findings of this study are included within the article.

Conflicts of Interest: The authors declare no conflicts of interests.

References

1. Solar Venero, E.C.; Galeano, M.B.; Luqman, A.; Ricardi, M.M.; Serral, F.; Fernandez Do Porto, D.; Robaldi, S.A.; Ashari, B.A.Z.; Munif, T.H.; Egoburo, D.E.; et al. Fever-like Temperature Impacts on *Staphylococcus aureus* and *Pseudomonas aeruginosa* Interaction, Physiology, and Virulence Both in Vitro and in Vivo. *BMC Biol.* **2024**, *22*, 27. [CrossRef] [PubMed]
2. Alexander, E.H.; Bento, J.L.; Hughes, F.M.; Marriott, I.; Hudson, M.C.; Bost, K.L. *Staphylococcus aureus* and *Salmonella enterica* Serovar Dublin Induce Tumor Necrosis Factor-Related Apoptosis-Inducing Ligand Expression by Normal Mouse and Human Osteoblasts. *Infect. Immun.* **2001**, *69*, 1581–1586. [CrossRef]
3. DeLeon, S.; Clinton, A.; Fowler, H.; Everett, J.; Horswill, A.R.; Rumbaugh, K.P. Synergistic Interactions of *Pseudomonas aeruginosa* and *Staphylococcus aureus* in an In Vitro Wound Model. *Infect. Immun.* **2014**, *82*, 4718–4728. [CrossRef]
4. Jfoster, T. Staphylococcus aureus. In *Molecular Medical Microbiology*; Elsevier: Amsterdam, The Netherlands, 2002; Volume 2, pp. 839–888, ISBN 978-0-12-677530-3.
5. Lowy, F.D. *Staphylococcus aureus* Infections. *N. Engl. J. Med.* **1998**, *339*, 520–532. [CrossRef] [PubMed]
6. Musher, D.M.; Lamm, N.; Darouiche, R.O.; Young, E.J.; Hamill, R.J.; Landon, G.C. The Current Spectrum of *Staphylococcus aureus* Infection in a Tertiary Care Hospital. *Medicine* **1994**, *73*, 186–208. [CrossRef] [PubMed]
7. Taylor, T.A.; Unakal, C.G. Staphylococcus aureus Infection. In *StatPearls*; StatPearls Publishing: Treasure Island, FL, USA, 2023.
8. Diggle, S.P.; Whiteley, M. Microbe Profile: *Pseudomonas aeruginosa*: Opportunistic Pathogen and Lab Rat. *Microbiology* **2020**, *166*, 30–33. [CrossRef]
9. Percival, S.L.; Williams, D.W. Chapter Ten—Salmonella. In *Microbiology of Waterborne Diseases*, 2nd ed.; Percival, S.L., Yates, M.V., Williams, D.W., Chalmers, R.M., Gray, N.F., Eds.; Academic Press: London, UK, 2014; pp. 209–222, ISBN 978-0-12-415846-7.
10. Ethelberg, S.; Mølbak, K.; Josefsen, M.H. Bacteria: *Salmonella* Non-Typhi. In *Encyclopedia of Food Safety*; Motarjemi, Y., Ed.; Academic Press: Waltham, MA, USA, 2014; pp. 501–514, ISBN 978-0-12-378613-5.
11. Dawoud, T.M.; Shi, Z.; Kwon, Y.M.; Ricke, S.C. Overview of Salmonellosis and Food-Borne *Salmonella*. In *Producing Safe Eggs*; Elsevier: Amsterdam, The Netherlands, 2017; pp. 113–138, ISBN 978-0-12-802582-6.
12. Mahmoud, K.H. Synthesis, Characterization, Optical and Antimicrobial Studies of Polyvinyl Alcohol–Silver Nanocomposites. *Spectrochim. Acta Part A Mol. Biomol. Spectrosc.* **2015**, *138*, 434–440. [CrossRef]
13. Jin, T.; Sun, D.; Su, J.Y.; Zhang, H.; Sue, H.-J. Antimicrobial Efficacy of Zinc Oxide Quantum Dots against *Listeria monocytogenes*, *Salmonella* Enteritidis, and *Escherichia coli* O157:H7. *J. Food Sci.* **2009**, *74*, M46–M52. [CrossRef]
14. Prestinaci, F.; Pezzotti, P.; Pantosti, A. Antimicrobial Resistance: A Global Multifaceted Phenomenon. *Pathog. Glob. Health* **2015**, *109*, 309–318. [CrossRef]
15. Nanda, A.; Saravanan, M. Biosynthesis of Silver Nanoparticles from *Staphylococcus aureus* and Its Antimicrobial Activity against MRSA and MRSE. *Nanomed. Nanotechnol. Biol. Med.* **2009**, *5*, 452–456. [CrossRef]
16. Li, W.-R.; Xie, X.-B.; Shi, Q.-S.; Duan, S.-S.; Ouyang, Y.-S.; Chen, Y.-B. Antibacterial Effect of Silver Nanoparticles on *Staphylococcus aureus*. *Biometals* **2011**, *24*, 135–141. [CrossRef] [PubMed]
17. Mirzajani, F.; Ghassempour, A.; Aliahmadi, A.; Esmaeili, M.A. Antibacterial Effect of Silver Nanoparticles on *Staphylococcus aureus*. *Res. Microbiol.* **2011**, *162*, 542–549. [CrossRef] [PubMed]
18. Shahverdi, A.R.; Fakhimi, A.; Shahverdi, H.R.; Minaian, S. Synthesis and Effect of Silver Nanoparticles on the Antibacterial Activity of Different Antibiotics against *Staphylococcus aureus* and *Escherichia coli*. *Nanomed. Nanotechnol. Biol. Med.* **2007**, *3*, 168–171. [CrossRef] [PubMed]
19. Arakha, M.; Pal, S.; Samantarrai, D.; Panigrahi, T.K.; Mallick, B.C.; Pramanik, K.; Mallick, B.; Jha, S. Antimicrobial Activity of Iron Oxide Nanoparticle upon Modulation of Nanoparticle-Bacteria Interface. *Sci. Rep.* **2015**, *5*, 14813. [CrossRef] [PubMed]
20. Borcherding, J.; Baltrusaitis, J.; Chen, H.; Stebounova, L.; Wu, C.-M.; Rubasinghege, G.; Mudunkotuwa, I.A.; Carlos Caraballo, J.; Zabner, J.H.; Grassian, V.; et al. Iron Oxide Nanoparticles Induce *Pseudomonas aeruginosa* Growth, Induce Biofilm Formation, and Inhibit Antimicrobial Peptide Function. *Environ. Sci. Nano* **2014**, *1*, 123–132. [CrossRef] [PubMed]
21. Zúñiga-Miranda, J.; Guerra, J.; Mueller, A.; Mayorga-Ramos, A.; Carrera-Pacheco, S.E.; Barba-Ostria, C.; Heredia-Moya, J.; Guamán, L.P. Iron Oxide Nanoparticles: Green Synthesis and Their Antimicrobial Activity. *Nanomaterials* **2023**, *13*, 2919. [CrossRef] [PubMed]

22. Ramola, B.; Joshi, N.C. Green Synthesis, Characterisations and Antimicrobial Activities of CaO Nanoparticles. *Orient. J. Chem.* **2019**, *35*, 1154–1157. [CrossRef]
23. Azam, A.; Ahmed, A.S.; Oves, M.; Khan, M.S.; Habib, S.S.; Memic, A. Antimicrobial Activity of Metal Oxide Nanoparticles against Gram-Positive and Gram-Negative Bacteria: A Comparative Study. *Int. J. Nanomed.* **2012**, *7*, 6003–6009. [CrossRef] [PubMed]
24. Duffy, L.L.; Osmond-McLeod, M.J.; Judy, J.; King, T. Investigation into the Antibacterial Activity of Silver, Zinc Oxide and Copper Oxide Nanoparticles against Poultry-Relevant Isolates of *Salmonella* and *Campylobacter*. *Food Control* **2018**, *92*, 293–300. [CrossRef]
25. Álvarez-Paino, M.; Muñoz-Bonilla, A.; Fernández-García, M. Antimicrobial Polymers in the Nano-World. *Nanomaterials* **2017**, *7*, 48. [CrossRef]
26. Zheng, K.; Setyawati, M.I.; Leong, D.T.; Xie, J. Antimicrobial Silver Nanomaterials. *Coord. Chem. Rev.* **2018**, *357*, 1–17. [CrossRef]
27. Mandapalli, P.K.; Labala, S.; Chawla, S.; Janupally, R.; Sriram, D.; Venuganti, V.V.K. Polymer–Gold Nanoparticle Composite Films for Topical Application: Evaluation of Physical Properties and Antibacterial Activity. *Polym. Compos.* **2017**, *38*, 2829–2840. [CrossRef]
28. Zhang, X.-F.; Liu, Z.-G.; Shen, W.; Gurunathan, S. Silver Nanoparticles: Synthesis, Characterization, Properties, Applications, and Therapeutic Approaches. *Int. J. Mol. Sci.* **2016**, *17*, 1534. [CrossRef]
29. Nguyen, N.P.U.; Dang, N.T.; Doan, L.; Nguyen, T.T.H. Synthesis of Silver Nanoparticles: From Conventional to 'Modern' Methods—A Review. *Processes* **2023**, *11*, 2617. [CrossRef]
30. Abdelghany, A.; Abelaziz, M.; Hezma, A.; Elashmawi, I. Spectroscopic and Antibacterial Speculation of Silver Nanoparticles Modified Chitosan/Polyvinyle Alcohol Polymer Blend. *Results Phys.* **2016**, in press. [CrossRef]
31. Wang, L.; Periyasami, G.; Aldalbahi, A.; Fogliano, V. The Antimicrobial Activity of Silver Nanoparticles Biocomposite Films Depends on the Silver Ions Release Behaviour. *Food Chem.* **2021**, *359*, 129859. [CrossRef] [PubMed]
32. Kalaivani, R.; Maruthupandy, M.; Muneeswaran, T.; Hameedha Beevi, A.; Anand, M.; Ramakritinan, C.M.; Kumaraguru, A.K. Synthesis of Chitosan Mediated Silver Nanoparticles (Ag NPs) for Potential Antimicrobial Applications. *Front. Lab. Med.* **2018**, *2*, 30–35. [CrossRef]
33. Abdallah, O.M.; EL-Baghdady, K.Z.; Khalil, M.M.H.; El Borhamy, M.I.; Meligi, G.A. Antibacterial, Antibiofilm and Cytotoxic Activities of Biogenic Polyvinyl Alcohol-Silver and Chitosan-Silver Nanocomposites. *J. Polym. Res.* **2020**, *27*, 74. [CrossRef]
34. Tripathi, S.; Mehrotra, G.K.; Dutta, P.K. Chitosan–Silver Oxide Nanocomposite Film: Preparation and Antimicrobial Activity. *Bull. Mater. Sci.* **2011**, *34*, 29–35. [CrossRef]
35. Wang, B.-L.; Liu, X.-S.; Ji, Y.; Ren, K.-F.; Ji, J. Fast and Long-Acting Antibacterial Properties of Chitosan-Ag/Polyvinylpyrrolidone Nanocomposite Films. *Carbohydr. Polym.* **2012**, *90*, 8–15. [CrossRef]
36. Khan, B.; Nawaz, M.; Hussain, R.; Price, G.J.; Warsi, M.F.; Waseem, M. Enhanced Antibacterial Activity of Size-Controlled Silver and Polyethylene Glycol Functionalized Silver Nanoparticles. *Chem. Pap.* **2021**, *75*, 743–752. [CrossRef]
37. Fahmy, A.; El-Zomrawy, A.; Saeed, A.M.; Sayed, A.Z.; El-Arab, M.A.E.; Shehata, H.A.; Friedrich, J. One-Step Synthesis of Silver Nanoparticles Embedded with Polyethylene Glycol as Thin Films. *J. Adhes. Sci. Technol.* **2017**, *31*, 1422–1440. [CrossRef]
38. El Hotaby, W.; Sherif, H.H.A.; Hemdan, B.A.; Khalil, W.A.; Khalil, S.K.H. Assessment of in Situ-Prepared Polyvinylpyrrolidone-Silver Nanocomposite for Antimicrobial Applications. *Acta Phys. Pol. A* **2017**, *131*, 1554–1560. [CrossRef]
39. Krishna Rao, K.S.V.; Ramasubba Reddy, P.; Lee, Y.-I.; Kim, C. Synthesis and Characterization of Chitosan–PEG–Ag Nanocomposites for Antimicrobial Application. *Carbohydr. Polym.* **2012**, *87*, 920–925. [CrossRef] [PubMed]
40. Mishra, S.K.; Raveendran, S.; Ferreira, J.M.F.; Kannan, S. In Situ Impregnation of Silver Nanoclusters in Microporous Chitosan-PEG Membranes as an Antibacterial and Drug Delivery Percutaneous Device. *Langmuir* **2016**, *32*, 10305–10316. [CrossRef] [PubMed]
41. Abdelgawad, A.M.; Hudson, S.M.; Rojas, O.J. Antimicrobial Wound Dressing Nanofiber Mats from Multicomponent (Chitosan/Silver-NPs/Polyvinyl Alcohol) Systems. *Carbohydr. Polym.* **2014**, *100*, 166–178. [CrossRef] [PubMed]
42. Pencheva, D.; Bryaskova, R.; Kantardjiev, T. Polyvinyl Alcohol/Silver Nanoparticles (PVA/AgNps) as a Model for Testing the Biological Activity of Hybrid Materials with Included Silver Nanoparticles. *Mater. Sci. Eng. C* **2012**, *32*, 2048–2051. [CrossRef] [PubMed]
43. Doan, L.; Tran, K. Relationship between the Polymer Blend Using Chitosan, Polyethylene Glycol, Polyvinyl Alcohol, Polyvinylpyrrolidone, and Antimicrobial Activities against *Staphylococcus aureus*. *Pharmaceutics* **2023**, *15*, 2453. [CrossRef] [PubMed]
44. Wiegand, I.; Hilpert, K.; Hancock, R.E.W. Agar and Broth Dilution Methods to Determine the Minimal Inhibitory Concentration (MIC) of Antimicrobial Substances. *Nat. Protoc.* **2008**, *3*, 163–175. [CrossRef]
45. El-Rafie, M.H.; Shaheen, T.I.; Mohamed, A.A.; Hebeish, A. Bio-Synthesis and Applications of Silver Nanoparticles onto Cotton Fabrics. *Carbohydr. Polym.* **2012**, *90*, 915–920. [CrossRef]
46. Alginate-Mediated Synthesis of Hetero-Shaped Silver Nanoparticles and Their Hydrogen Peroxide Sensing Ability—PubMed. Available online: https://pubmed.ncbi.nlm.nih.gov/31972997/ (accessed on 15 April 2024).
47. Monshi, A.; Foroughi, M.R.; Monshi, M.R. Modified Scherrer Equation to Estimate More Accurately Nano-Crystallite Size Using XRD. *WJNSE* **2012**, *2*, 154–160. [CrossRef]
48. El Farissi, H.; Lakhmiri, R.; Albourine, A.; Safi, M.; Cherkaoui, O. Removal of RR-23 Dye from Industrial Textile Wastewater by Adsorption on Cistus Ladaniferus Seeds and Their Biochar. *J. Environ. Earth Sci.* **2017**, *7*.

49. Rafienia, M.; Zarinmehr, B.; Poursamar, S.A.; Bonakdar, S.; Ghavami, M.; Janmaleki, M. Coated Urinary Catheter by PEG/PVA/Gentamicin with Drug Delivery Capability against Hospital Infection. *Iran Polym. J.* **2013**, *22*, 75–83. [CrossRef]
50. Barrias, C.C.; Martins, M.C.L.; Almeida-Porada, G.; Barbosa, M.A.; Granja, P.L. The Correlation between the Adsorption of Adhesive Proteins and Cell Behaviour on Hydroxyl-Methyl Mixed Self-Assembled Monolayers. *Biomaterials* **2009**, *30*, 307–316. [CrossRef] [PubMed]
51. Yin, I.X.; Zhang, J.; Zhao, I.S.; Mei, M.L.; Li, Q.; Chu, C.H. The Antibacterial Mechanism of Silver Nanoparticles and Its Application in Dentistry. *Int. J. Nanomed.* **2020**, *15*, 2555–2562. [CrossRef]
52. Luo, L.; Huang, W.; Zhang, J.; Yu, Y.; Sun, T. Metal-Based Nanoparticles as Antimicrobial Agents: A Review. *ACS Appl. Nano Mater.* **2024**, *7*, 2529–2545. [CrossRef]
53. Meikle, T.G.; Dyett, B.P.; Strachan, J.B.; White, J.; Drummond, C.J.; Conn, C.E. Preparation, Characterization, and Antimicrobial Activity of Cubosome Encapsulated Metal Nanocrystals. *ACS Appl. Mater. Interfaces* **2020**, *12*, 6944–6954. [CrossRef]

Disclaimer/Publisher's Note: The statements, opinions and data contained in all publications are solely those of the individual author(s) and contributor(s) and not of MDPI and/or the editor(s). MDPI and/or the editor(s) disclaim responsibility for any injury to people or property resulting from any ideas, methods, instructions or products referred to in the content.

Article

Effect of Chitosan–Riboflavin Bioconjugate on Green Mold Caused by *Penicillium digitatum* in Lemon Fruit

Brenda M. Ipinza-Concha [1], Luciano Dibona-Villanueva [2], Denis Fuentealba [2], Alexander Pinilla-Quispe [2], Daniel Schwantes [1,2,3], María A. Garzón-Nivia [1], Mario A. Herrera-Défaz [1] and Héctor A. Valdés-Gómez [1,*]

1. Facultad de Agronomía y Sistemas Naturales, Pontificia Universidad Católica de Chile, Santiago 7820436, Chile; bmipinza@uc.cl (B.M.I.-C.)
2. Facultad de Química y de Farmacia, Pontificia Universidad Católica de Chile, Santiago 7820436, Chile
3. Facultad de Medicina, Pontificia Universidad Católica de Chile, Santiago 7820436, Chile
* Correspondence: hevaldes@uc.cl

Abstract: *Penicillium digitatum* is the causal agent of green mold, a primary postharvest disease of citrus fruits. This study evaluated the efficacy of a novel photoactive chitosan–riboflavin bioconjugate (CH-RF) to control green mold in vitro and in lemon fruit. The results showed total inhibition of *P. digitatum* growth on APDA supplemented with CH-RF at 0.5% (*w*/*v*) and a significant reduction of 84.8% at 0.25% (*w*/*v*). Lemons treated with CH-RF and kept under controlled conditions (20 °C and 90–95% relative humidity) exhibited a noteworthy reduction in green mold incidence four days post-inoculation. Notably, these effects persisted, with all treatments remaining significantly distinct from the control group until day 14. Furthermore, CH-RF showed high control of green mold in lemons after 20 days of cold storage (5 ± 1 °C). The disease incidence five days after cold storage indicated significant differences from the values observed in the control. Most CH-RF treatments showed enhanced control of green mold when riboflavin was activated by white-light exposure. These findings suggest that this novel fungicide could be a viable alternative to conventional synthetic fungicides, allowing more sustainable management of lemon fruit diseases.

Keywords: antifungal; *Penicillium digitatum*; biofungicide; chitosan; riboflavin; photoactive; citrus

Citation: Ipinza-Concha, B.M.; Dibona-Villanueva, L.; Fuentealba, D.; Pinilla-Quispe, A.; Schwantes, D.; Garzón-Nivia, M.A.; Herrera-Défaz, M.A.; Valdés-Gómez, H.A. Effect of Chitosan–Riboflavin Bioconjugate on Green Mold Caused by *Penicillium digitatum* in Lemon Fruit. *Polymers* **2024**, *16*, 884. https://doi.org/10.3390/polym16070884

Academic Editors: William Facchinatto and Sérgio Paulo Campana-Filho

Received: 13 February 2024
Revised: 11 March 2024
Accepted: 14 March 2024
Published: 23 March 2024

Copyright: © 2024 by the authors. Licensee MDPI, Basel, Switzerland. This article is an open access article distributed under the terms and conditions of the Creative Commons Attribution (CC BY) license (https://creativecommons.org/licenses/by/4.0/).

1. Introduction

Penicillium digitatum is a phytopathogenic fungus of the Ascomycota phylum that causes green mold in citrus fruits. Along with *Penicillium italicum* (blue mold), these are the primary sources of decay loss in citrus production worldwide [1]. *P. digitatum* is a pathogen that infects fruit through wounds that can occur accidentally in the field during handling, due to insects or thorns, or in transportation [2]. This pathogen is primarily controlled by synthetic fungicides used before cold storage [3], which are applied as a wax coating, dip, or drench [4]. Despite their extensive use, conventional fungicides are under critical scrutiny due to their adverse environmental effects and potential risks to human health. These effects include the contamination of surface water sources, leading to toxicity to different species, such as invertebrates, vertebrates, and microorganisms [5]. In terms of human health, there is evidence of toxic effects on human cells [6] and neurological disruption in neonates [7]. Furthermore, the extensive and frequent use of fungicides has led to the emergence of resistant strains of many *Penicillium* species [8]. These factors have promoted research into alternative biodegradable and less toxic fungicides.

Chitosan is the deacetylated form of chitin, one of the most abundant polymers in nature. It is a constituent of the exoskeleton of crustaceans and insects, but it can also be found as a component of fungal cells [9]. Chitosan has been extensively studied due to its antifungal and resistance-inducing properties. In this context, assays have been conducted under preharvest and postharvest conditions to control fungal decay in fruits [10]. Chitosan alone, associated or in combination with other compounds, has been tested on postharvest fungi such as *Botrytis cinerea* [11,12], *Colletotrichum gloeosporioides* [13], and *Penicillium italicum* [14]. Chitosan has also been tested to control *P. digitatum* decay in several citrus species, with variable results. In grapefruit, a reduction in severity was observed, but no reduction in incidence six days after inoculation [15]. In assays in mandarins, a reduction in incidence and severity was observed seven days after inoculation [16]. Similar results were observed on cold storage [17], while other authors report no reduction in incidence four days after inoculation [18].

Photodynamic inactivation (PDI) is a promising alternative method that has been studied due to its capacity to treat human diseases. This technique involves the use of a photosensitizer at a specific wavelength of light in the presence of oxygen to generate reactive oxygen compounds that trigger cell damage and death [19]. PDI has recently been tested for the inactivation of human pathogenic fungal species such as *Trichophyton rubrum*, *Candida albicans*, and *Tinea* [20]. Studies on the control of phytopathogenic fungi are very limited. Some of the photosensitizers studied in plant disease control include curcumin for *B. cinerea* [21] and *P. expansum* [22], harmol for *P. digitatum* and *B. cinerea* [23], and vitamin K3 in *P. digitatum* [24].

In this context, a cutting-edge photoactive fungicide crafted from the combination of chitosan and riboflavin (vitamin B2) has been developed [25]. The fungicidal effect of this molecule (CH-RF) is driven by exposure to white light. This molecule has been shown to significantly inhibit the growth of *B. cinerea* in a culture medium and significantly reduce its incidence in inoculated grapes in a concentration-dependent manner [12]. Preliminary results have shown that CH-RF inhibits the mycelial growth of *P. digitatum* at low doses under in vitro conditions [25,26]. Nevertheless, its efficacy in preventing *P. digitatum* infection in citrus fruits and its performance under postharvest conditions remain unexplored.

Thus, this study aims to evaluate the efficacy of the photoactive molecule composed of chitosan–riboflavin (CH-RF) in controlling the damage caused by *Penicillium digitatum*. Efficacy was assessed under three conditions: in vitro in Petri dishes in laboratory biotests, in lemon fruits under a high-susceptibility environment (20 °C; 90–95% RH), and in microscale cold storage followed by shelf-life postharvest conditions.

2. Materials and Methods

2.1. Plant Material

Commercially mature lemons (*Citrus limon* L. Burm. f. cv. Eureka frost) without recent fungicide applications were harvested from an orchard in the Melipilla area, Chile. The absence of recent fungicide applications ensures the conditions for green mold growth. Fruits were selected based on uniformity, ripeness, and absence of mechanical damage.

2.2. Fungal Isolates

The *Penicillium digitatum* strain used in the experiments was obtained from the Fruit Pathology Laboratory of the Agricultural and Natural Systems Faculty of the Pontificia Universidad Católica de Chile. This pathogen was obtained from diseased fruits and maintained in a 20% glycerol solution at -20 °C until use. Conidial suspension for the different experiments was obtained from 7-day-old *P. digitatum* growth on acidified potato dextrose agar (APDA) at 20 ± 1 °C. A 1×1 cm portion of the recent sporulated fungal growth was scraped from the culture medium and mixed with 0.05% (v/v) Tween 80 detergent to prepare the pathogen suspension. This solution was poured into a glass beaker through a sterilized double-gauze layer to remove the mycelia. An aliquot of 5 µL of the suspension was extracted and placed on a hemocytometer (BOECO, Hamburg, Germany) to count the conidia under a light microscope (Olympus CX31, Tokyo, Japan). Concentration was adjusted using Tween 80® solution. For in vitro assays, the suspension was adjusted to 1×10^5 conidia mL^{-1}; for the in vivo assays, the concentration was 2×10^5 conidia mL^{-1} [13,17].

2.3. CH-RF Biofungicide

CH-RF biofungicide was provided by the Supramolecular Chemistry and Photobiology Laboratory of the Chemistry and Pharmacy Faculty of the Pontificia Universidad Católica de Chile. The protocol developed by Dibona-Villanueva and Fuentealba (2021) was implemented to elaborate the conjugate. Briefly, 2 mg of N-hydroxysuccinimide (NHS), 1 mL of thioglycolic acid, and 3.5 mg of 1-ethyl-3-(3-dimethylaminopropyl) carbodiimide (EDC) were dissolved in 2 mL of dimethylformamide (DMF) and stirred overnight at room temperature. Simultaneously, a 2.5% (w/v) chitosan solution (90% degree of deacetylation, ≤3.0 kDa, Chitolytic, Toronto, ON, Canada) was prepared in HCl 0.2 M and adjusted to pH 5 by adding 10 M NaOH. The activated NHS-ester obtained in the previous step was added to the chitosan solution and stirred for 24 h at room temperature. The thiolated chitosan (CH-SH) was recovered through dialysis separation, lyophilized, and stored in a cold and dark environment. A total of 500 mg of CH-SH was dissolved in 20 mL of water and adjusted to pH 6. Simultaneously, 19 mg of RF-PMPI previously synthetized using the protocols depicted by Dibona-Villanueva and Fuentealba (2021) was dissolved in a small amount of DMSO and added to the thiolated chitosan solution under constant stirring. The CH-SH-and-RF-PMPI mixture was stirred at room temperature under a N_2 atmosphere and protected from light overnight. The CH-RF conjugate was recovered by ethanol precipitation and washed several times with ethanol and cold water until no color was observed in the filtrate. The product was then dried and stored at 4 °C while protected from light. The conjugate was characterized by the FTIR and MS techniques detailed in the previous report. The final structure of the conjugate as well as the percentual components are depicted in Figure 1.

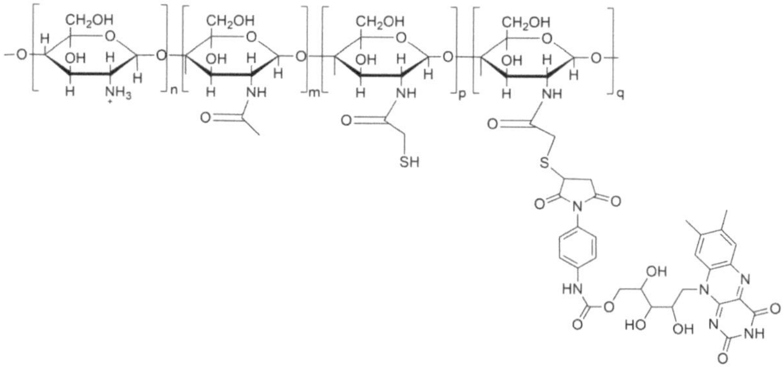

Figure 1. CH-RF chemical structure. The subscripts n, m, p and q depict the recurrent unit composition of free amino, acetylated, thiolated and RF containing glucosamine groups, respectively. Substitution degree n = 70%, m = 14%, p = 15% and q = 1%.

2.4. In Vitro Tests

CH-RF's efficacy in controling *P. digitatum* growth was assessed by a five-treatment in vitro assay. Each treatment was mixed with APDA containing 3.2 g potato puree, 3.2 g dextrose, 4 g agar–agar, and 80 µL lactic acid for a 160 mL solution. APDA was diluted in 160 mL of distilled water for control treatment and autoclaved at 121 °C for 15 min. For fludioxonil treatment (FLU 0.1), 160 µL of commercial fungicide Shield Brite FDL 230SC (Anasac Chile S.A, Lampa, Chile) was mixed with the 160 mL autoclaved APDA to reach a concentration of 0.1% (v/v) (0.023% w/v fludioxonil). APDA was mixed with 80 mL of a 1% (w/v) stock solution to obtain a concentration of 0.5% (v/v) (CH-RF0.5). The same procedure was used to prepare CH-RF0.25 and CH-RF0.1. pH of CH-RF treatments ranged between 4.3 and 4.5. A total of 12 mL of the different culture media was poured into Petri dishes and left to solidify. An amount of 5 µL of conidial suspension of *P. digitatum* was placed at the center of each Petri dish. Each treatment consisted of five replicates of three plates. The treatments were maintained at 20 ± 1 °C under a light/dark cycle of 12:12 h.

2.5. Biotests in Lemon Fruit in Temperate and Cold Storage

Fruits were superficially disinfected with 2% v/v sodium hypochlorite solution for 2–3 min and washed twice with sterilized distilled water. Subsequently, lemons were air-dried in a laminar flow cabinet for 1–2 h. Lemons were wounded at the equatorial region using a scalpel, creating 2 × 2 mm lesions in depth and length. An aliquot of 10 μL of a *P. digitatum* conidial suspension (2×10^5 conidia mL^{-1}) was placed in the wound and left to dry for 5 h. Treatments were performed by dipping lemons in sterile water (control), Fludioxonil (FLU) 0.2% (v/v) (Shield Brite FDL 230SC, Anasac Chile S.A, Lampa, Chile), and CH-RF 2% (w/v) exposed for 0 min (CH-RF), 6 min (CH-RF6), 15 min (CH-RF15), and 30 min (Ch-RF30) to white LED with irradiance of 17–23 W/m^2, and continuously rotated to illuminate the whole lemon surface. The control and FLU treatment groups were not exposed to light. An illumination step was required to activate riboflavin in the conjugate. Fruits were left to dry completely in a laminar flow cabinet for 1–2 h.

For temperate- and humid-condition assays, individual lemons that were inoculated and treated as indicated above were placed inside a sealed plastic container with a height of 15 cm and a diameter of 10 cm. The humidity level in the box was maintained at 90–95% by pouring 12 mL of sterilized distilled water into a paper towel at the bottom of the box. The fruit was placed above a rack and was not in direct contact with the wet bottom of the box. Plastic containers were stored at 20 ± 1 °C. Each treatment had four replicates of three boxes, with one lemon each.

For cold storage, lemons were arranged in cardboard packaging with fruit trays inside cardboard boxes to simulate the postharvest conditions of citrus fruits. Lemons were stored for 20 days at 5 ± 1 °C, and then, exposed to room temperature (20 ± 1 °C) for 9 days. The same treatments used for the humid box assays were implemented, and each treatment had four replicates of 16 fruits each. Experiments with lemon fruits in humid boxes and cold storage conditions were conducted twice.

2.6. Disease Evaluation

For the in vitro tests, the diameter of mycelium growth (mm) was measured daily until day 10, and the mean of the cross-measurements of the colony was calculated. Mycelial growth inhibition was calculated using the following equation:

$$\text{MGI (\%)} = ((Dc - Dt)/Dc) \times 100 \qquad (1)$$

where MGI is mycelial growth inhibition, Dc is the diameter of mycelial growth in the control (mm), and Dt is the diameter of the treatment.

For both in vivo assays, the incidence (lemons with evident green mold infection) was calculated as the mean of the diseased lemons in the four replicates. The severity was assessed as the percentage of lemon surface with green mold growth [27]. This methodology was used to avoid cross-infection for manipulation and because of the round surface of the lemons. A lemon with half of its surface exhibiting green mold growth was estimated to have a severity of 50%. Efficacy in controlling the disease was evaluated using the following equation:

$$\text{Efficacy (\%)} = ((Ci - Ti)/Ci) \times 100 \qquad (2)$$

where Ci is the incidence of the control, and Ti is the incidence of the treatment in evaluation.

Evaluation of lemons infected with green mold in humid boxes was carried out 4, 7, 11, and 14 days after inoculation. For the postharvest simulation tests, green mold infection was evaluated during cold storage on days 7 and 15 and at room temperature on days 0, 5, 7, and 9.

2.7. Statistical Analysis

Disease development in the different assays was analyzed using an ANOVA (Infostat 2020 version statistical software, Córdoba, Argentina). When significant effects were observed, differences between the means were determined using Tukey's test (HSD) ($p < 0.05$). Data as percentages were arcsine angular transformed before running the analyses.

3. Results

3.1. In Vitro Performance of CH-RF

The results showed inhibition of *P. digitatum* growth in most CH-RF treatments (Figure 2). Green mold mycelia were observed on day 3 of evaluation for the control and CH-RF0.1 and on day 5 in the CH-RF0.25 treatment (Figure 2b). On day 10 of evaluation, there was no visible growth of *P. digitatum* on the CH-RF0.5 and FLU treatments (Figure 2a,b), while the control and CH-RF0.1 showed 68 mm and 67.8 mm growth, respectively. Mycelial growth was completely inhibited on the CH-RF0.5 and FLU treatments, while CH-RF0.25 showed a value of 84.5%. Non-significant differences ($p < 0.05$) were observed between the control and CH-RF0.1 (Figure 2c).

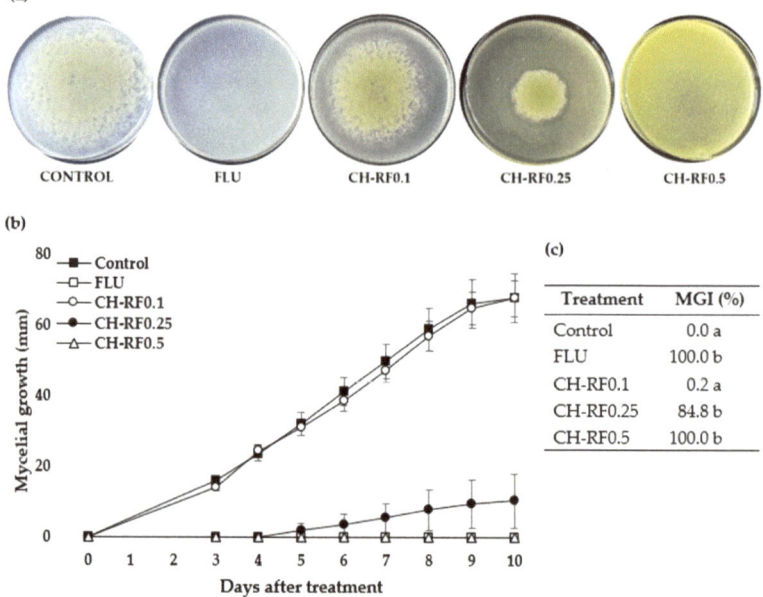

Figure 2. Effect of chitosan–riboflavin conjugate (CH-RF) on in vitro mycelial growth of *Penicillium digitatum*. (a) Growth of *P. digitatum* in the different treatments on day 10 post-sowing. (b) Mycelial growth (mm) from days 0 to 10 and (c) mycelial growth inhibition (MGI) (%) on day 10. Vertical bars represent the standard error of the mean. Different letters indicate statistical differences according to Tukey's test ($p < 0.05$).

3.2. In Vivo Performance of CH-RF on Lemon Fruit in Temperate Condition

The incidence and severity of green mold were evaluated in lemons under optimal growth conditions for the pathogen (20 ± 1 °C 90–95% RH). Previous studies using CH-RF in lemons showed a lower efficacy in the control of green mold at concentrations of 0.7% and 1.4% (Supplementary Figures S1 and S2). It was then decided to use a concentration of 2% w/v of CH-FR. At this concentration, a significant reduction in green mold incidence was observed in inoculated lemons across all treatments compared to the control. Statistical differences were observed on days 4, 7, 11, and 14 post-inoculations (Figure 3a). All treatments showed an incidence below 8.5% and 21% for days 4 and 7, respectively, while the control incidence was 42% and 92% for the same days. No differences between the FLU and CH-RF treatments were observed for days 4 and 7. After that, all treatments remained significantly different from the control, but the lowest incidence was observed for FLU, CH-RF6, and CH-RF30 until day 14 of evaluation.

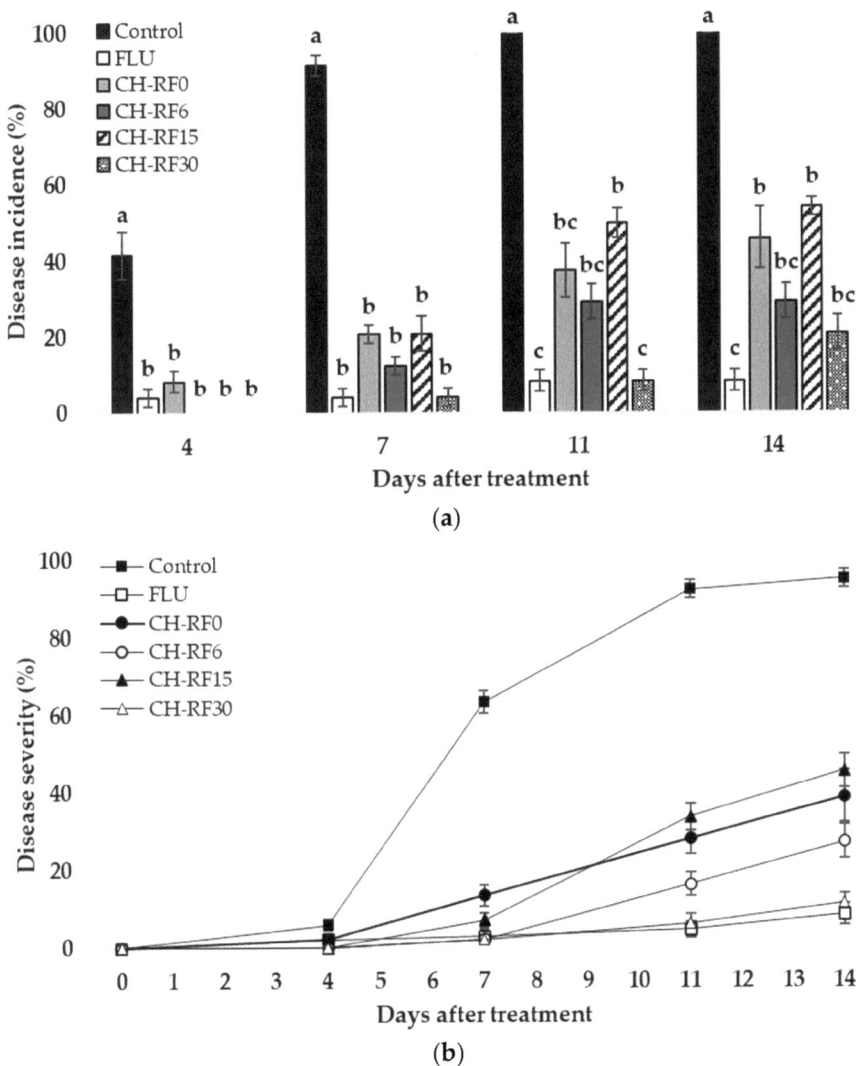

Figure 3. Cont.

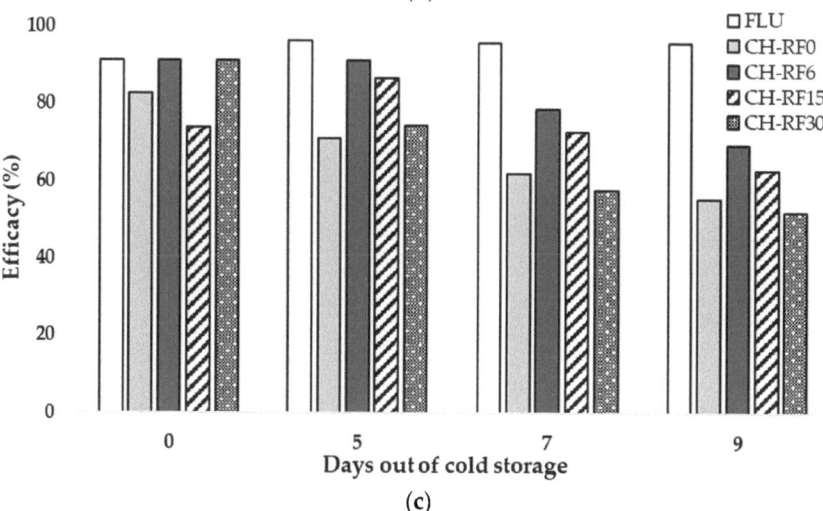

Figure 3. Effect of chitosan–riboflavin conjugate (CH-RF) on incidence (%) (**a**), mean severity (%) (**b**), and efficacy of control (%) (**c**) of green mold disease in lemon fruit inoculated, treated, and evaluated after 4, 7, 11, and 14 days in 20 ± 1 °C and 90–95% RH. Treatments consisted of sterile distilled water (control), fludioxonil 0.2% (FLU), or CH-RF 2% without (CH-RF0) or with 6 (CH-RF6), 15 (CH-RF15), and 30 min (CH-RF30) of white LED light exposure. The experiments were conducted twice. Vertical bars indicate standard errors of the mean. Different letters indicate statistical differences according to Tukey's test ($p < 0.05$).

Disease severity was also evaluated in lemons on days 4, 7, 11, and 14 (Figure 3b), and their trends closely mirrored those observed for the incidence. A low severity (<6%) was observed in all treatments on day 4. On day 7, all treatments showed green mold growth, with the control exhibiting an increase in severity, reaching 63.3%, followed by CH-RF0 at 13.4%. On day 11, the control displayed a disease severity of 92.3%, followed by CH-RF15 with 33.5%. On the last evaluation day, the control reached 95% severity, and CH-RF15 reached 45.4%. In this assessment, all treatments demonstrated statistical differences from the control; however, the treatments with the lowest severity values were consistently observed in FLU, CH-RF6, and CH-RF30.

All treatments showed substantial efficacy in the control of green mold (Figure 3c). Treatments CH-RF6, CH-RF15, and CH-RF30 showed 100% efficacy on day 4 of the evaluation. FLU remained the most effective treatment until day 14 (92%), followed by CH-RF30 (79%), and CH-RF6 (71%).

3.3. In Vivo Performance of CH-RF in Lemon Fruit in Cold Storage

Inoculated lemons were treated and kept for 20 days in cold storage (5 ± 1 °C), and then, placed at room temperature (20 ± 1 °C). Evaluations were made on days 0, 5, 7, and 9. The disease incidence results showed effective control of CH-RF over green mold throughout all evaluation days (Figure 4a). After 14 days of cold storage, the highest level of incidence was observed in the control, with 17.9% of lemons exhibiting green mold. On the fifth day post-release from cold storage, a notable increase in the occurrence of moldy fruits was observed. All CH-RF treatments differed from the control (89.1%), and the lowest levels of diseased lemons were observed in FLU, CH-RF6, and CH-RF15. Day 7 followed the trend of day 5, with FLU and CH-RF6 showing the best performance, with disease incidences of 3.9% and 19.5%, respectively. On day 9 of evaluation, the control showed a 94.5% incidence, while all CH-RF treatments were significantly different, with an incidence of 42.2% or below. In addition, all CH-RF treatments were different from FLU treatments.

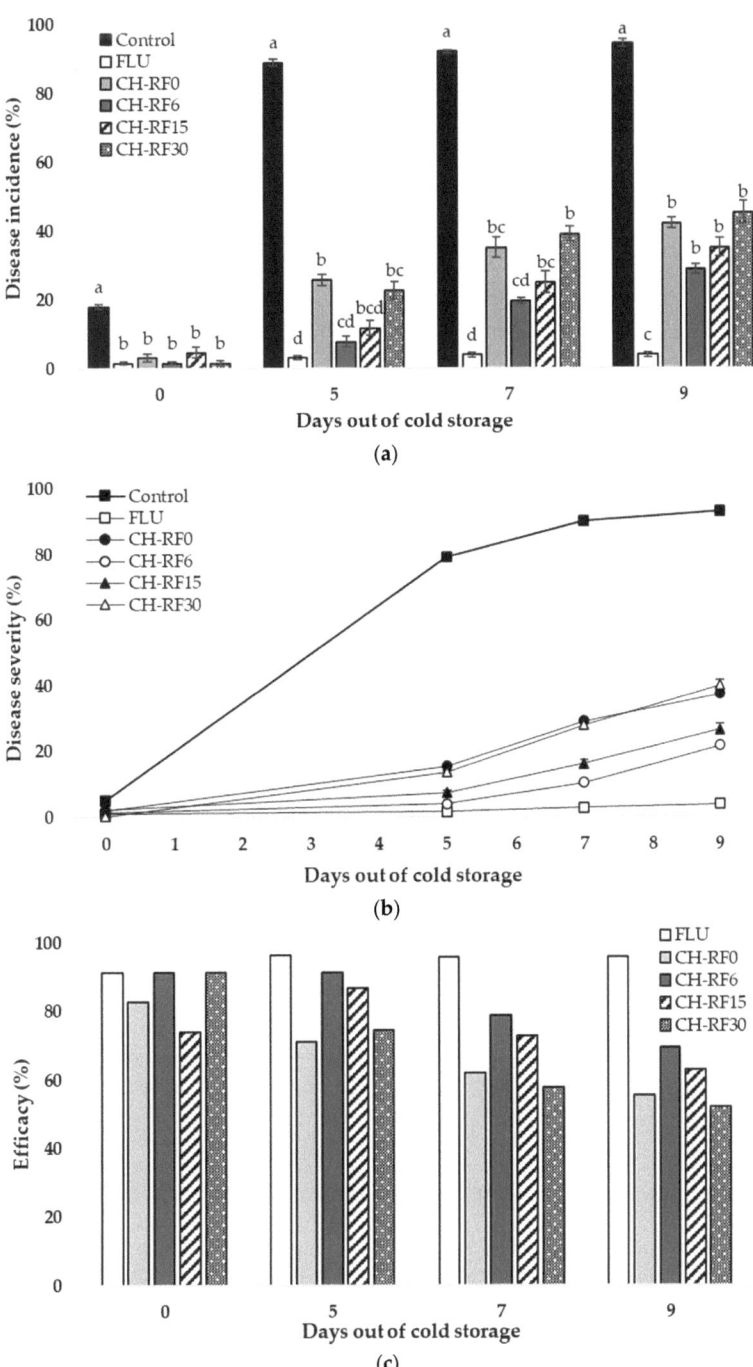

Figure 4. Effect of chitosan–riboflavin conjugate (CH-RF) on incidence (**a**), mean severity (**b**), and efficacy of control (**c**) of green mold disease in lemon fruit inoculated, treated, and maintained in cold storage for 14 days and evaluated after 0, 5, 7, and 9 days at room temperature Treatments consisted

of sterile distilled water (control), fludioxonil 0.2% (FLU), or CH-RF 2% without (CH-RF0) or with 6 (CH-RF6), 15 (CH-RF15), and 30 min (CH-RF30) of white LED light exposure. The experiments were conducted twice. Vertical bars indicate standard errors of the mean. Different letters indicate statistical differences according to Tukey's test ($p < 0.05$).

The control reached 92.9% severity on day 9 of evaluation (Figure 4b), the highest value compared to the other treatments. The lowest severity was measured in FLU at 3.8% and CH-RF6 at 21.7%, both showing differences from the rest of the treatments.

The efficacy in controlling green mold observed in the cold storage tests, followed by subsequent maintenance at room temperature, was similar to that in the tests conducted in temperate temperatures and at high RH, showing a decline from day 0 to day 9 (Figure 4c). The conventional commercial fungicide FLU achieved 95.9% efficacy. Greater efficacy of the CH-RF treatments was observed for CH-RF6 (69.4%).

4. Discussion

Chitosan is a natural biopolymer with proven antifungal effects against various fungi. The control of green mold with chitosan has been previously evaluated in vitro and in different citrus species, such as lemons, oranges, and grapefruit [16–18,28,29]. In addition, there are increasing studies using photoactive compounds to inhibit or reduce the growth of several postharvest fungi. Examples include harmol [23] and vitamin K3 [24], both exhibiting promising potential control of *P. digitatum*. In this study, we evaluated the fungicidal effect of CH-RF, a photoactive conjugate made up of riboflavin (1%) and chitosan (99%), on *P. digitatum* in vitro tests and using infected lemon fruits. As demonstrated in the in vitro evaluations, the growth of *P. digitatum* was completely decreased by 0.5% w/v CH-RF. This trend is consistent with previous studies showing that chitosan at a concentration of 0.5% inhibits entirely or reduces the growth of the fungus by more than 90% [16]. Our findings reveal a heightened inhibitory impact resulting from the incorporation of riboflavin into chitosan. Notably, a 0.25% concentration of CH-RF effectively regulates the growth of *P. digitatum*, underscoring the substantial inhibitory potential of this composite. While other studies have demonstrated nearly complete control of pathogen growth using 0.25% chitosan alone, it is important to note that these investigations employed significantly lower spore concentrations compared to those utilized in our study [29].

Efficacy of CH-RF on lemons under optimal growth conditions for the pathogen (20 °C; 90–95%RH): The in vivo experiments demonstrated the need to increase the doses that were effective in controlling the pathogen under in vitro conditions. The preliminary experiments with CH-RF in lemons in humid boxes showed lower antifungal effects at concentrations of 0.7% w/v and 1.4% w/v, with an efficacy of nearly 100% on day 5, which decreased on day 7 of evaluation (see Supplementary Materials). These results reaffirmed the use of 2% w/v CH-RF as the dose that showed a more extended inhibitory period. This result is consistent with previous reports about the use of chitosan in grapefruits, demonstrating nearly no control of the pathogen with 1% w/v chitosan [15] and no significant differences in the incidence on day 4 of evaluation compared to the control using chitosan even at 3% w/v [18]. Our results showed an increased and significant inhibitory effect on incidence and severity using 2% w/v CH-RF. The enhanced effect could be explained by the activation of the riboflavin portion of the conjugate. We can observe this effect by comparing the reduction in pathogen damage at seven days by chitosan alone with that of the CH-RF molecule (see Supplementary Materials).

Efficacy of CH-RF under cold storage and shelf display conditions: Cold storage is a commonly used method that extends the commercial life of citrus fruits and reduces the growth of several fungus species [30]. For this reason, the use of CH-RF for lemons, and the simulation of commercial cold storage and shelf-life conditions, were tested. Studies on mandarins treated with 1% w/v chitosan for 17 days in cold storage reported a 40% decay incidence due to *P. digitatum* [17]. Other reports using a coating of chitosan solution conducted with oranges showed no incidence of *P. digitatum* in cold storage for up to three or four weeks [29]. Also, using 2% w/v chitosan coating on mandarins, and then, storing them for 15 days at 5 °C and 5 days at 25 °C [18] found no reduction in the incidence of green mold but a reduction in decay severity. The differences across studies regarding cultivar susceptibility, inoculum concentration, or strain virulence may account for the inconsistent results. In our study, green mold was scarcely present in the evaluations during the 20 days of cold storage. Disease incidence was observed only when the fruits were exposed to temperate temperature conditions (day 0). On day 5, it was below 25% for all treatments with CH-RF and light exposure. This difference could be explained by the enhanced fungicidal mechanism shown by the riboflavin portion of the conjugate, suggesting greater performance of the CH-RF conjugate above CH by itself.

Influence of light exposure on CH-RF efficacy: Li et al. [24] observed increased inhibition of the spore germination of a photoactive vitamin K3 analog when exposed to UV and sunlight compared to its use in the dark. Dibona-Villanueva and Fuentealba [25] conducted pioneering in vitro experiments using CH-RF at a 0.5% w/v concentration. Their findings revealed the remarkable complete inhibition of *P. digitatum* growth under visible light irradiation, showcasing the powerful antifungal effect of CH-RF. In contrast, the inhibition rate dropped to approximately 25% when the experiment was conducted in dark conditions, underscoring the crucial role of light in enhancing the inhibitory efficacy against *P. digitatum*. In our experiments, we consistently observe enhanced efficacy in pathogen control when the CH-RF molecule is irradiated for 6 min, as opposed to situations where no initial irradiation is applied. For alternative irradiation durations (15 or 30 min), an improvement in control is not consistently observed. The observed variations in the latter results can be attributed to the following: (i) prolonged LED light exposure, resulting in faster riboflavin photodecomposition [31], and consequently, the efficacy diminishes over time; (ii) in the fruits where CH-RF was applied without initial LED light exposure, they were kept in a natural-light environment in the laboratory, rather than complete darkness. This result suggests the possibility of some activation of riboflavin taking place. Nonetheless, it has to be mentioned that the trend shown in Figure 4 of large reductions at low light exposure and a reduced effect at higher light exposure aligns with the general trends in the literature [32]. The effect of delivering a fixed amount of energy through either a low-power source over extended periods or a high-power source over shorter durations and the biological implications of both remain a topic of ongoing inquiry.

Potential mechanisms explaining CH-RF efficacy: The precise mechanisms responsible for CH-RF's antifungal activity remain incompletely elucidated; however, several potential pathways have been suggested. Previous studies have shed light on potential pathways for the effectiveness of photodynamic inactivation (PDI), particularly in disrupting microbial cell membranes [33]. Furthermore, the critical role of reactive oxygen species (ROS) in causing oxidative stress within the cell has been highlighted [34]. Notably, the involvement of ergosterol degradation in the PDI mechanism has been a focal point of recent investigations. Ergosterol, a vital component of fungal cell membranes, is identified as a target during PDI, with studies demonstrating its degradation as a crucial step in the photodynamic inactivation process [35]. Dibona-Villanueva and Fuentealba [25,36] described the importance of the localization and accumulation of photosensitizers in fungal spores' superficial cell structures in photoinactivation extent. Additionally, there is evidence that supports the idea that the photoinactivation effect strongly depends on the interaction of photosensitizers and fungal cell envelopes [37].

These findings deepen our comprehension of the mechanisms underlying the efficacy of the chitosan–riboflavin bioconjugate, providing valuable insights for fine-tuning protocols and applications against a range of fungal pathogens. Furthermore, these results suggest that this novel fungicide could be used as a promising alternative to traditional synthetic fungicides, enabling a more sustainable approach to the management of lemon fruit diseases.

5. Conclusions

The present study demonstrated the antifungal effect of a novel chitosan–riboflavin bioconjugate (CH-RF) on the growth of *P. digitatum*, one of the most relevant postharvest pathogens in citrus fruit. The results indicate better performance of the bioconjugate compared to the use of chitosan by itself. Moreover, most CH-RF treatments showed enhanced control of green mold when riboflavin was activated by white-light exposure. Our study suggests that this conjugate could be an excellent natural fungicide for the biological control of green mold and could thus be used as an alternative to synthetic fungicides. Further investigations focused on evaluating the bioconjugate in different pathogens and fruit species are currently being pursued.

6. Patents

Patent application PCT/CL2020/050154 "photoactive biofungicide" USA: No. 17/784,797 (13 June 2022). European Patent Office 20899708.0 (11 July 2022). Brazil: No. BR 11 2022 011326 (9 June 2022). Patent application INAPI No: 201903654, Chile.

Supplementary Materials: The following supporting information can be downloaded at: https://www.mdpi.com/article/10.3390/polym16070884/s1, Figure S1: Disease incidence (%) of green mold on lemons inoculated and treated with sterile water (control); chitosan–riboflavin conjugate at 0.7% with 15 (CH-RF0.7% 15) and 30 min (CH-RF0.7% 30) of LED exposure or CH-RF at 2% with 15 (CH-RF2% 15) and 30 min (CH-RF2% 30) of LED exposure; Figure S2: Disease incidence (%) of green mold on lemons inoculated and treated with sterile water without (control0) and with 15 min of white LED exposure (control15); chitosan at 1.4% (CH1.4%); chitosan–riboflavin conjugate at 0.7% with 30 min of LED exposure (CH-RF0.7% 30) and CH-RF at 1.4% with 15 min of LED exposure (CH-RF1.4% 15). Figure S3: Disease severity (%) of green mold on lemons inoculated and treated with sterile water (Control); Chitosan-Riboflavin conjugate at 0.7% with 15 (CH-RF0.7% 15) and 30 min (CH-RF0.7% 30) of LED exposure or CH-RF at 2% with 15 (CH-RF2% 15) and 30 min (CH-RF2% 30) of LED. Figure S4: Disease severity (%) of green mold on lemons inoculated and treated with sterile water without (Control0) and with 15 min of white LED exposure (Control15); Chitosan at 1.4% (CH1.4%); Chitosan-Riboflavin conjugate at 0.7% with 30 min of LED exposure (CH-RF0.7% 30) and CH-RF at 1.4% with 15 min of LED exposure (CH-RF1.4% 15).

Author Contributions: Conceptualization, B.M.I.-C.; methodology, L.D.-V., D.F., A.P.-Q., H.A.V.-G. and B.M.I.-C.; validation, L.D.-V., D.F., D.S., H.A.V.-G. and B.M.I.-C.; formal analysis, B.M.I.-C.; investigation, M.A.G.-N., M.A.H.-D. and B.M.I.-C.; resources, D.F. and H.A.V.-G.; data curation, B.M.I.-C.; writing—original draft preparation, B.M.I.-C.; writing—review and editing, L.D.-V., D.F., H.A.V.-G. and B.M.I.-C.; visualization, H.A.V.-G. and B.M.I.-C.; supervision, D.F., D.S. and H.A.V.-G.; project administration, D.F. and H.A.V.-G.; funding acquisition, L.D.-V., D.F., D.S. and H.A.V.-G. All authors have read and agreed to the published version of the manuscript.

Funding: This research was supported by the Chilean project COPEC-UC 2019.J.1273 and by the Pontificia Universidad Católica Vicerrectoría de investigación (VRI) Doctoral Fellowship 2023.

Institutional Review Board Statement: Not applicable.

Data Availability Statement: Data will be made available upon reasonable request.

Acknowledgments: We are grateful to Julio Cornejo and Johanna Mártiz for their technical support and for the fruit material used in the in vivo and cold storage assays. We are also very grateful to the Fruit Pathology Laboratory staff—B. Puebla and L. Silva—for their significant scientific and/or technical support, notably concerning the laboratory tests.

Conflicts of Interest: The authors declare no conflicts of interest.

References

1. Costa, J.H.; Bazioli, J.M.; de Moraes Pontes, J.G.; Fill, T.P. Penicillium Digitatum Infection Mechanisms in Citrus: What Do We Know so Far? *Fungal Biol.* **2019**, *123*, 584–593. [CrossRef] [PubMed]
2. Palou, L. Penicillium Digitatum, Penicillium Italicum (Green Mold, Blue Mold). In *Postharvest Decay*; Elsevier: Amsterdam, The Netherlands, 2014; pp. 45–102. ISBN 978-0-12-411552-1.

3. Ladaniya, M. Postharvest Disease Management with Fungicides. In *Citrus Fruit*; Elsevier: Amsterdam, The Netherlands, 2023; pp. 563–594, ISBN 978-0-323-99306-7.
4. Kellerman, M.; Liebenberg, E.; Njombolwana, N.; Erasmus, A.; Fourie, P.H. Postharvest Dip, Drench and Wax Coating Application of Pyrimethanil on Citrus Fruit: Residue Loading and Green Mould Control. *Crop Prot.* 2018, *103*, 115–129. [CrossRef]
5. Zubrod, J.P.; Bundschuh, M.; Arts, G.; Brühl, C.A.; Imfeld, G.; Knäbel, A.; Payraudeau, S.; Rasmussen, J.J.; Rohr, J.; Scharmüller, A.; et al. Fungicides: An Overlooked Pesticide Class? *Environ. Sci. Technol.* 2019, *53*, 3347–3365. [CrossRef]
6. Xu, J.; Xiong, H.; Zhang, X.; Muhayimana, S.; Liu, X.; Xue, Y.; Huang, Q. Comparative Cytotoxic Effects of Five Commonly Used Triazole Alcohol Fungicides on Human Cells of Different Tissue Types. *J. Environ. Sci. Health Part B* 2020, *55*, 438–446. [CrossRef] [PubMed]
7. Wang, Y.; Lafon, P.-A.; Salvador-Prince, L.; Gines, A.R.; Trousse, F.; Torrent, J.; Prevostel, C.; Crozet, C.; Liu, J.; Perrier, V. Prenatal Exposure to Low Doses of Fungicides Corrupts Neurogenesis in Neonates. *Environ. Res.* 2021, *195*, 110829. [CrossRef]
8. Oiki, S.; Yaguchi, T.; Urayama, S.; Hagiwara, D. Wide Distribution of Resistance to the Fungicides Fludioxonil and Iprodione in Penicillium Species. *PLoS ONE* 2022, *17*, e0262521. [CrossRef]
9. Muanprasat, C.; Chatsudthipong, V. Chitosan Oligosaccharide: Biological Activities and Potential Therapeutic Applications. *Pharmacol. Ther.* 2017, *170*, 80–97. [CrossRef]
10. Romanazzi, G.; Feliziani, E.; Baños, S.B.; Sivakumar, D. Shelf Life Extension of Fresh Fruit and Vegetables by Chitosan Treatment. *Crit. Rev. Food Sci. Nutr.* 2017, *57*, 579–601. [CrossRef]
11. Hua, C.; Li, Y.; Wang, X.; Kai, K.; Su, M.; Shi, W.; Zhang, D.; Liu, Y. The Effect of Low and High Molecular Weight Chitosan on the Control of Gray Mold (*Botrytis Cinerea*) on Kiwifruit and Host Response. *Sci. Hortic.* 2019, *246*, 700–709. [CrossRef]
12. Herrera-Défaz, M.; Fuentealba, D.; Dibona-Villanueva, L.; Schwantes, D.; Jiménez, B.; Ipinza, B.; Latorre, B.; Valdés-Gómez, H.; Fermaud, M. Biocontrol of Botrytis Cinerea on Grape Berries in Chile: Use of Registered Biofungicides and a New Chitosan-Based Fungicide. *Horticulturae* 2023, *9*, 746. [CrossRef]
13. Zhao, Y.; Deng, L.; Zhou, Y.; Ming, J.; Yao, S.; Zeng, K. Wound Healing in Citrus Fruit Is Promoted by Chitosan and Pichia Membranaefaciens as a Resistance Mechanism against Colletotrichum Gloeosporioides. *Postharvest Biol. Technol.* 2018, *145*, 134–143. [CrossRef]
14. García-Bramasco, C.A.; Blancas-Benitez, F.J.; Montaño-Leyva, B.; Medrano-Castellón, L.M.; Gutierrez-Martinez, P.; González-Estrada, R.R. Influence of Marine Yeast Debaryomyces Hansenii on Antifungal and Physicochemical Properties of Chitosan-Based Films. *J. Fungi* 2022, *8*, 369. [CrossRef] [PubMed]
15. Shi, Z.; Wang, F.; Lu, Y.; Deng, J. Combination of Chitosan and Salicylic Acid to Control Postharvest Green Mold Caused by Penicillium Digitatum in Grapefruit Fruit. *Sci. Hortic.* 2018, *233*, 54–60. [CrossRef]
16. Waewthongrak, W.; Pisuchpen, S.; Leelasuphakul, W. Effect of *Bacillus Subtilis* and Chitosan Applications on Green Mold (*Penicilium Digitatum* Sacc.) Decay in Citrus Fruit. *Postharvest Biol. Technol.* 2015, *99*, 44–49. [CrossRef]
17. Shao, X.; Cao, B.; Xu, F.; Xie, S.; Yu, D.; Wang, H. Effect of Postharvest Application of Chitosan Combined with Clove Oil against Citrus Green Mold. *Postharvest Biol. Technol.* 2015, *99*, 37–43. [CrossRef]
18. Da Silva, Y.C.R.; Alves, R.M.; Da Silva, B.M.P.; Bron, I.U.; Cia, P. Chitosan and Hot Water Treatments Reduce Postharvest Green Mould in 'Murcott' Tangor. *J. Phytopathol.* 2020, *168*, 542–550. [CrossRef]
19. Correia, J.H.; Rodrigues, J.A.; Pimenta, S.; Dong, T.; Yang, Z. Photodynamic Therapy Review: Principles, Photosensitizers, Applications, and Future Directions. *Pharmaceutics* 2021, *13*, 1332. [CrossRef]
20. Wu, X.; Hu, Y. Photodynamic Therapy for the Treatment of Fungal Infections. *Infect. Drug Resist.* 2022, *15*, 3251–3266. [CrossRef]
21. Seididamyeh, M.; Netzel, M.E.; Mereddy, R.; Harmer, J.R.; Sultanbawa, Y. Photodynamic Inactivation of Botrytis Cinerea Spores by Curcumin—Effect of Treatment Factors and Characterization of Photo-Generated Reactive Oxygen Species. *Food Bioprocess Technol.* 2023, *17*, 670–685. [CrossRef]
22. Song, L.; Zhang, F.; Yu, J.; Wei, C.; Han, Q.; Meng, X. Antifungal Effect and Possible Mechanism of Curcumin Mediated Photodynamic Technology against Penicillium Expansum. *Postharvest Biol. Technol.* 2020, *167*, 111234. [CrossRef]
23. Olmedo, G.M.; Cerioni, L.; González, M.M.; Cabrerizo, F.M.; Volentini, S.I.; Rapisarda, V.A. UVA Photoactivation of Harmol Enhances Its Antifungal Activity against the Phytopathogens Penicillium Digitatum and Botrytis Cinerea. *Front. Microbiol.* 2017, *8*, 347. [CrossRef]
24. Li, X.; Sheng, L.; Sbodio, A.O.; Zhang, Z.; Sun, G.; Blanco-Ulate, B.; Wang, L. Photodynamic Control of Fungicide-Resistant Penicillium Digitatum by Vitamin K3 Water-Soluble Analogue. *Food Control* 2022, *135*, 108807. [CrossRef]
25. Dibona-Villanueva, L.; Fuentealba, D. Novel Chitosan-Riboflavin Conjugate with Visible Light-Enhanced Antifungal Properties against Penicillium Digitatum. *J. Agric. Food Chem.* 2021, *69*, 945–954. [CrossRef]
26. Jiménez Jiménez, B. Evaluación de un Biofungicida Fotoactivo en Hongos Patógenos de Frutales. Master's Thesis, Pontificia Universidad Católica de Chile, Santiago, Chile, 2021.
27. Pérez-Alfonso, C.O.; Martínez-Romero, D.; Zapata, P.J.; Serrano, M.; Valero, D.; Castillo, S. The Effects of Essential Oils Carvacrol and Thymol on Growth of Penicillium Digitatum and P. Italicum Involved in Lemon Decay. *Int. J. Food Microbiol.* 2012, *158*, 101–106. [CrossRef]
28. Bhatta, U.K. Alternative Management Approaches of Citrus Diseases Caused by Penicillium Digitatum (Green Mold) and Penicillium Italicum (Blue Mold). *Front. Plant Sci.* 2022, *12*, 833328. [CrossRef]

29. Khalil Bagy, H.M.M.; Ibtesam, B.F.M.; Abou-Zaid, E.A.A.; Sabah, B.M.; Nashwa, S.M.A. Control of Green Mold Disease Using Chitosan and Its Effect on Orange Properties during Cold Storage. *Arch. Phytopathol. Plant Prot.* **2021**, *54*, 570–585. [CrossRef]
30. Strano, M.C.; Altieri, G.; Admane, N.; Genovese, F.; Di Renzo, G.C. Advance in Citrus Postharvest Management: Diseases, Cold Storage and Quality Evaluation. In *Citrus Pathology*; Gill, H., Garg, H., Eds.; InTech: London, UK, 2017; ISBN 978-953-51-3071-0.
31. Crocker, L.B.; Lee, J.H.; Mital, S.; Mills, G.C.; Schack, S.; Bistrović-Popov, A.; Franck, C.O.; Mela, I.; Kaminski, C.F.; Christie, G.; et al. Tuning Riboflavin Derivatives for Photodynamic Inactivation of Pathogens. *Sci. Rep.* **2022**, *12*, 6580. [CrossRef] [PubMed]
32. Piksa, M.; Lian, C.; Samuel, I.C.; Pawlik, K.J.; Samuel, I.D.W.; Matczyszyn, K. The Role of the Light Source in Antimicrobial Photodynamic Therapy. *Chem. Soc. Rev.* **2023**, *52*, 1697–1722. [CrossRef] [PubMed]
33. Maisch, T.; Baier, J.; Franz, B.; Maier, M.; Landthaler, M.; Szeimies, R.-M.; Bäumler, W. The Role of Singlet Oxygen and Oxygen Concentration in Photodynamic Inactivation of Bacteria. *Proc. Natl. Acad. Sci. USA* **2007**, *104*, 7223–7228. [CrossRef] [PubMed]
34. Jori, G.; Magaraggia, M.; Fabris, C.; Soncin, M.; Camerin, M.; Tallandini, L.; Coppelotti, O.; Guidolin, L. Photodynamic Inactivation of Microbial Pathogens: Disinfection of Water and Prevention of Water-Borne Diseases. *J. Environ. Pathol. Toxicol. Oncol.* **2011**, *30*, 261–271. [CrossRef] [PubMed]
35. Böcking, T.; Barrow, K.D.; Netting, A.G.; Chilcott, T.C.; Coster, H.G.L.; Höfer, M. Effects of Singlet Oxygen on Membrane Sterols in the Yeast *Saccharomyces cerevisiae*. *Eur. J. Biochem.* **2000**, *267*, 1607–1618. [CrossRef] [PubMed]
36. Dibona-Villanueva, L.; Fuentealba, D. Protoporphyrin IX–Chitosan Oligosaccharide Conjugate with Potent Antifungal Photodynamic Activity. *J. Agric. Food Chem.* **2022**, *70*, 9276–9282. [CrossRef] [PubMed]
37. Dai, T.; Fuchs, B.B.; Coleman, J.J.; Prates, R.A.; Astrakas, C.; St. Denis, T.G.; Ribeiro, M.S.; Mylonakis, E.; Hamblin, M.R.; Tegos, G.P. Concepts and Principles of Photodynamic Therapy as an Alternative Antifungal Discovery Platform. *Front. Microbiol.* **2012**, *3*, 120. [CrossRef] [PubMed]

Disclaimer/Publisher's Note: The statements, opinions and data contained in all publications are solely those of the individual author(s) and contributor(s) and not of MDPI and/or the editor(s). MDPI and/or the editor(s) disclaim responsibility for any injury to people or property resulting from any ideas, methods, instructions or products referred to in the content.

Article

Rhodamine B-Containing Chitosan-Based Films: Preparation, Luminescent, Antibacterial, and Antioxidant Properties

Omar M. Khubiev [1], Anton R. Egorov [1], Daria I. Semenkova [1,2], Darina S. Salokho [1,2], Roman A. Golubev [1,2], Nkumbu D. Sikaona [1], Nikolai N. Lobanov [1], Ilya S. Kritchenkov [2,3], Alexander G. Tskhovrebov [1], Anatoly A. Kirichuk [1], Victor N. Khrustalev [1] and Andreii S. Kritchenkov [1,2,*]

1. Department of Human Ecology and Bioelementology, RUDN University, Miklukho-Maklaya St. 6, Moscow 117198, Russia; ihubievomar1@gmail.com (O.M.K.); sab.icex@mail.ru (A.R.E.); darya.semenkova02@mail.ru (D.I.S.); dalialesma01@gmail.com (D.S.S.); asdfdss.asdasf@yandex.ru (R.A.G.); nkumbusikaona534@gmail.com (N.D.S.); lobanov-nn@rudn.ru (N.N.L.); alexander.tskhovrebov@gmail.com (A.G.T.); kirichuk-aa@rudn.ru (A.A.K.); vnkhrustalev@gmail.com (V.N.K.)
2. Metal Physics Laboratory, Institute of Technical Acoustics NAS of Belarus, General Lyudnikov Ave. 13, 210009 Vitebsk, Belarus; ilya.kritchenkov@gmail.com
3. Department of General and Inorganic Chemistry, St. Petersburg State University, Universitetskaya Embankment 7–9, St. Petersburg 199034, Russia
* Correspondence: kritchenkov-as@rudn.ru

Citation: Khubiev, O.M.; Egorov, A.R.; Semenkova, D.I.; Salokho, D.S.; Golubev, R.A.; Sikaona, N.D.; Lobanov, N.N.; Kritchenkov, I.S.; Tskhovrebov, A.G.; Kirichuk, A.A.; et al. Rhodamine B-Containing Chitosan-Based Films: Preparation, Luminescent, Antibacterial, and Antioxidant Properties. *Polymers* **2024**, *16*, 755. https://doi.org/10.3390/polym16060755

Academic Editors: William Facchinatto and Sérgio Paulo Campana-Filho

Received: 29 January 2024
Revised: 1 March 2024
Accepted: 7 March 2024
Published: 9 March 2024

Copyright: © 2024 by the authors. Licensee MDPI, Basel, Switzerland. This article is an open access article distributed under the terms and conditions of the Creative Commons Attribution (CC BY) license (https://creativecommons.org/licenses/by/4.0/).

Abstract: In this study, Rhodamine B-containing chitosan-based films were prepared and characterized using their mechanical, photophysical, and antibacterial properties. The films were synthesized using the casting method and their mechanical properties, such as tensile strength and elongation at break, were found to be dependent on the chemical composition and drying process. Infrared spectroscopy and X-ray diffraction analysis were used to examine the chemical structure and degree of structural perfection of the films. The photophysical properties of the films, including absorption spectra, fluorescence detection, emission quantum yields, and lifetimes of excited states, were studied in detail. Rhodamine B-containing films exhibited higher temperature sensitivity and showed potential as fluorescent temperature sensors in the physiological range. The antibacterial activity of the films was tested against Gram-positive bacteria *S. aureus* and Gram-negative bacteria *E. coli*, with Rhodamine B-containing films demonstrating more pronounced antibacterial activity compared to blank films. The findings suggest that the elaborated chitosan-based films, particularly those containing Rhodamine B can be of interest for further research regarding their application in various fields such as clinical practice, the food industry, and agriculture due to their mechanical, photophysical, and antibacterial properties.

Keywords: chitosan; rhodamine B; films; photophysical properties; antibacterial activity

1. Introduction

Chitosan, a biopolymer derived from chitin, has garnered significant attention in the past few decades due to its attractive characteristics such as biocompatibility, biodegradability, antioxidant effect, and antimicrobial activity [1–4]. Chitosan-based films have been extensively studied for a wide range of applications, including wound dressings, drug delivery, and food packaging [5,6]. In many instances, properties of chitosan films can be enhanced through the addition of dyes or pigments, to impart new attractive functionalities [7,8]. In the literature, there are numerous examples of chitosan films containing various dyes, such as natural dyes like curcumin and betalains, as well as synthetic dyes like methylene blue and crystal violet [9,10]. The importance and relevance of the development of chitosan-based films has been repeatedly emphasized in recent research papers and reviews [5,11–13].

However, among the wide range of chitosan films reported in the literature, there is a growing interest in films containing fluorescent dyes, which have attractive photophysical properties that can open up new possibilities for applications in bioimaging, sensing, and theranostics [14]. While there are several examples of chitosan films incorporated with fluorescent dyes, such as fluorescein and coumarins, there is limited literature on chitosan films containing Rhodamine B, a widely used fluorescent dye [15,16]. Rhodamine B has been extensively studied for its photophysical properties, including its high fluorescence quantum yield and photostability, making it a promising candidate for various applications [17,18]. Therefore, the combination of chitosan films with Rhodamine B holds great potential for developing novel functional materials with enhanced properties (and not necessarily only fluorescent properties, since the new film can demonstrate, for example, improved mechanical parameters, an enhanced antimicrobial effect, and other, sometimes unexpected, characteristics).

In this study, we aimed to investigate chitosan-based films containing Rhodamine B, filling the gap in the literature, and exploring the preparative, photophysical, and biological properties of these films. To the best of our knowledge, there are limited examples of chitosan films incorporating Rhodamine B, highlighting the uniqueness and novelty of our research. We prepared chitosan films containing Rhodamine B using a simple and reproducible method, and characterized the films using various techniques, including mechanical tests, UV/Vis spectroscopy, fluorescence spectroscopy, and X-ray diffraction (XRD). The photophysical properties of the films, such as fluorescence intensity and lifetime, were evaluated. Additionally, we conducted in vitro studies to assess the potential antimicrobial properties of the chitosan films containing Rhodamine B and their antioxidant activity.

The combination of chitosan and Rhodamine B in a film format presents a promising platform for developing multifunctional materials with applications in diverse fields, such as bioimaging, sensing, and antimicrobial and antioxidant coatings (Figure 1). The results of this study, their discussion, and perspectives are presented in the sections that follow below.

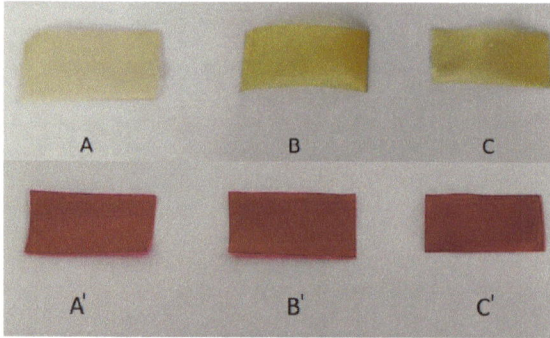

Figure 1. Blank (**A–C**) films and Rhodamine B-containing (**A′–C′**) films.

2. Materials and Methods

Chitosan 40 kDa was purchased from Bioprogress (Russia, Moscow) and Rhodamine B was obtained from Aldrich. Other chemicals and solvents were obtained from commercial sources and were used as received.

Preparation of films: Rhodamine B-containing films and blank films were prepared as follows. An amount of 1.0 g of chitosan was dissolved in 40 mL of 1% aqueous acetic acid, then 0.5 mL of glycerol and 0.5 mL of 0.1% aqueous Rhodamine B were added (no Rhodamine B was added in the case of the blank films). The resultant solutions were cast on Petri dishes and dried at 60 °C (A and A′) and 90 °C (B and B′). C and C′ were prepared as A and A′, followed by 20 min treatment of the film using 20% NH_3 in EtOH solution and drying at room temperature to constant mass.

Tensile strength and elongation at break were measured using a RAM.1.1.A_RA apparatus (Dongguan, China) at 22 °C (samples 6.0 cm long and 2.0 cm wide).

IR spectra were recorded using a Shimadzu IRSpirit (Kyoto, Japan) at 4700 to 350 cm^{-1} (10 mg of sample without any specified sample preparation).

Differential thermal analysis (DTA) and thermogravimetric analysis (TGA) were recorded using a SDT Q600 (New Castle, USA) using heating rate 5 °C/min in the temperature range from 40 °C to 600 °C.

X-ray diffraction analysis was carried out on a Dron-7 X-ray diffractometer (Saint Petersburg, Russia). A 2θ angle interval from 7° to 40° with scanning step Δ2θ = 0.02° and exposure of 7 s per point were used. Cu Kα radiation (Ni filter) was used, which was subsequently decomposed into Kα1 and Kα2 components during processing of the spectra [19].

Photophysical experiments: UV/Vis light absorption spectra were recorded using a spectrophotometer UV-1800 (Shimadzu, Kyoto, Japan). The emission spectra were registered using an Avantes AvaSpec-2048 × 64 spectrometer (Avantes, Apeldoorn, The Netherlands). The absolute emission quantum yield was determined using an integrating sphere AvaSphere-50 (Avantes, Apeldoorn, The Netherlands). An LED (365 nm) (Ocean Optics, Largo, FL, USA) was applied for pumping. A pulse laser LDH-P-C-405 (wavelength 405 nm, pulse width 50 ps, repetition frequency 10 MHz) (PicoQuant, Berlin, Germany), a photon counting head H10682-01 (Hamamatsu, Hamamatsu, Japan), a multiple-event time digitizer MCS6A1T4 (FAST ComTec, Oberhaching, Germany), and a monochromator Monoscan-2000 (interval of wavelengths 1 nm) (Ocean Optics, Largo, FL, USA) were used for lifetime measurements. Temperature control was performed by using a cuvette sample compartment qpod-2e (Quantum Northwest Inc., Liberty Lake, WA, USA).

Antimicrobial activity (in vitro) was evaluated completely as previously described by some of us [20–22]. The DPPH• scavenging effect was evaluated according to the published procedure [23].

3. Results and Discussion

3.1. Preparation of Rhodamine B-Containing Films

The chitosan-based Rhodamine B-containing films were prepared using the conventional solution casting method. The chitosan which was used in the current study (MW = 40 kDa) is not water-soluble. We dissolved the chitosan in 1% acetic acid solution. Acetic acid protonates primary amino groups of the chitosan, thus, destroying the native interchained hydrogen bonds system. This, in turn, results in the complete dissolution of chitosan.

To improve the flexibility and mechanical properties of the resultant films, we used glycerol as a common plasticizer for polysaccharide-based films. The volume of glycerol added to the chitosan solution was determined based on the literature data to achieve the desired film properties [20]. To the chitosan solution containing the plasticizer, we added a solution of Rhodamine B under vigorous stirring. The initial colorless chitosan/plasticizer solution immediately turned purple. One notable observation was that no heterogeneity was observed in the solution even at a temperature of 80 °C, indicating the uniform distribution of Rhodamine B in the chitosan matrix. This could be attributed to the strong stirring during the mixing process, which ensured a homogeneous dispersion of the dye in the solution.

In the same manner described above, we prepared the blank films, which were obtained by the same procedures except the addition of the dye Rhodamine B. The resultant solutions were cast in plastic dishes and dried under different conditions which are presented schematically as follows:

1. Dried at 60 °C for 24 h (films A and A')*;
2. Dried at 60 °C for 24 h and then 90 °C for 2 h (films B and B');
3. Dried at 60 °C for 24 h, then treatment of the film using 20% NH$_3$ in EtOH solution and drying at room temperature (films C and C');
4. *—the abbreviations A, B, and C belong to the blank films while A', B', and C' belong to the rhodamine-containing films.

Thus, films A and A′ were dried at 60 °C for 24 h, films B and B′ were dried at 60 °C for 24 h, followed by drying at 90 °C for 2 h, and films C and C′ were dried at 60 °C for 24 h, followed by treatment with a 20% solution of ammonia in ethanol (transferring the film to the base form) and air drying to constant weight.

The resultant dye-containing films were transparent purple, while the blank films were practically colorless (Figure 1).

3.2. Mechanical Properties of the Films

The most important mechanical parameters for films include tensile strength and elongation at break, which strongly depend on the chemical structure of the components of the elaborated film. Tensile strength is the maximum stress that a material can withstand while being stretched or pulled before breaking. Elongation at break is the ratio of the initial and final lengths of the film before it breaks. Thus, tensile strength is a measure of film strength, while elongation at break characterizes the ductility of the film material. Pure chitosan films are characterized by high strength but very low ductility, so they are dramatically brittle. To avoid this drawback, plasticizers such as glycerin are used. Figures 2 and 3 demonstrate the results of the mechanical tests of the elaborated films.

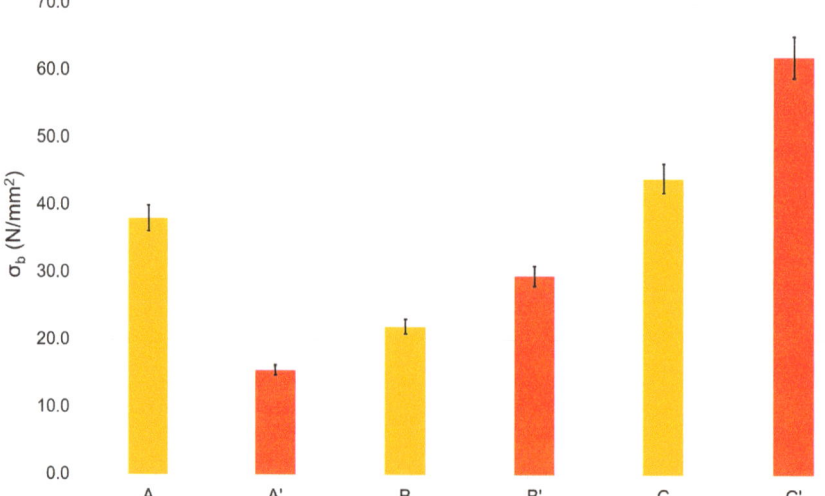

Figure 2. Tensile strength of the blank (A–C) films and Rhodamine B-containing (A′–C′) films.

In many instances, the heating of chitosan films or their transferring into their base form (see Section 3.1, issue 2 and 3) results in changes in the mechanical characteristics of the films, and this is usually explained by different packaging of chitosan macromolecular coils in the films, different water contents, and differences in the hydrogen bonds networks [24]. Thus, heating a blank film for 2 h at 90 °C (A → B) leads to an almost 1.5-fold decrease in strength, and transferring the film to the base form (A → C) results in an approximate 15% increase in strength. In opposition, the heating of the Rhodamine B-containing film (A′ → B′) results in a 2-fold increase in the film strength. Similarly, transferring the Rhodamine B-containing film to its base form (A′ → C′) furnishes a dramatic rise in its strength.

In general, the ductility varies in the opposite direction of strength. Both heating of rhodamine-containing films (A′ → B′) and their transfer to their base form (A′ → C′) reduce ductility by more than five times. Heating the blank films (A → B) strongly increases their ductility. The transfer of blank films to their base form (A → C) decreases their ductility, but only slightly.

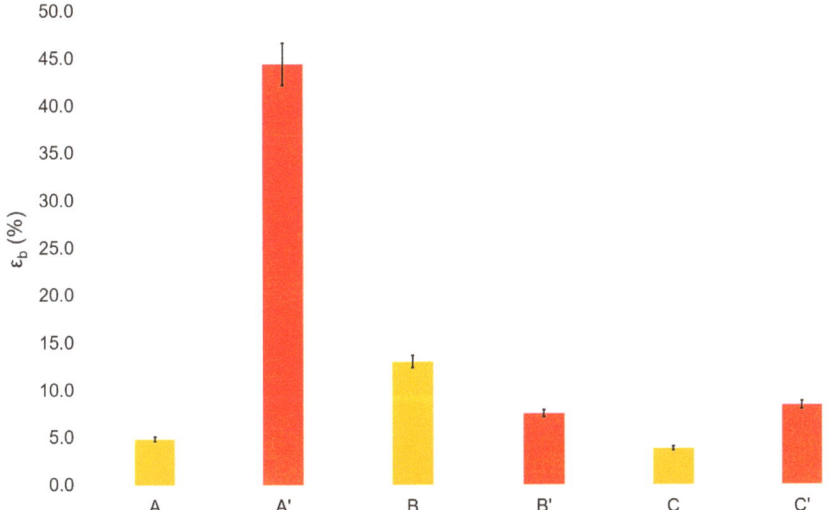

Figure 3. Elongation at break of the blank (A–C) films and Rhodamine B-containing (A'–C') films.

Low ductility is an important problem in the mechanics of chitosan-based films. In this regard, we considered film A' to be the film with the best mechanical characteristics, i.e., the film with the best ductility (the highest elongation at break). The tensile strength of this film is quite sufficient for practical applications in both medicine and the food industry and it is comparable to the strength of similar polyethylene films [25].

3.3. Infrared Spectroscopy

IR spectroscopy is based on the absorption of infrared light by a substance, and it is commonly used to estimate the chemical structure of films. In this study, we employed this method to reveal the basic or salt forms of chitosan, as well as to identify a Rhodamine B presence in the elaborated films.

Thus, in all prepared films we were able to identify only the starting chitosan. The spectra of the films, A, B, C, A', B', and C' exhibited stretching vibration bands characteristic for chitosan, i.e., wide bands of O–H and N–H stretching (3440–3100 cm^{-1}), C–H stretching (2870 cm^{-1}) and bending (1460, 1420, and 1380 cm^{-1}) vibrations, and N–H deformation vibrations (1590–1650 cm^{-1}). Absorption bands in the range of 900–1200 cm^{-1} are due to C–O–C, C–C, and N–H deformation vibrations. Spectra of samples with the salt form of chitosan (A, A' B, B', C') also show bands characteristic of the protonated NH_3^+ group and CH_3COO^- (1300–1640 cm^{-1}) (Figures 4 and 5) [26].

Based on these spectra, it is easy to identify the salt or base form of chitosan by the characteristic structure of the specific vibration bands (1250–1750 cm^{-1}), as seen by the moieties highlighted in the corresponding area (Figures 4 and 5). Thus, samples of starting chitosan and C are in base form. Chitosan in the sample C' was likely partially transferred to its base form. This is apparently because of the significantly lower availability of NH_3^+ groups due to the greater compaction of macromolecular coils in the rhodamine-containing film compared to the corresponding blank film. However, the salt form of chitosan in the C' sample is unambiguously prevailing. Spectra of A', B', and C' do not show Rhodamine B characteristic bands [27] due to its extremely small concentration.

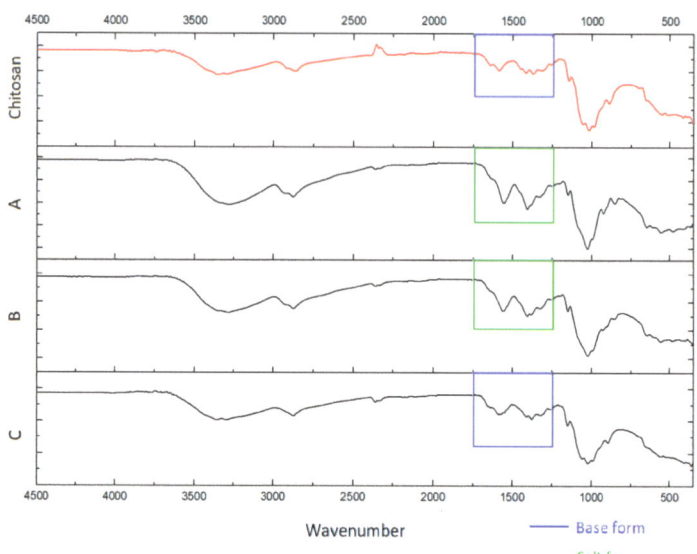

Figure 4. IR spectra of chitosan and blank films A, B, and C.

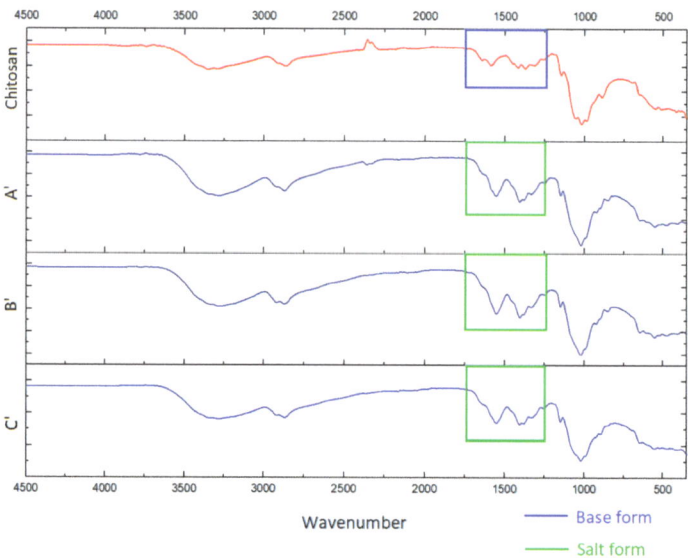

Figure 5. IR spectrum of chitosan and Rhodamine B-containing films A', B', and C'.

3.4. Photophysical Properties of the Films

For all films under study, we investigated in detail the photophysical properties, including the measurement of absorption spectra in the ultraviolet, visible, and near-infrared regions, the detection of fluorescence, the determination of emission quantum yields, and the lifetimes of excited states in different conditions. The results of these measurements are shown below in Figures 6 and 7 and are summarized in Table 1.

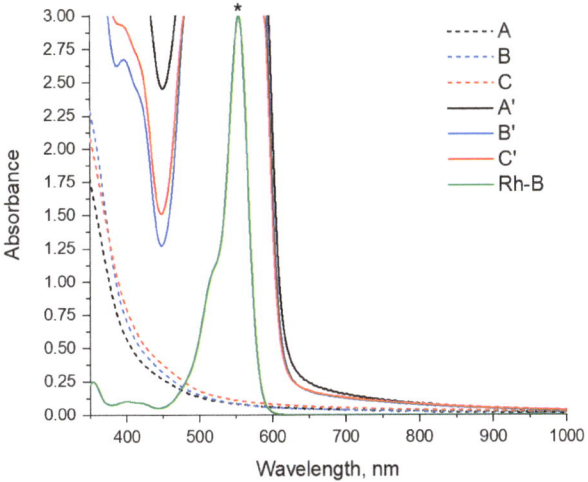

Figure 6. UV/Vis light absorption spectra of films A′–C′, A–C, and Rhodamine B (Rh-B, measured in water solution, C = 1 μM). Temperature 298 K. * Absorbance of Rh-B is normalized to the maximum for the purpose of comparison (absorption data with values greater than 3 are not valid and, therefore, are not displayed).

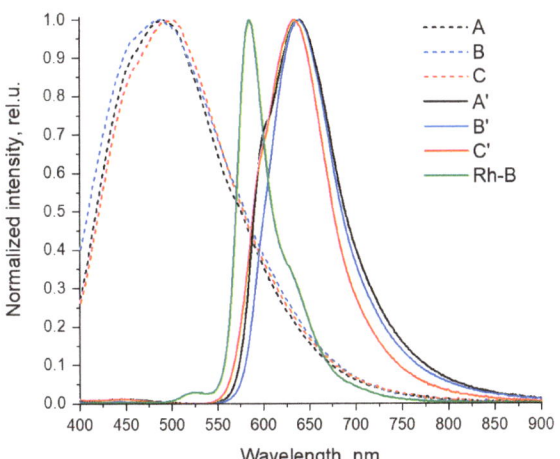

Figure 7. Normalized emission spectra of films A′–C′, A–C, and Rhodamine B (Rh-B, measured in water solution, C = 1 μM). Temperature 298 K. Excitation at 365 nm.

Table 1. Photophysical data for films A–C, A′–C′, and Rhodamine B (Rh-B, * measured in water solution, C = 1 μM). Temperature 298 K (unless otherwise specified).

	A	B	C	A′	B′	C′	Rh-B *
λ_{em}, nm [a]	452(sh); 490	454(sh); 487	454(sh); 498	601(sh); 639	598(sh); 639	596(sh); 633	585; 630(sh)
QY, % [a]	4.16	3.91	3.83	1.85	3.34	2.45	31.1
τ, ns (32 °C) [b]	4.09 [c]	3.58 [c]	3.52 [c]	3.37 [d]	3.62 [d]	3.02 [d]	1.42 [e]
τ, ns (42 °C) [b]	3.75 [c]	3.49 [c]	3.38 [c]	2.78 [d]	3.01 [d]	2.47 [d]	1.12 [e]
$\Delta\tau/(\tau\cdot\Delta T)$,%/K	0.87	0.26	0.43	1.92	1.82	1.98	2.38

[a]—excitation at 365 nm, [b]—excitation at 355 nm, [c]—emission at 495 nm, [d]—emission at 635 nm, [e]—emission at 585 nm.

The obtained films A′–C′ have an intense color due to the strong absorption of the visible light (more than 90%) with wavelengths less than 600 nm (see Figure 6). Thus, these films transmit only the red part of the visible spectrum (more than 600 nm), which determines their intense red color. This absorption is mostly due to the introduction of Rhodamine B in the composition of these materials (for illustration, Figure 6 shows the spectrum of Rhodamine B in a dilute aqueous solution).

Blank films A–C exhibit strong absorption (over 90% of the light) in the ultraviolet region of the spectrum (less than 380 nm, Figure 6), as well as moderate absorption up to the 500 nm region, which determines that they have only a weak yellow color.

All the films obtained show a noticeable photoluminescence (Figure 7). The quantum yields of this fluorescence are moderate and are approximately 4% for blank A–C films and 2–3% for A′–C′ rhodamine-containing films (Table 1).

Blank films A–C exhibit fluorescence in the blue–green region of the visible spectrum, with band maxima in the region of 490–500 nm. Moreover, the transition from acid films A and B to the basic film C is characterized by an increase in the emission wavelength by 10 nm.

Films A′–C′ containing Rhodamine B show photoluminescence with maxima in the region of 630–640 nm, which is noticeably greater than for Rhodamine B in solution (585 nm, spectra are compared in Figure 7 and Table 1). This phenomenon can be explained by the strong self-absorption of light with wavelengths less than 600 nm in the film with the Rhodamine B molecules [28]. The result of such strong self-absorption is much lower (2–3%) fluorescence quantum yields of loaded films A′–C′ than of free Rhodamine B in solution (31%), despite the fact that the lifetime of the excited state of Rhodamine B in films is longer than in solution (on the order of 4 ns versus 1.6 ns). An additional factor that reduces the quantum yield of the loaded films is the noticeable absorption of light in the ultraviolet region by the film material itself.

The observed hypsochromic shift in the emission maximum of film C′ (633 nm) compared to its analogs A′ and B′ (both 639 nm) is most likely associated with the somewhat lower fluorescence self-quenching in this sample, due to a decrease in absorption with an increase in the pH of the Rhodamine B environment. Nevertheless, this does not lead to an increase in the quantum yield of the film C′, since an increase in pH also leads to a decrease in the luminescence intensity of Rhodamine B [29–32].

The observed structure of the emission bands of the Rhodamine B-containing films A′–C′ is characterized by the presence, in addition to the main maximum (633–639 nm), of shorter wavelength shoulders in the region of 600 nm, which gives an energy difference between these bands of approximately 1050 cm^{-1}, which is in good agreement with the vibration frequencies of aromatic systems in chromophore, and, thus, can be attributed to the vibrational structure of this spectrum. Similarly, a shoulder is observed in the spectrum of free Rhodamine B, but in a longer wavelength region (approximately 630 nm) and with a similar difference in the energies of these bands (approximately 1200 cm^{-1}).

Rhodamine B exhibits a significant dependence of the fluorescence intensity and lifetime on temperature changes and is being studied as a temperature sensor [33–39]. Therefore, it was of great interest to study the dependence of the lifetime of an excited state (a parameter largely independent of concentration, in contrast to intensity) on temperature. As the range of interest, we chose that in the physiological region, i.e., 37 ± 5 °C (see Table 1).

It turned out that blank A–C films show an insignificant response of this parameter to temperature variations (lifetime changes were approximately 0.3–0.9% per 1 °C). In turn, the films loaded with Rhodamine B showed a noticeably higher temperature sensitivity, in the region of 2%. This value is somewhat smaller than for free Rhodamine B in solution (2.4%), which can be explained by the greater rigidity of the chromophore environment in the film matrix.

It should be noted that the decay curves of the fluorescence intensity (which, upon processing, give the values of the excited state lifetimes) in the case of films A'–C' have the form of a biexponential (or multiexponential) dependence, in contrast to the strictly monoexponential decay of luminescence in the case of free Rhodamine B in solution. Most likely, this fact is explained by the heterogeneity of the state of the Rhodamine B chromophore in the film, its different local surrounding, as well as by possible π–π aggregation, which results in differences in the rates of radiative and nonradiative relaxation [40–42].

The emission band maximum in film is longer because of the self-absorption of Rhodamine B (since the Rhodamine B concentration in film is high). As a result, the shorter wavelength part of the emission band of rhodamine in film is partially absorbed by rhodamine itself. The quantum yield of Rhodamine B in film is lower than in solution because of the mentioned above self-absorption of fluorescence shorter than 600 nm by Rhodamine B itself. But the lifetime in this case is independent from absorption (since it is not FRET in its nature). Even vice versa, the lifetime in film is higher because of the rigidity raise due to the insertion of rhodamine in the polymer matrix.

As a result, we demonstrated that the obtained films can be used as luminescent materials and, in the case of films A'–C' containing Rhodamine B, as effective fluorescent temperature sensors in the physiological range. Therefore, further study and optimization of such composites looks very promising.

3.5. X-ray Diffraction Study

The X-ray phase analysis of these samples was carried out on a DRON-7 automatic X-ray diffractometer for polycrystalline materials in the step-by-step scanning mode. A 2θ angle interval from 5° to 45° with scanning step Δ2θ = 0.02° and 5 s exposure per point were used. Cu Kα radiation was used, which was subsequently decomposed into Kα1 and Kα2 components during the processing of the spectra.

Figures 8 and 9 show the diffraction patterns of the studied samples.

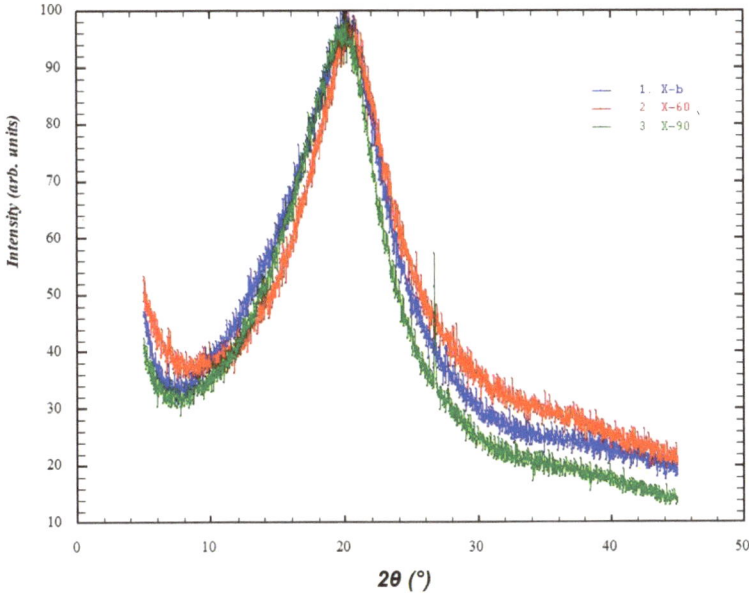

Figure 8. X-ray diffraction patterns of blank films (red—A', green—B', blue—C').

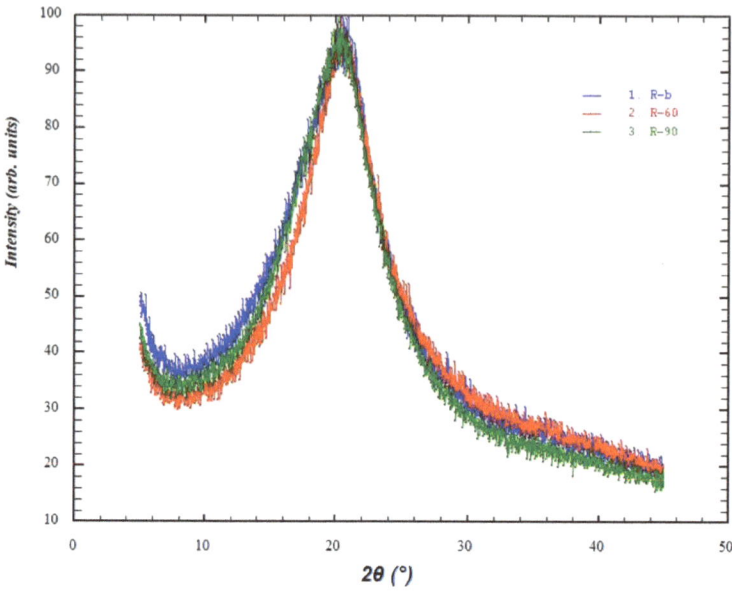

Figure 9. X-ray diffraction patterns of Rhodamine B-containing films (red—A, green—B, blue—C).

Based on the results of the X-ray diffraction studies, we assessed the degree of perfection of the structure of blank and Rhodamine B-containing films. The X-ray diffraction profiles of the amorphous peaks were approximated using the Pseudo-Voigt function. We also refined the peak position, intensity, half-width, and integral peak width. Since the films containing chitosan in its basic form were very hard and had an uneven surface, the peak broadening also occurred due to defocusing of the X-ray beam, so the samples of films in the basic form were excluded from the calculations. As is known, integral broadening is related to the degree of perfection of the structure or broadening due to the size of micro- or nanoblocks. Less integral broadening is evidence of a more perfect structure. Table 2 shows the characteristics of the integral broadening obtained from the results of refinement of the profile of the amorphous peak of the studied samples. For example, in the case of rhodamine films, the integral broadening for films dried at 90 °C (B) was less than for films dried at 60 °C (A). This indicates a more perfect structure of film B and is consistent with its increased strength (see Figures 2 and 3). The same patterns were observed for Rhodamine B-containing films. Moreover, Table 2 clearly shows that the introduction of Rhodamine B into the chitosan-based films increases the perfection of their structure.

Table 2. Integral broadening of the amorphous peak for blank and rhodamine films dried at 60° and heated at 90°.

Sample	Integral Broadening
A	9.76
B	9.51
A'	10.22
B'	10.18

3.6. Antimicrobial Activity of the Films

Antimicrobial films are of interest in clinical practice and pharmacology [43], in food industry for prolongation of food products shelf-life [12], and in agriculture as plant protecting systems [44]. The elaborated films were tested in vitro as microbial systems toward Gram-positive bacteria *S. aureus* and Gram-negative bacteria *E. coli*, and also toward fungi *A. fumigatus* and *G. candidum*. The results of the biological experiment are presented in Table 3.

Table 3. Antimicrobial effect of the elaborated films.

Sample	Inhibition Zone, mm *			
	S. aureus	E. coli	A. fumigatus	G. candidum
A	12.7 ± 0.3	9.2 ± 0.1	11.7 ± 0.2	9.8 ± 0.1
B	12.4 ± 0.2	9.0 ± 0.2	11.8 ± 0.1	9.8 ± 0.3
C	10.3 ± 0.3	7.8 ± 0.2	10.0 ± 0.1	8.6 ± 0.2
A′	14.2 ± 0.1	9.2 ± 0.2	17.4 ± 0.3	16.1 ± 0.1
B′	14.4 ± 0.1	9.2 ± 0.3	16.2 ± 0.2	13.9 ± 0.3
C′	13.7 ± 0.1	8.6 ± 0.2	13.7 ± 0.3	12.7 ± 0.1

* mean value ± S.D.

Antimicrobial activity of the blank films in salt form A and B exceeds the activity of the blank film in the base form C. This fact can be explained by the enhanced cationic density of chitosan in its salt form. The salt form of chitosan provides its polycationic nature. The chitosan polycation effectively interacts with the anionic regions of the bacterial cell membrane. This interaction results in a cascade of events unfavorable for the bacterium, including disruption of ion pumps, osmotic imbalance, and membrane rupture. All this ultimately leads to the inevitable death of the bacterial cell [45]. Thus, an increase in the cationic density of chitosan entails an increase in its antibacterial effect [26].

Antibacterial activity of the Rhodamine B-containing films, generally, is more pronounced than that of the blank films. This fact is likely due to the presence of Rhodamine B which is characterized by its strong antimicrobial effect [46]. The slightly reduced antibacterial activity of film C can be explained by the fact that this film has partially passed into the base form. The antifungal activity of Rhodamine B-containing films is much more pronounced than their antibacterial activity. The antifungal effect of the blank films is significantly less, therefore, the antifungal effect of films A′–C′ is due to Rhodamine B. The most active antifungal film is A′, which is approximately 50% more effective in comparison with the corresponding blank film. Since it is fungal spoilage that is the main part of microbial spoilage of products, the most antifungal film A′ seems to be the most promising, for example, for protecting food products.

3.7. Antioxidant Activity of the Films

On the one hand, one of the reasons for reducing the shelf life of food is oxidative spoilage, so antioxidant active food packaging is of paramount importance in the food industry [47]. On the other hand, pathological processes in wounds and burns are accompanied by oxidative stress. The use of antioxidant systems leads to a decrease in oxidative stress, a decrease in the production of cytokines and inflammatory mediators, and has a beneficial effect on regeneration processes. Therefore, antioxidation films are important in biomedical applications such as wound and burn coatings [48].

In this work, we assessed the antioxidant activity of the prepared films and compared this with a reference highly active antioxidant, i.e., ascorbic acid. The conventional approach to evaluate antioxidant activity is to estimate the capacity to trap reactive DPPH• free radical [49] (Figure 10). The best antioxidant effect is demonstrated by ascorbic acid; at a concentration 1 mg/ml, it scavenges 100% of free reactive radicals DPPH•. The lowest antioxidant effect is characteristic of blank films A–C since they are capable of binding only approximately 40% of DPPH•. The prepared Rhodamine B-containing films A′–C′ display a more pronounced antioxidant activity; at the same concentration, films A′–C′ trap approximately 80% DPPH•. It should be especially noted that within each series A–C and A′–C′, the antioxidant effect does not depend on the method of film processing (drying at 60 °C, or the same followed by 90 °C, or transferring the film to its basic form). In addition, Figure 10 demonstrates that antioxidant activity has a strong concentration dependence;

as the amount of any of the tested films in the system decreases, the antioxidant activity decreases. Of course, the increased antioxidant activity of Rhodamine B-containing films is explained by the presence of Rhodamine B in them. The literature data indicate that the capacity to bind reactive DPPH• free radical usually is provided by the H-atom of phenol or aromatic amine functionalities [50,51]. The Rhodamine B molecule contains primary aromatic amino groups, and this explains the increased antioxidant activity of the corresponding films A'–C'.

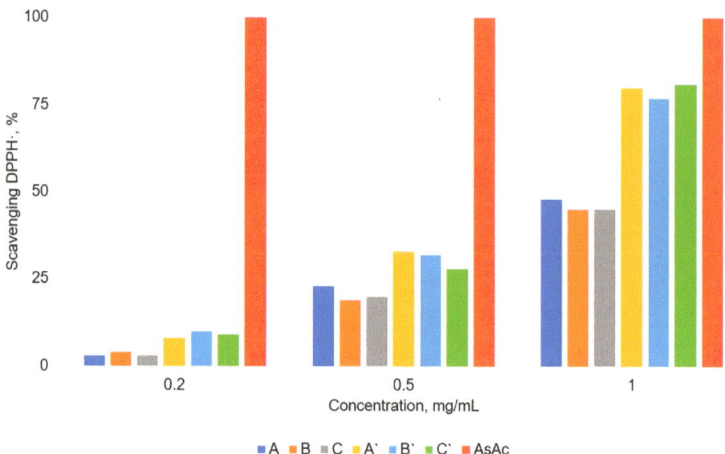

Figure 10. Antioxidant activity of the films (AsAc—ascorbic acid).

4. Conclusions

In this study, we successfully prepared chitosan-based films and films containing Rhodamine B using the casting method. The films were characterized by their mechanical properties, chemical structures, photophysical properties, and antimicrobial and antioxidant activities. The mechanical properties of the films were significantly influenced by the chemical composition and the drying process. Infrared spectroscopy confirmed the presence of chitosan in all samples, as well as the salt or base form of chitosan.

The photophysical properties of the films were studied, and it was found that Rhodamine B-containing films showed strong absorption in the visible light range, giving them an intense red color, while blank films exhibited strong absorption in the ultraviolet region. All films displayed noticeable photoluminescence. The quantum yields of fluorescence were moderate. Rhodamine B-containing films exhibited a higher temperature sensitivity in comparison to blank films, making them promising candidates for use as fluorescent temperature sensors in the physiological range.

The X-ray diffraction study of the films revealed a correlation between integral broadening and the mechanical properties of the films. The antimicrobial activity of the films was found to be higher for Rhodamine B-containing films, which is likely due to the presence of Rhodamine B and its antibacterial and especially antifungal effect. The films in salt form exhibited higher antimicrobial activity compared to those in their base form. Moreover, Rhodamine B-containing films are characterized by significantly improved antioxidant activity compared with the corresponding blank films.

These results indicate that chitosan-based films, especially those containing Rhodamine B, have promising applications in various fields such as clinical practice, food industry, and agriculture, thanks to their mechanical, photophysical, antibacterial, and antioxidant properties. Further research and optimization of these composites are warranted to enhance their potential uses.

Author Contributions: Conceptualization, O.M.K. and A.S.K.; methodology, O.M.K. and R.A.G.; software, A.R.E. and D.S.S.; validation, A.A.K., A.G.T. and A.S.K.; formal analysis, N.N.L.; investigation, O.M.K. and N.D.S.; resources, D.I.S.; data curation, A.R.E.; writing—original draft preparation, I.S.K. and O.M.K.; writing—review and editing, A.G.T.; visualization, V.N.K.; supervision, A.S.K.; project administration, V.N.K. and A.S.K.; funding acquisition, A.A.K. and A.S.K. All authors have read and agreed to the published version of the manuscript.

Funding: This paper was supported by the RUDN University Strategic Academic Leadership Program (recipient: A.A. Kirichuk, award no. 202717-0-000 "Development of a scientifically based methodology for the ecological adaptation of foreign students to new environmental conditions"). The work was carried out with the financial support of separate project of fundamental and applied scientific research "Ultrasonic synthesis of layered double hydroxides for medical purposes" (Republic of Belarus).

Institutional Review Board Statement: Not applicable.

Data Availability Statement: Data are contained within the article.

Acknowledgments: In commemoration of the 300th anniversary of St Petersburg State University's founding.

Conflicts of Interest: The authors declare no conflicts of interest.

References

1. Kumar, S.; Shukla, A.; Baul, P.P.; Mitra, A.; Halder, D. Biodegradable hybrid nanocomposites of chitosan/gelatin and silver nanoparticles for active food packaging applications. *Food Packag. Shelf Life* **2018**, *16*, 178–184. [CrossRef]
2. Rinaudo, M. Chitin and chitosan: Properties and applications. *Prog. Polym. Sci.* **2006**, *31*, 603–632. [CrossRef]
3. Lun'kov, A.P.; Shagdarova, B.T.; Zhuikova, Y.V.; Il'ina, A.V.; Varlamov, V.P. Properties of Functional Films Based on Chitosan Derivative with Gallic Acid. *Appl. Biochem. Microbiol.* **2018**, *54*, 484–490. [CrossRef]
4. Varlamov, V.P.; Il'ina, A.V.; Shagdarova, B.T.; Lunkov, A.P.; Mysyakina, I.S. Chitin/Chitosan and Its Derivatives: Fundamental Problems and Practical Approaches. *Biochemistry* **2020**, *85*, 154–176. [CrossRef] [PubMed]
5. Jayakumar, R.; Prabaharan, M.; Muzzarelli, R.A. *Chitosan for Biomaterials I*; Springer: Berlin/Heidelberg, Germany, 2011; Volume 243.
6. Kumar, S.; Ye, F.; Dobretsov, S.; Dutta, J. Chitosan nanocomposite coatings for food, paints, and water treatment applications. *Appl. Sci.* **2019**, *9*, 2409. [CrossRef]
7. Tardajos, M.G.; Cama, G.; Dash, M.; Misseeuw, L.; Gheysens, T.; Gorzelanny, C.; Coenye, T.; Dubruel, P. Chitosan functionalized poly-ε-caprolactone electrospun fibers and 3D printed scaffolds as antibacterial materials for tissue engineering applications. *Carbohydr. Polym.* **2018**, *191*, 127–135. [CrossRef]
8. Zhang, C.; Yang, X.; Li, Y.; Qiao, C.; Wang, S.; Wang, X.; Xu, C.; Yang, H.; Li, T. Enhancement of a zwitterionic chitosan derivative on mechanical properties and antibacterial activity of carboxymethyl cellulose-based films. *Int. J. Biol. Macromol.* **2020**, *159*, 1197–1205. [CrossRef]
9. Goy, R.C.; Britto, D.d.; Assis, O.B.G. A review of the antimicrobial activity of chitosan. *Polímeros* **2009**, *19*, 241–247. [CrossRef]
10. Albadarin, A.B.; Collins, M.N.; Naushad, M.; Shirazian, S.; Walker, G.; Mangwandi, C. Activated lignin-chitosan extruded blends for efficient adsorption of methylene blue. *Chem. Eng. J.* **2017**, *307*, 264–272. [CrossRef]
11. Elsabee, M.Z.; Abdou, E.S. Chitosan based edible films and coatings: A review. *Mater. Sci. Eng. C* **2013**, *33*, 1819–1841. [CrossRef] [PubMed]
12. Kumar, S.; Mukherjee, A.; Dutta, J. Chitosan based nanocomposite films and coatings: Emerging antimicrobial food packaging alternatives. *Trends Food Sci. Technol.* **2020**, *97*, 196–209. [CrossRef]
13. Yuan, G.; Chen, X.; Li, D. Chitosan films and coatings containing essential oils: The antioxidant and antimicrobial activity, and application in food systems. *Food Res. Int.* **2016**, *89*, 117–128. [CrossRef]
14. Zheng, L.; Hu, X.; Wu, H.; Mo, L.; Xie, S.; Li, J.; Peng, C.; Xu, S.; Qiu, L.; Tan, W. In Vivo Monocyte/Macrophage-Hitchhiked Intratumoral Accumulation of Nanomedicines for Enhanced Tumor Therapy. *J. Am. Chem. Soc.* **2020**, *142*, 382–391. [CrossRef]
15. Vanamudan, A.; Pamidimukkala, P. Chitosan, nanoclay and chitosan-nanoclay composite as adsorbents for Rhodamine-6G and the resulting optical properties. *Int. J. Biol. Macromol.* **2015**, *74*, 127–135. [CrossRef]
16. Benbettaïeb, N.; Chambin, O.; Assifaoui, A.; Al-Assaf, S.; Karbowiak, T.; Debeaufort, F. Release of coumarin incorporated into chitosan-gelatin irradiated films. *Food Hydrocoll.* **2016**, *56*, 266–276. [CrossRef]
17. Setiawan, D.; Kazaryan, A.; Martoprawiro, M.A.; Filatov, M. A first principles study of fluorescence quenching in rhodamine B dimers: How can quenching occur in dimeric species? *Phys. Chem. Chem. Phys.* **2010**, *12*, 11238–11244. [CrossRef] [PubMed]
18. Chen, X.; Jia, J.; Ma, H.; Wang, S.; Wang, X. Characterization of rhodamine B hydroxylamide as a highly selective and sensitive fluorescence probe for copper(II). *Anal. Chim. Acta* **2009**, *632*, 9–14. [CrossRef] [PubMed]

19. Kritchenkov, A.S.; Luzyanin, K.V.; Bokach, N.A.; Kuznetsov, M.L.; Gurzhiy, V.V.; Kukushkin, V.Y. Selective Nucleophilic Oxygenation of Palladium-Bound Isocyanide Ligands: Route to Imine Complexes That Serve as Efficient Catalysts for Copper-/Phosphine-Free Sonogashira Reactions. *Organometallics* **2013**, *32*, 1979–1987. [CrossRef]
20. Kritchenkov, A.S.; Egorov, A.R.; Volkova, O.V.; Zabodalova, L.A.; Suchkova, E.P.; Yagafarov, N.Z.; Kurasova, M.N.; Dysin, A.P.; Kurliuk, A.V.; Shakola, T.V.; et al. Active antibacterial food coatings based on blends of succinyl chitosan and triazole betaine chitosan derivatives. *Food Packag. Shelf Life* **2020**, *25*, 100534. [CrossRef]
21. Kritchenkov, A.S.; Zhaliazniak, N.V.; Egorov, A.R.; Lobanov, N.N.; Volkova, O.V.; Zabodalova, L.A.; Suchkova, E.P.; Kurliuk, A.V.; Shakola, T.V.; Rubanik, V.V.; et al. Chitosan derivatives and their based nanoparticles: Ultrasonic approach to the synthesis, antimicrobial and transfection properties. *Carbohydr. Polym.* **2020**, *242*, 116478. [CrossRef] [PubMed]
22. Kritchenkov, A.S.; Egorov, A.R.; Volkova, O.V.; Kritchenkov, I.S.; Kurliuk, A.V.; Shakola, T.V.; Khrustalev, V.N. Ultrasound-assisted catalyst-free phenol-yne reaction for the synthesis of new water-soluble chitosan derivatives and their nanoparticles with enhanced antibacterial properties. *Int. J. Biol. Macromol.* **2019**, *139*, 103–113. [CrossRef] [PubMed]
23. Shi, M.-J.; Wei, X.; Xu, J.; Chen, B.-J.; Zhao, D.-Y.; Cui, S.; Zhou, T. Carboxymethylated degraded polysaccharides from Enteromorpha prolifera: Preparation and in vitro antioxidant activity. *Food Chem.* **2017**, *215*, 76–83. [CrossRef] [PubMed]
24. van den Broek, L.A.M.; Knoop, R.J.I.; Kappen, F.H.J.; Boeriu, C.G. Chitosan films and blends for packaging material. *Carbohydr. Polym.* **2015**, *116*, 237–242. [CrossRef] [PubMed]
25. Gerassimidou, S.; Geueke, B.; Groh, K.J.; Muncke, J.; Hahladakis, J.N.; Martin, O.V.; Iacovidou, E. Unpacking the complexity of the polyethylene food contact articles value chain: A chemicals perspective. *J. Hazard. Mater.* **2023**, *454*, 131422. [CrossRef] [PubMed]
26. Khubiev, O.M.; Esakova, V.E.; Egorov, A.R.; Bely, A.E.; Golubev, R.A.; Tachaev, M.V.; Kirichuk, A.A.; Lobanov, N.N.; Tskhovrebov, A.G.; Kritchenkov, A.S. Novel Non-Toxic Highly Antibacterial Chitosan/Fe(III)-Based Nanoparticles That Contain a Deferoxamine—Trojan Horse Ligands: Combined Synthetic and Biological Studies. *Processes* **2023**, *11*, 870.
27. Li, Y.L.; Wang, W.X.; Wang, Y.; Zhang, W.B.; Gong, H.M.; Liu, M.X. Synthesis and Characterization of Rhodamine B-ethylenediamine-hyaluronan Acid as Potential Biological Functional Materials. *IOP Conf. Ser. Mater. Sci. Eng.* **2018**, *359*, 012040. [CrossRef]
28. Arbeloa, F.L.; Ojeda, P.R.; Arbeloa, I.L. Flourescence self-quenching of the molecular forms of Rhodamine B in aqueous and ethanolic solutions. *J. Lumin.* **1989**, *44*, 105–112. [CrossRef]
29. Knauer, K.-H.; Gleiter, R. Photochromism of Rhodamine Derivatives. *Angew. Chem. Int. Ed. Engl.* **1977**, *16*, 113. [CrossRef]
30. Yang, Y.; Zhao, Q.; Feng, W.; Li, F. Luminescent Chemodosimeters for Bioimaging. *Chem. Rev.* **2013**, *113*, 192–270. [CrossRef]
31. Srivastava, P.; Fürstenwerth, P.C.; Witte, J.F.; Resch-Genger, U. Synthesis and spectroscopic characterization of a fluorescent phenanthrene-rhodamine dyad for ratiometric measurements of acid pH values. *New J. Chem.* **2021**, *45*, 13755–13762. [CrossRef]
32. Semenov, K.N.; Charykov, N.A.; Keskinov, V.A.; Kritchenkov, A.S.; Murin, I.V. Fullerenol-d Solubility in Fullerenol-d-Inorganic Salt-Water Ternary Systems at 25 degrees C. *Ind. Eng. Chem. Res.* **2013**, *52*, 16095–16100. [CrossRef]
33. Kitamura, N.; Hosoda, Y.; Iwasaki, C.; Ueno, K.; Kim, H.-B. Thermal Phase Transition of an Aqueous Poly(N-isopropylacrylamide) Solution in a Polymer Microchannel-Microheater Chip. *Langmuir* **2003**, *19*, 8484–8489. [CrossRef]
34. Benninger, R.K.P.; Koç, Y.; Hofmann, O.; Requejo-Isidro, J.; Neil, M.A.A.; French, P.M.W.; deMello, A.J. Quantitative 3D Mapping of Fluidic Temperatures within Microchannel Networks Using Fluorescence Lifetime Imaging. *Anal. Chem.* **2006**, *78*, 2272–2278. [CrossRef]
35. Müller, C.B.; Weiß, K.; Loman, A.; Enderlein, J.; Richtering, W. Remote temperature measurements in femto-liter volumes using dual-focus-Fluorescence Correlation Spectroscopy. *Lab Chip* **2009**, *9*, 1248–1253. [CrossRef]
36. Mercadé-Prieto, R.; Rodriguez-Rivera, L.; Chen, X.D. Fluorescence lifetime of Rhodamine B in aqueous solutions of polysaccharides and proteins as a function of viscosity and temperature. *Photochem. Photobiol. Sci.* **2017**, *16*, 1727–1734. [CrossRef]
37. Kitchenkov, I.S.; Melnikov, A.S.; Serdobintsev, P.S.; Khodorkovskii, M.A.; Pavlovskii, V.V.; Porsev, V.V.; Tunik, S.P. Energy Transfer Processes in the Excited States of an {[Ir(N C)2(N N)]+-Rhodamine} Dyad: An Experimental and Theoretical Study. *ChemPhotoChem* **2022**, *6*, e202200048. [CrossRef]
38. Kritchenkov, A.S.; Bokach, N.A.; Starova, G.L.; Kukushkin, V.Y. A palladium(II) center activates nitrile ligands toward 1,3-dipolar cycloaddition of nitrones substantially more than the corresponding platinum(II) center. *Inorg. Chem.* **2012**, *51*, 11971–11979. [CrossRef]
39. Tskhovrebov, A.G.; Novikov, A.S.; Tupertsev, B.S.; Nazarov, A.A.; Antonets, A.A.; Astafiev, A.A.; Kritchenkov, A.S.; Kubasov, A.S.; Nenajdenko, V.G.; Khrustalev, V.N. Azoimidazole gold(III) complexes: Synthesis, structural characterization and self-assembly in the solid state. *Inorg. Chim. Acta* **2021**, *522*, 120373. [CrossRef]
40. López Arbeloa, I.; Ruiz Ojeda, P. Dimeric states of rhodamine B. *Chem. Phys. Lett.* **1982**, *87*, 556–560. [CrossRef]
41. Ilich, P.; Mishra, P.K.; Macura, S.; Burghardt, T.P. Direct observation of rhodamine dimer structures in water. *Spectrochim. Acta A Mol. Biomol. Spectrosc.* **1996**, *52*, 1323–1330. [CrossRef]
42. McHedlov-Petrosyan, N.O.; Kholin, Y.V. Aggregation of Rhodamine B in Water. *Russ. J. Appl. Chem.* **2004**, *77*, 414–422. [CrossRef]
43. Wang, P.; Yin, B.; Dong, H.; Zhang, Y.; Zhang, Y.; Chen, R.; Yang, Z.; Huang, C.; Jiang, Q. Coupling Biocompatible Au Nanoclusters and Cellulose Nanofibrils to Prepare the Antibacterial Nanocomposite Films. *Front. Bioeng. Biotechnol.* **2020**, *8*, 986. [CrossRef]
44. Nazarov, P.A.; Baleev, D.N.; Ivanova, M.I.; Sokolova, L.M.; Karakozova, M.V. Infectious Plant Diseases: Etiology, Current Status, Problems and Prospects in Plant Protection. *Acta Nat.* **2020**, *12*, 46–59. [CrossRef]

45. Egorov, A.R.; Artemjev, A.A.; Kozyrev, V.A.; Sikaona, D.N.; Rubanik, V.V.; Rubanik, V.V., Jr.; Kritchenkov, I.S.; Yagafarov, N.Z.; Khubiev, O.M.; Tereshina, T.A.; et al. Synthesis of Selenium-Containing Chitosan Derivatives and Their Antibacterial Activity. *Appl. Biochem. Microbiol.* **2022**, *58*, 132–135. [CrossRef]
46. Bucki, R.; Pastore, J.J.; Randhawa, P.; Vegners, R.; Weiner, D.J.; Janmey, P.A. Antibacterial activities of rhodamine B-conjugated gelsolin-derived peptides compared to those of the antimicrobial peptides cathelicidin LL37, magainin II, and melittin. *Antimicrob. Agents Chemother.* **2004**, *48*, 1526–1533. [CrossRef]
47. Rangaraj, V.M.; Rambabu, K.; Banat, F.; Mittal, V. Natural antioxidants-based edible active food packaging: An overview of current advancements. *Food Biosci.* **2021**, *43*, 101251. [CrossRef]
48. Fadilah, N.I.M.; Phang, S.J.; Kamaruzaman, N.; Salleh, A.; Zawani, M.; Sanyal, A.; Maarof, M.; Fauzi, M.B. Antioxidant Biomaterials in Cutaneous Wound Healing and Tissue Regeneration: A Critical Review. *Antioxidants* **2023**, *12*, 787. [CrossRef]
49. Xia, W.; Wei, X.-Y.; Xie, Y.-Y.; Zhou, T. A novel chitosan oligosaccharide derivative: Synthesis, antioxidant and antibacterial properties. *Carbohydr. Polym.* **2022**, *291*, 119608. [CrossRef] [PubMed]
50. Zeb, A. Concept, mechanism, and applications of phenolic antioxidants in foods. *J. Food Biochem.* **2020**, *44*, e13394. [CrossRef] [PubMed]
51. Liu, Y.; Li, X.; Pu, Q.; Fu, R.; Wang, Z.; Li, Y.; Li, X. Innovative screening for functional improved aromatic amine derivatives: Toxicokinetics, free radical oxidation pathway and carcinogenic adverse outcome pathway. *J. Hazard. Mater.* **2023**, *454*, 131541. [CrossRef] [PubMed]

Disclaimer/Publisher's Note: The statements, opinions and data contained in all publications are solely those of the individual author(s) and contributor(s) and not of MDPI and/or the editor(s). MDPI and/or the editor(s) disclaim responsibility for any injury to people or property resulting from any ideas, methods, instructions or products referred to in the content.

Article

Specific FRET Probes Sensitive to Chitosan-Based Polymeric Micelles Formation, Drug-Loading, and Fine Structural Features

Igor D. Zlotnikov, Ivan V. Savchenko and Elena V. Kudryashova *

Faculty of Chemistry, Lomonosov Moscow State University, Leninskie Gory, 1/3, 119991 Moscow, Russia; zlotnikovid@my.msu.ru (I.D.Z.)
* Correspondence: helenakoudriachova@yandex.ru

Abstract: Förster resonance energy transfer (FRET) probes are a promising tool for studying numerous biochemical processes. In this paper, we show the application of the FRET phenomenon to observe the micelle formation from surfactants, micelles self-assembling from chitosan grafted with fatty acid (oleic—OA, or lipoic—LA), cross-linking of SH groups in the micelle's core, and inclusion and release of the model drug cargo from the micelles. Using the carbodiimide approach, amphiphilic chitosan-based polymers with (1) SH groups, (2) crosslinked with S-S between polymer chains, and (3) without SH and S-S groups were synthesized, followed by characterization by FTIR and NMR spectroscopy. Two pairs of fluorophores were investigated: 4-methylumbelliferon-trimethylammoniocinnamate—rhodamine (MUTMAC–R6G) and fluorescein isothiocyanate—rhodamine (FITC–R6G). While FITC–R6G has been described before as an FRET-producing pair, for MUTMAC–R6G, this has not been described. R6G, in addition to being an acceptor fluorophore, also serves as a model cytostatic drug in drug-release experiments. As one could expect, in aqueous solution, FRET effect was poor, but when exposed to the micelles, both MUTMAC–R6G and FITC–R6G yielded a pronounced FRET effect. Most likely, the formation of micelles is accompanied by the forced convergence of fluorophores in the hydrophobic micelle core by a donor-to-acceptor distance (**r**) significantly closer than in the aqueous buffer solution, which was reflected in the increase in the FRET efficiency (**E**). Therefore, **r(E)** could be used as analytical signal of the micelle formation, including critical micelle concentration (CMC) and critical pre-micelle concentration (CPMC), yielding values in good agreement with the literature for similar systems. We found that the **r**-function provides analytically valuable information about the nature and mechanism of micelle formation. S-S crosslinking between polymer chains makes the micelle more compact and stable in the normal physiological conditions, but loosens in the glutathione-rich tumor microenvironment, which is considered as an efficient approach in targeted drug delivery. Indeed, we found that R6G, as a model cytostatic agent, is released from micelles with initial rate of 5%/h in a normal tissue microenvironment, but in a tumor microenvironment model (10 mM glutathione), the release of R6G from S-S stitched polymeric micelles increased up to 24%/h. Drug-loading capacity differed substantially: from 75–80% for nonstitched polymeric micelles to ~90% for S-S stitched micelles. Therefore, appropriate FRET probes can provide comprehensive information about the micellar system, thus helping to fine-tune the drug delivery system.

Keywords: FRET probes; rhodamine 6G; chitosan; polymeric micelles; surfactants; stimulus-sensitivity; tumor microenvironment

Citation: Zlotnikov, I.D.; Savchenko, I.V.; Kudryashova, E.V. Specific FRET Probes Sensitive to Chitosan-Based Polymeric Micelles Formation, Drug-Loading, and Fine Structural Features. *Polymers* **2024**, *16*, 739. https://doi.org/10.3390/polym16060739

Academic Editors: Luminita Marin, William Facchinatto and Sérgio Paulo Campana-Filho

Received: 18 January 2024
Revised: 28 February 2024
Accepted: 7 March 2024
Published: 8 March 2024

Copyright: © 2024 by the authors. Licensee MDPI, Basel, Switzerland. This article is an open access article distributed under the terms and conditions of the Creative Commons Attribution (CC BY) license (https://creativecommons.org/licenses/by/4.0/).

1. Introduction

In the last decade, Förster resonance energy transfer (FRET) between fluorophore molecules [1–4] has been actively developed as a quantitative approach to determine a number of biochemical parameters in real time [5]. This approach turned out to be advantageous, providing high sensitivity and selectivity, since the FRET effect reflects the specific molecular organization in the system. In addition, changes in the FRET signal can be monitored online upon the supramolecular assembly self-organization process [6], which is undoubtedly an advantage over other methods such as electron microscopy, radioactive tagging, and dynamic light scattering. It is worth noting that the FRET signal can be used to create powerful sensors in the biological research and medical applications.

FRET can be observed when the emission spectrum of the donor overlaps with the excitation spectrum of the acceptor, and the distance at which the energy transfer can occur is limited to ~10 nm. The quantum yield of this energy-transfer transition, FRET efficiency (E), is determined by the donor-to-acceptor distance r [7–10]:

$$E = 1/(1 + (r/R_0)^6), \qquad (1)$$

where R_0 is the Förster distance of the given pair donor–acceptor, which can range from 10 to 100 Å. The Förster radius R_0 for fluorescein isothiocyanate (FITC) and rhodamine 6G (R6G) is ~50 Å, and for—4-methylumbelliferyl p-trimethylammoniocinnamate chloride (MUTMAC) and R6G is about 60 Å.

The choice of fluorophores for FRET probes is also justified by their potential as potential medicines: Rhodamine 6G and its derivatives [11,12], as well as Coumarin and derivatives, are model cytostatics proposed for use as medicines [13] and as model fluorophores for studying the loading degree.

Since even small changes in the donor–acceptor distance (r/R_0) crucially affect FRET efficiency, the FRET-based approach can be considered as a powerful tool for the studies involving accurate estimation of the inter- and intramolecular distances, in the molecular dynamics assays, molecular interactions, and binding events. Interestingly, the FRET approach seems promising in the study of the formation and functional properties of polymeric micelles [14,15], the most popular drug delivery systems. Among advantages of polymeric micelles as drug carriers are their ability to encapsulate a wide array of hydrophobic and poorly soluble therapeutic agents, coupled with their propensity for prolonged circulation and passive tumor targeting through the enhanced permeability and retention (EPR) effect [6].

In this paper, we investigated the role of the micelles formation in FRET phenomenon, where the FRET effect is expected to be increased due to concentration and convergence of the donor–acceptor agents caused by its specific hydrophobic–hydrophilic phase distribution in micelles. Therefore, two main applications of FRET probes are considered: (1) Determination of the CMC and CPMC (critical micelle and pre-micelle concentrations, respectively) values, and (2) the study of the kinetics of the formation and destruction of S-S bonds in the tumor microenvironment using the example of stimulus-sensitive micelles from molecules of chitosan grafted with lipoic acid residues.

In the case of classical micelles, the spontaneous formation of spherical particles associated with surfactant molecules (SDS, Triton X-100, etc.) leads to the loading of the aromatic fluorophore molecules into the hydrophobic micelle core, which can be used in the observation of FRET during micelle formation [16]. The FRET phenomenon depends on the environment of the fluorophores (buffer, cationic/anionic/zwitterionic, or neutral surfactant): energy transfer on rhodamine is active in anionic/nonionic media [16].

We used surfactants (control systems with parameters (CMC) described in the literature) for validation of the FRET-based approach in order to proceed further in the investigation of grafted chitosan polymeric micelles. Recently, we suggested the approach where FRET was used as an effective tool for monitoring the formation of micro-/nanogels [5,17]. We showed that the formation of chitosan nanogels promotes the interaction of pyrene covalently attached to chitosan with added model drug molecules of tryptophan (biologically active substance), which is necessary for the appearance of the FRET effect and which is not observed in the solution before nanogel formation.

The study of micelle formation is important from the point of view of creating smart delivery systems for antibacterial and antitumor drugs [18–31]. FRET is applicable for studying the formation of various types of nanoparticles based on polymers (chitosan, chitosan-PEG) and proteins (ovalbumin, casein, etc.) [32]. Nanoparticles, along with micelles, deserve special attention as promising drug carriers [33–36]. Polymeric micelles are promising carriers of a wide range of drugs, since they have a number of properties [37–43]: (1) the external hydrophilic shell ensures the colloidal stability of the system; (2) the internal hydrophobic core is necessary for the solubilization of drugs, which are often poorly soluble (which limits their use in medicine); (3) thermodynamic stability; (4) the possibility of obtaining biocompatible micellar structures; (5) increased permeability of the drug to target cells due to fatty acids; (6) wide possibilities for creating stimulus-sensitive delivery systems, for example, in tumors. For the latter, chitosan demonstrated pH sensitivity to a slightly acidic environment (tumors), and lipoic acid residues with S-S bonds between various polymer chains provided glutathione sensitivity [44]. Here, we propose to study the mechanisms of formation of such micelles using the FRET probe technique based on changes in the FRET efficiency and the distance between fluorophores during aggregation and disaggregation of amphiphilic molecules.

2. Materials and Methods

2.1. Reagents

Surfactants SDS (sodium dodecyl sulfate), Triton X-100 and zephirol (N-benzoyl-N,N-dimethyldodecan-1-ammonium chloride) were purchased from Reachim (Moscow, Russia). The fluorophores rhodamine 6G (R6G), fluorescein isothiocyanate (FITC), 4-methylumbelliferyl p-trimethylammoniocinnamate chloride (MUTMAC), 4-methylumbelliferone (MUmb); chitosan oligosaccharide lactate 5 kDa (Chit5), lipoic acid (LA), 1-ethyl-3-(3-dimethylaminopropyl) carbodiimide (EDC), N-hydroxysuccinimide (NHS), 1 M 2,4,6-trinitrobenzenesulfonic acid, and the enzyme α-chymotrypsin from bovine pancreas (EC 3.4.21.1, ≥40 units/mg protein) were purchased from Sigma-Aldrich (St. Louis, MO, USA).

2.2. Synthesis of Chitosan Grafted with Lipoic Acid (Chit5-LA) and Oleic Acid (Chit5-OA)—Micelles Preparation

The synthesis of modified chitosan was carried out as described by us earlier with some modifications [45–47]. Chitosan was dissolved in 1 mM HCl solution (10 mg/mL) and then the pH was adjusted to 7.4 using 0.1 M phosphate buffer. Lipoic acid was dissolved in PBS/EtOH (50/50 v/v) to a concentration of 20 mg/mL. NHS and EDC were dissolved in EtOH (50 mg/mL). The crosslinking reaction was carried out using a carbodiimide approach, for which the solutions described above were mixed so as to obtain the Chit5/LA/EDC/NHS mass ratios = 1/0.33/3/1, for OA 1/0.35/3/1. The mixture was incubated for 6 h at a temperature of 50 °C. The product was then purified by three-stage dialysis against water (12 h × 3, cut-off 3.5 kDa). The polymer was freeze-dried at −70 °C.

Amphiphilic chitosan-based polymers (1 nM–50 μM) were mixed with FRET probes (1 μM) in PBS (0.01 M, pH 7.4), and the mixtures were then incubated at 37 °C for 1 h. Micelle samples were obtained by ultrasonic treatment of solutions (22 kHz) for 15 min with constant cooling in an ultrasonic device (Cole-Parmer, Vernon Hills, IL, USA). Micellar solutions were extruded (5-fold, 400 nm membrane, Avanti Polar Lipids). The free fluorophores were then separated by dialysis against PBS (with a cut-off mass of 8 kDa), and the degree of loading was then determined by fluorescence intensity: (1) For MUTMAC λ_{exci} = 360 nm, λ_{emi} = 450 nm; (2) for R6G λ_{exci} = 515 nm, λ_{emi} = 550 nm; (3) for FITC λ_{exci} = 490 nm, λ_{emi} = 520 nm were used.

2.3. Characterization of Chitosan Grafted with Lipoic Acid (Chit5-LA)

The characterization of chitosan grafted with lipoic acid (Chit5-LA) was carried out by the methods of FTIR, ^{1}H NMR spectroscopy, atomic force microscopy, and circular dichroism spectroscopy.

FTIR spectra of Chit5, LA, OA, Chit5-LA, and Chit5-OA were recorded using an FTIR microscope MICRAN-3 and Bruker Tensor 27 spectrometer equipped with a liquid-nitrogen-cooled MCT (mercury cadmium telluride) detector, as described earlier [45,48].

^{1}H NMR spectra of samples (7–10 mg/mL in D$_2$O) were recorded on a Bruker Avance 400 spectrometer (Germany, 400 MHz). FTIR and NMR spectroscopy was used to calculate the modification degree of chitosan.

Circular dichroism spectroscopy (Jasco J-815 CD Spectrometer, Tokyo, Japan) were used to estimate the deacetylation degree in Chitosan, which amounted to (92 ± 3)%.

Atomic force microscopy (AFM microscope NTEGRA II) was used to visualize polymeric micelles based on grafted chitosan and compare it in terms of shape and size with nonmodified chitosan.

The degree of chitosan modification by fatty acid residues was determined by a well-proven method of spectrophotometric titration of amino groups using 2,4,6-trinitrobenzenesulfonic acid forming colored adduct with amino groups (absorption at 420 nm). To 300 µL of solutions of modified and unmodified chitosan (0.03–0.2 mg/mL) in 0.02 M Na-borate buffer (pH 9.2), 3 µL of 1 M solution of trinitrobenzenesulfonic acid (TNBS) was added, and then kinetic curves at 420 nm (A420) were recorded for an hour. The grafting degree was calculated from the change in A420 relative to unmodified chitosan.

Hemolytic activity and thrombogenicity are the primary parameters for evaluating the safety of medical formulations. For chitosan and polymer micelles in concentrations up to 1 mg/mL, the values of hemolytic activity and thrombogenicity did not exceed 1–2%.

2.4. FRET Probes for Determination of CMC for Micelles Formed from Surfactants and Chit5-LA

2.4.1. Determination of CMC for Micelles Formed from Surfactants

FRET probes are two pairs of fluorophores FITC–R6G and MUTMAC–R6G, where for both, R6G is the acceptor. We chose surfactants zephirol, Triton X-100, and SDS as amphiphilic compounds for studying micelles formation.

The excitation and emission spectra of fluorescence were recorded on the device Varian Cary Eclipse fluorescence spectrometer (Agilent Technologies, Santa Clara, CA, USA). For FRET probe 1 (MUTMAC + R6G), λ_{exci} = 360 nm, λ_{emi} = 450 nm (donor), and 550 nm (acceptor) were used. For FRET probe 2 (FITC + R6G), λ_{exci} = 460 nm, λ_{emi} = 520 nm (donor), and 550 nm (acceptor) were used.

The final concentration of fluorophores was 1 µg/mL. Fluorophore emission and excitation spectra were recorded for each separately and for a donor–acceptor mixture in a buffer solution (PBS 0.01 M, pH 7.4) in the absence of surfactants and in its presence of various amounts.

FRET efficiency E was calculated as

$$E = 1 - F_{DA}/F_D \qquad (2)$$

for MUTMAC + R6G pair and as

$$E = F_{AD}/F_A - 1 \qquad (3)$$

for FITC + R6G pair. Where F_{DA} and F_D—the intensities of donor fluorescence in the presence and absence of the acceptor, respectively; F_{AD} and F_A—the intensities of acceptor fluorescence in the presence and absence of the donor, respectively.

The ratio r/R_0 was calculated as an analytical signal of micelle formation;

$$r/R_0 = (1/E - 1)^{(1/6)} \qquad (4)$$

where r is the distance between donor and acceptor and R_0 is Förster radius. Förster distance was calculated based on an assumption that orientation factor (κ^2) is 0.667. Critical micelle concentration (CMC and CPMC) was estimated using x-coordinate of a point on the right branch of the graph (r/R_0 versus surfactant concentration) with the value r/R_0 equal to the initial one (for a pair of fluorophores in a buffer solution without surfactants).

2.4.2. Determination of CMC for Polymeric Micelles

We chose chitosan grafted with lipoic acid (Chit5-LA) that formed S-S bonds, and as a control, chitosan grafted with oleic acid (Chit5-OA), as amphiphilic compounds for studying micelles formation.

The formation of S-S bonds between Chit5-LA polymeric chains was studied using FRET probes, registering their fluorescence as described above. First, dithiothreitol was added to the self-assembled Chit5-LA samples (0.05 mg/mL) to the final concentration of 0.2 mg/mL, and incubated for 30 min at 37 °C, followed by oxidized glutathione GSSG addition to the final concentration of 2–5 mg/mL. The fluorescence values were recorded before and after the addition of each component.

2.4.3. Flow Cytometry for Micelle Formation Study

A CytoFLEX S flow cytometer (Beckman Coulter) was used to study micelles with fluorophore (R6G). The polymers (Chit5-LA, Chit5-(oleic acid)) were incubated with pure rhodamine 6G (5 µg/mL) for 15 min, then treated with ultrasound. We used a 488 nm laser for excitation. The fluorescence emissions were collected using a 585/42 nm bandpass filter for 30,000 micelles for each sample. The collected data were then analyzed using CytExpert software (v. 2.0).

2.4.4. Release of R6G from Micelles by Addition of Reduced Glutathione as Thiol-Disulfide Exchange Agent (Tumor Microenvironment Model)

R6G-loaded micelles formed from self-assembled polymers (Chit5-LA and control Chit5-(oleic acid) without S-H bonds) were prepared in PBS (pH = 7.4, 0.01 M) after ultrasound treatment of 1 mL of each sample: polymer solution (2 mg/mL) + R6G solution (0.1 mg/mL). Further, reduced glutathione (as thiol-disulfide exchange agent) was added to the samples to destroy S-S bonds in micelles at concentrations of 0, 0.2, and 3 mg/mL. Release of R6G from micelles was studied using dialysis technique (6–8 kDa cut-off, 150 rpm) to external 10 mL PBS buffer solution at 37 °C. R6G in external solution was detected by absorption at 515 nm and fluorescence intensity at λ_{exci} = 515, λ_{emi} = 550 nm.

2.5. Enzyme Activity Studies for Determination of the Fluorophore Inclusion Degree in Micelles

The catalytic activity of α–chymotrypsin was determined fluorometrically on the device Varian Cary Eclipse fluorescence spectrometer (Agilent Technologies, Santa Clara, CA, USA). The reaction rate was measured at λ_{exci} = 360 nm, λ_{emi} = 450 nm, and T = 37 °C in PBS (0.01 M, pH 7.4) by the accumulation of the fluorescent product (MUTMAC --> MUmb): specific parameters are indicated in the captions of the tables and figures. The concentration of chymotrypsin was optimized as follows: We varied the concentration of chymotrypsin in the range of 0.05–3 µM, and chose the optimal concentration of 0.4 µM so that the initial section of the kinetic curve was linear for at least 1–2 min and the substrate was consumed within about 2–5 min, and not instantly. This approach allowed the determination of the concentrations of the MUTMAC fluorophore substrate from 0.01 mM to 1 mM.

3. Results

3.1. Article Design

The present work is aimed at studying the applications of the FRET effect as a selective indicator of surfactant molecules aggregation and as a tool for studying the promising drug carriers—polymeric micelles formed from chitosan-fatty acid conjugates. The first stage of the work is the validation of the FRET probe technique with classical surfactant micelles: we study the effect of charge, size, geometry, and degree of surfactants aggregation in micelles on the FRET effectiveness, and its correlations with micelle formation (CMC, CPMC). On the basis of this, the developed FRET technique, using chosen donor–acceptor pairs, was used to study the mechanisms of formation of stimulus-sensitive polymer micelles (with S-S bonds) to the tumor microenvironment (low pH and increased concentrations of glutathione, GSH). We considered the formation of micelles from surfactants of different structure (cationic Zephirol, anionic SDS, and neutral Triton X-100) and modified polymers (chitosan grafted with fatty acid), where FRET occurs between two fluorophores pairs (R6G with MUTMAC or FITC) due to their convergence in the core of the micelle. Objects of research (Figure 1): (i) Two pairs of fluorophores, FITC–R6G and MUTMAC–R6G; (ii) surfactants zephirol, Triton X-100, and SDS; (iii) chitosan grafted with lipoic (with S-S bond forming function) and oleic (without S-S bond) acid.

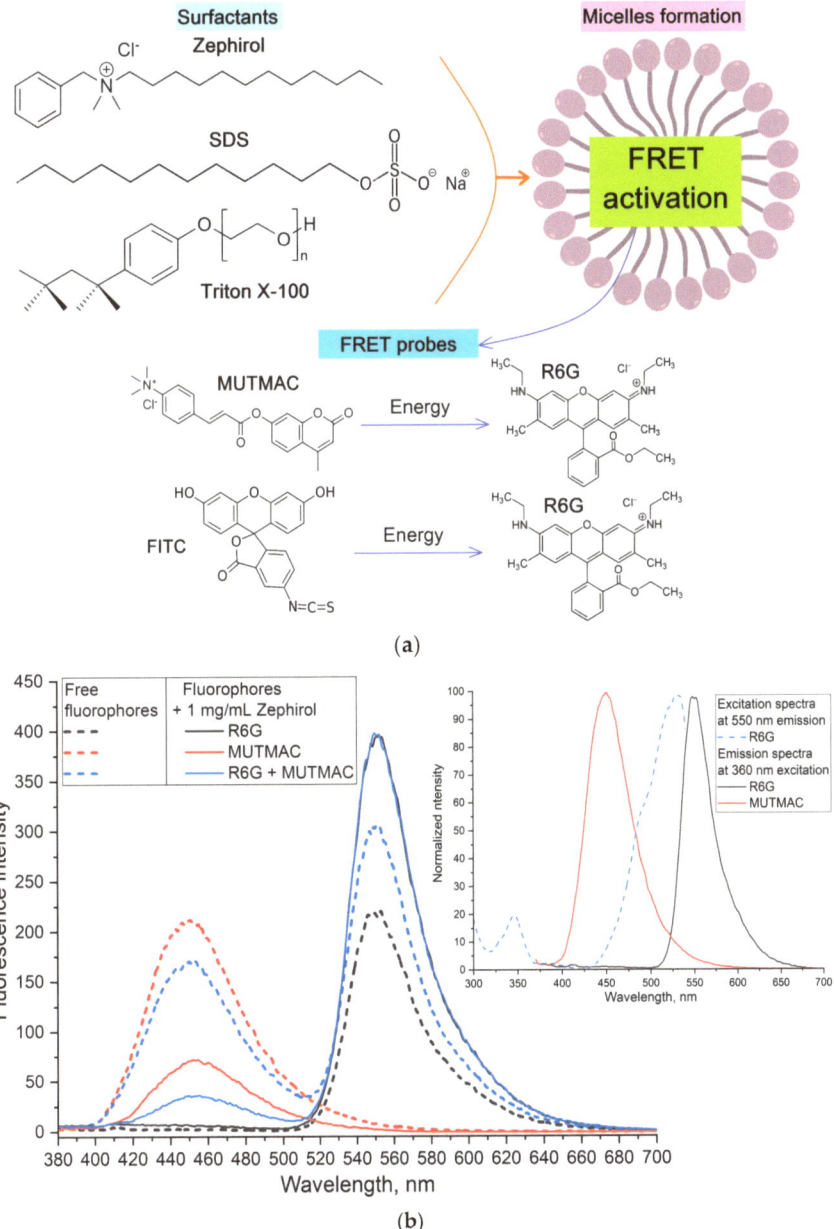

Figure 1. Cont.

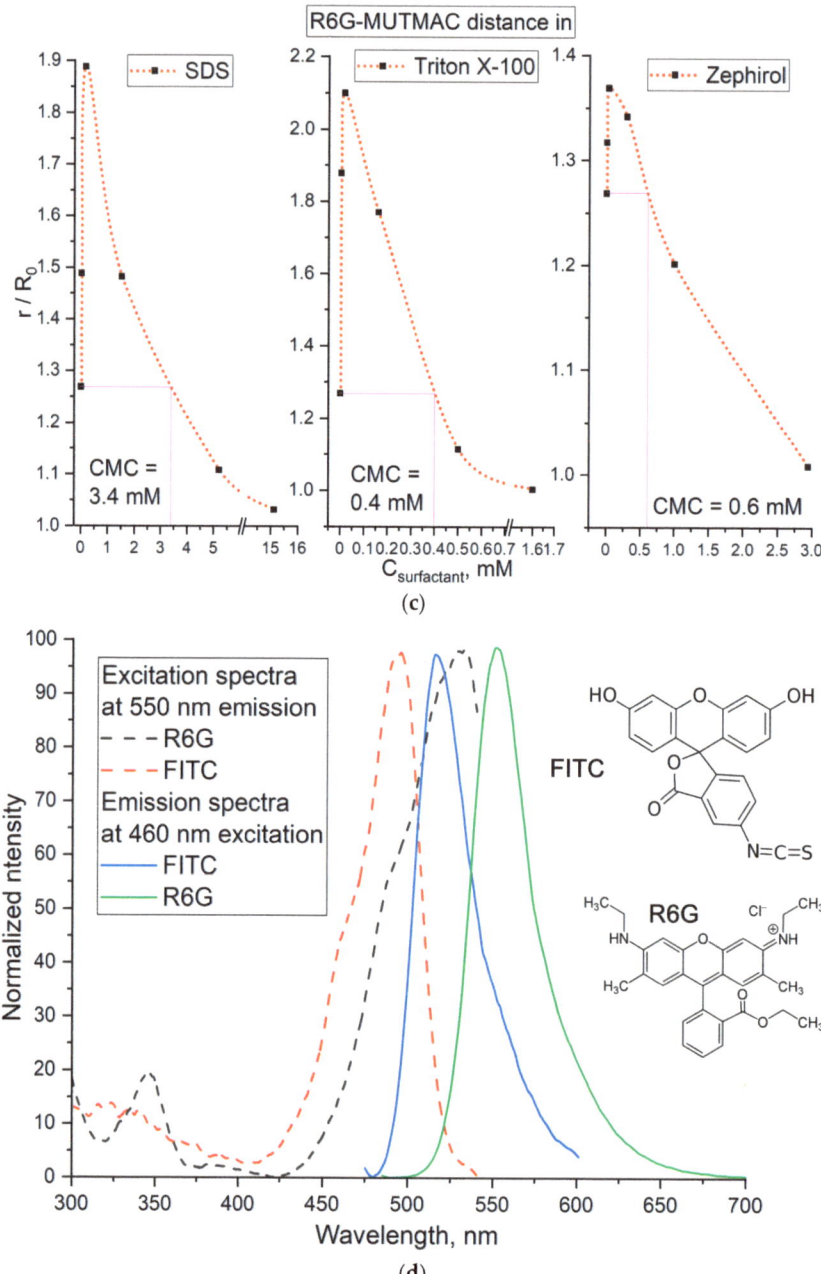

Figure 1. *Cont.*

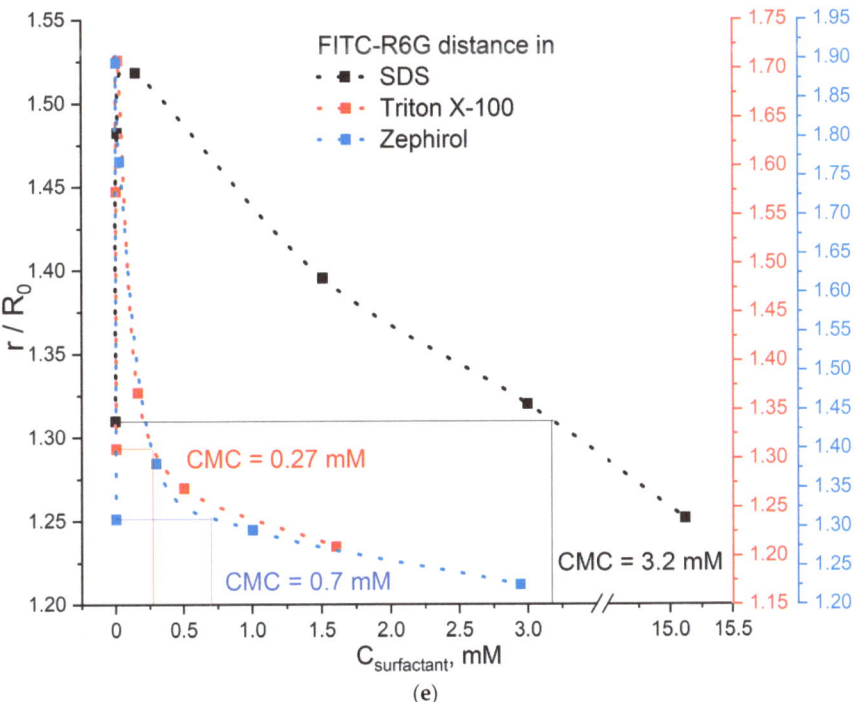

(e)

Figure 1. (a) Experiment design: FRET as an indicator of micelle formation from surfactants. (b) Emission fluorescence spectra of MUTMAC, R6G alone, and its mixtures 1 to 1 (1 µM/1 µM) in PBS buffer solution (0.01 M, pH 7.4) and in the presence of 1 mg/mL of the surfactant zephirol. The excitation wavelength is 360 nm. The insert shows the excitation and emission spectra of these fluorophores in PBS. (c) The dependences of r/R_0 (MUTMAC–R6G) on the surfactants' concentration; r—the distance between donor and acceptor, and R_0 is Förster radius. (d) The excitation and emission spectra of FITC and R6G fluorophores in PBS at excitation wavelength 460 nm. (e) The dependence of r/R_0 (MUTMAC–R6G) on the surfactants' concentration (r is the distance between donor and acceptor and R_0 is Förster radius). T = 22 °C.

3.2. FRET as an Indicator of Micelle Formation in Surfactants Solution

As pairs of fluorophores with the FRET function, we chose MUTMAC–R6G and FITC–R6G (Figure 1a). The first pair is appropriate in terms of the ratio of the fluorescence intensities of the donor and acceptor (approximately 1 to 1), as well as the visual separation of emission peaks. The second pair: visually, the fluorescence peaks are not well resolved into components due to the close location of the bands of donor emission and acceptor absorption; however, this determines the high efficiency of FRET (E value Equation (1)). Such variability (spatial resolution)/(FRET efficiency) was studied here to select the optimal pair of fluorophores with the FRET function.

3.2.1. MUTMAC–R6G Pair

Figure 1b shows the excitation and emission fluorescence spectra of MUTMAC (donor) and R6G (acceptor). The main components are the fluorescence peak of the donor at 450 nm and the acceptor—at 550 nm. The excitation wavelength was 360 nm, such that both MUTMAC and partially R6G would be excited, which makes it possible to monitor the fluorescence of both fluorophores.

The degree of fluorophore loading was controlled by changing the fluorescence intensity from the concentration of the added surfactant (Figure S1). For MUTMAC, an increase in emission intensity was observed during the formation of pre-micelles, and during the formation of micelles, fluorescence quenching occurred—a marker of the loading degree. At a concentration of surfactants of the order of 1 mg/mL, the degree of MUTMAC loading is 75–80%. In the case of rhodamine 6G, fluorescence ignition is mainly observed during the formation of pre-micelles and slight quenching during the formation of micelles based on Triton X-100 and Zephirol, and quenching during the formation of micelles of anionic SDS. The degree of loading of R6G at a concentration of surfactants of the order of 1 mg/mL can be estimated as 80–85%. The fluorescence emission spectra of R6G in free and micellar form are shown in Figure S1e. A shift of the maximum position to 5–10 nm is observed due to the inclusion of fluorophores in the micelles hydrophobic areas.

To quantify the formation of micelles from surfactants, it is necessary to select the target signal: the most pronounced is FRET efficiency (E value—Equations (1)–(3)) and the ratio r/R_0 (Equation (4)), characterizing the distance between two molecules of the fluorophore: the donor and acceptor. r/R_0 is directly related to the formation/disaggregation of micelles: (1) The addition of small amounts of surfactant to the system leads to the increasing the donor–acceptor molecules distance (Figure 1c); (2) The formation of micellar structures is reflected in the convergence of fluorophores due to its incorporation into the hydrophobic core of micelles, enhancing with the increase in surfactant concentrations. Therefore, the dependences of r/R_0 on surfactants' concentrations has a maximum, which means the initial process of the surfactant molecule aggregation (pre-micelles). The critical pre-micelle concentration (CPMC) can be determined from the position of the maximum curve. However, another analytically significant parameter is the critical micelle concentration (CMC). In this case, the CMC corresponds to a point on the right branch of the graph with the value r/R_0 equal to the initial one (for a pair of fluorophores in a buffer solution without surfactants)—Figure 1c.

3.2.2. FITC–R6G Pair

Figure 1d shows the excitation and emission fluorescence spectra of FITC (donor) and R6G (acceptor) separately from each other in a buffer solution. The main components are the fluorescence maximum for the donor observed at 520 nm and at 550 nm for the acceptor. The excitation wavelength was 460 nm for the selective observation of the FITC emission peak.

The degree of FITC and R6G loading was controlled by changing the fluorescence intensity from the concentration of the added surfactant (Figure S1d). For R6G, the observations are described above. In the case of FITC, an interesting fact is observed: the dependence of the fluorescence intensity on the concentration of surfactant is a curve with a minimum corresponding to the formation of pre-micelles and a right shoulder corresponding to the compactization of surfactant molecules into micelles.

In this system, it is most informative to determine the r/R_0 ratio by the igniting of the acceptor (R6G) fluorescence intensity. Similarly to the MUTMAC–R6G pair considered above, the dependences of r/R_0 on $C_{surfactant}$ with a maximum are obtained for the FITC–R6G pair. Graphically, the points corresponding to CMC are marked in Figure 1e. The MUTMAC–R6G pair is more sensitive than FITC–R6G to the formation of micelles from charged surfactants, since the value of r/R_0 changed significantly, and, in addition, the visual separation of the peaks of fluorescence emission makes it possible to estimate the values of CMC, CPMC, etc., much more accurately.

3.2.3. Comparison of CMC Values Obtained Using Two FRET Probes and the Literature Data

Based on the plots given in Figure 1b,d (distances between the fluorophore donor and acceptor plotted on the concentration of surfactants), the CMC values were graphically determined (the results are presented in Table 1). The data obtained using two FRET probes coincide within the margin of error and satisfy the literature data obtained using fast titration method with ionic organic dyes. This means that the technique of FRET probes allows us to study the mechanisms of micelle formation and determine not only CMC, but also CPMC, which was previously available only by indirect methods.

The effect of surfactants charge on FRET efficiency. The largest maximum on the graph of r/R_0 as a function of surfactant concentration (Figure 1c) is achieved for neutral Triton X-100 (2.1 units) and decreases for anionic SDS (1.9 units) and cationic Zephirol (1.37 units). This difference between anionic and cationic surfactants can be attributed to the positive charge on the FRET probe itself; therefore, in the case of «+»–charged surfactants, a convergence of fluorophores is observed due to repulsion from charged surfactant groups (Figure 1). By the magnitude of the maximum on the r/R_0 curve, it is possible to judge the charge effect of surfactants on the micelles formation semi-quantitatively.

The effect of micellar size on FRET efficiency. Micellar size affects the distance between fluorophores, and it is therefore important to monitor the right branch of graphs of r/R_0 versus surfactant concentration (Figure 1c). For classical surfactants, the curves exceed $r/R_0 \approx 1$, whereas for large chitosan polymer micelles (size 100–200 nm), the r/R_0 is significantly larger than 1 (shown in Section 3.4.2). The highest CMC value is typical for surfactants with a low molecular weight—SDS, an order of magnitude lower CMC values are typical for surfactants with a high molecular weight such as Triton X-100 and Zephirol—due to multipoint interactions. The effect of the surfactants charge on the CMC is rather pronounced: the smallest CMC values are characteristic for uncharged surfactants. At the same time, Triton X-100 is characterized by a higher aggregation degree of 143 versus 50 for SDS [49,50]. This is reflected as an in increase in the sharpness of the peak r/R_0 vs. $C_{surfactant}$, which indicates the sensitivity of the presented FRET probes. Due to the charged groups in R6G and MUTMAC, these fluorophores can interact with anionic groups in surfactants, and specifically with the sulfogroup in SDS. This affects the observed FRET: the r/R_0 parameter varies from 1.3 to 2.0 units; and in the case of cationic zephirol, r/R_0 varies only slightly from 1.27 to 1.37 units. Thus, using the FRET technique, it is possible to judge the degree of aggregation and the size of micelles.

Additionally, we showed the *specificity of the probes to different types of micelles in terms of the FRET signal*. The **MUTMAC–R6G pair** is more specific than FITC–R6G to the formation of micelles from charged surfactants, due to the visual separation of fluorescence peaks. In addition, MUTMAC is a fluorescent substrate and can be used to study enzyme activity using the FRET phenomenon in the micellar systems or even in the living cells.

At the same time, the **FITC–R6G pair** is characterized by a higher degree of inclusion in the core of micelles >80–85% (for surfactant micelles at concentration higher than 1 mg/mL), and for the more hydrophilic MUTMAC, this value is about 75% (as can be judged from the fluorescence data). The difference in the inclusion degrees of fluorophores affects the sensitivity of the FRET probe and the spike in the analytical signal r/R_0 (Figure 1).

Table 1. Critical micelle concentrations (CMCs) and critical pre-micelle concentrations (CPMCs) for surfactants determined using FRET probes in comparison with the literature data. PBS (0.01 M, pH 7.4). T = 22 °C.

Surfactant	CPMC, µM	CMC, mM		
		FRET Probe 1: MUTMAC + R6G	FRET Probe 2: FITC + R6G	Literature Data
SDS (sodium dodecyl sulfate)	15 ± 4	3.4 ± 0.1	3.2 ± 0.3	3.32 ± 0.01 [51]
Triton X-100	16 ± 3	0.39 ± 0.05	0.27 ± 0.08	0.3 ± 0.01 [51]
Zephirol (benzalkonium chloride)	4 ± 1	0.6 ± 0.1	0.7 ± 0.2	0.6 mM [52]

3.3. Determination of the Fluorophore Inclusion Degree in Micelles by Enzymatic Activity

A complementary approach to the FRET probes technique to determine the fluorophore loading degree in the micelles and the micelle formation (CMC) is the use of enzyme catalytic activity. α-Chymotrypsin (proteinase) catalyzes the hydrolysis reaction of MUTMAC to 4-methylumbelliferon (MUmb) (Figure 2) accompanied by the ignition of fluorescence at 450 nm (MUmb fluorescence). Upon formation of the micellar structures from Zephirol, MUTMAC enters the hydrophobic core; therefore, it becomes inaccessible for enzymatic reactions. Thus, with an increase in the surfactant concentration, there would be a decrease in the apparent reaction rate due to a decrease in the effective concentration of the substrate in aqueous phase. This experiment was specially designed so that the micelles were loaded with fluorophore, but not separated by dialysis, so that part of the fluorophore would not be in the micelles. This is to show the applicability of the method for determining fluorophore loading by enzymatic reactions based on the amount of fluorophore remaining outside the micellar structure.

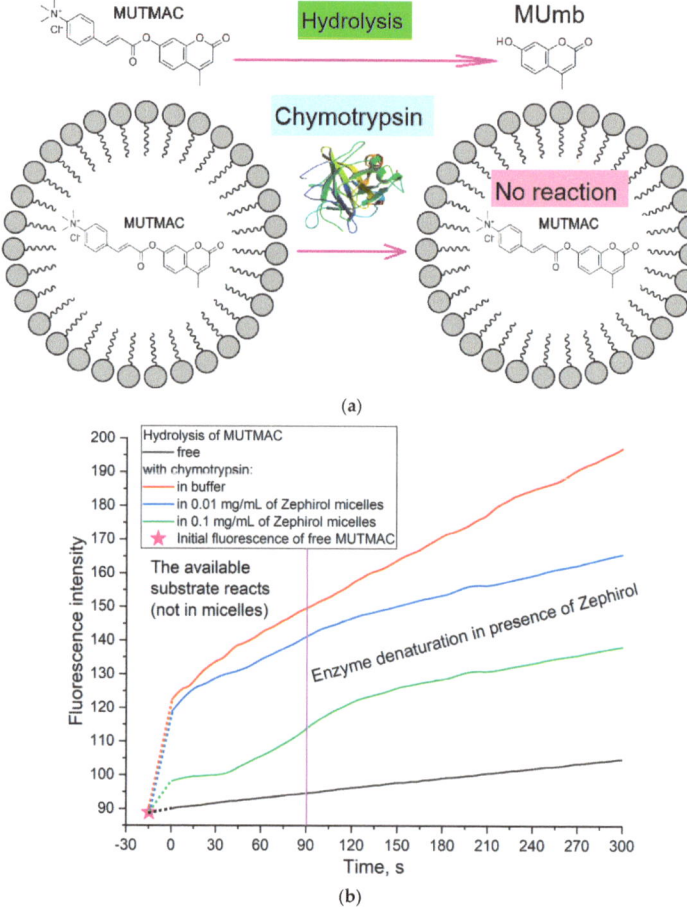

Figure 2. (**a**) Experiment design: determination of the fluorophore inclusion degree in micelles by enzymatic activity. (**b**) Kinetic curves of MUTMAC (0.1 mM) hydrolysis in the presence/absence of chymotrypsin (0.4 μM) and various concentrations of zephirol. λ_{exci} = 360 nm, λ_{emi} = 450 nm. PBS (0.01 M, pH 7.4). T = 37 °C. The reaction rate was determined by the initial spike in the fluorescence intensity of the product, and not by the tangent of the tilt angle, since the enzyme is partially denatured. The purple vertical line indicates >10% denaturation of the enzyme.

According to the values of the fluorescence intensity changes at 450 nm (initial splash, which corresponds to the release MUmb product), and in comparison, with an aqueous solution, it is possible to judge the amount of fluorophore loaded into the micelle core: 20% of the fluorophore was screened by the surfactant at $C_{zephirol}$ 0.01 mg/mL, and 65% was loaded in the micelles at $C_{zephirol}$ = 0.1 mg/mL. The estimated CMC value calculated using the enzyme technique is 0.25 mM (=0.1 mg/mL), close to those given in Table 1 obtained using FRET probes approach (R6G with MUTMAC and FITC). Given that the surfactant can cause denaturation of the enzyme after 1–2 min (for chymotripsin) (Figure 2), the initial reaction rate should be used as a relevant analytical signal. The results obtained using enzymatic techniques for loading fluorophores into micelles are in good agreement with those obtained using FRET probes (section above). However, using chymotrypsin, it is possible to more accurately determine the distribution of fluorophores in the micellar system.

3.4. Formation of Polymeric Micelles as Assessed by FRET Probes

3.4.1. Self-Assembled Amphiphilic Chitosan Grafted with Lipoic and Oleic Acid Residues

The first part of the work was devoted to the validation and optimization of the FRET technique for studying micelle formation. During the validation of the technique, a more sensitive MUTMAC–R6G FRET pair was selected. The main practical interest is, rather, polymeric micelles that are widely used for drug delivery. Accordingly, with regard to self-assembled amphiphilic chitosan grafted with lipoic and oleic acid residues, we aimed to study the mechanism of formation of the micelles, as well as to study the subtle nature of stimulus sensitivity due to loosening of the 3D structure of chitosan in a weakly acidic medium and the reducing of S-S bonds to S-H in the presence of glutathione as a model tumor microenvironment [53–55].

The synthesis of chitosan grafted with fatty acids was carried out using the carbodiimide approach described earlier [45,47]. The Chit5 and lipoic acid (LA) or oleic acid (OA) (Figure 3) chemical conjugation was confirmed by a significant decrease in intensity of the absorption band of carboxylic acid group (1730–1700 cm^{-1}) of lipoic acid, and the appearance of characteristic peaks of ν(C=O) at 1630 cm^{-1} and δ(N–H) at 1560 cm^{-1} oscillation in amide bond between chitosan and acid residues. Conjugate formation is also confirmed by a decrease in the intensity of ν(N–H) at 3500–3300 cm^{-1} in NH$_2$ groups of chitosan, since they are modified into amide. Grafted chitosan is characterized by three peaks of characteristic oscillation bands ν(C–H) in fatty acid residues at 3000–2850 cm^{-1}. Interestingly, the structure of the C-O-C bond oscillation band (1200–1000 cm^{-1}) in chitosan changes from two-component to multicomponent after modification with lipoic acid. This occurs due to the formation of micelles and various variants of the microenvironment of glucosamine fragments of chitosan.

^{1}H NMR spectra of polymers with S-H or S-S bonds loaded with R6G are presented in Figure S2. Chit5-LA was studied as a self-assembled polymer with S-S bonds between chains or non-stitched S-H bonds. As a control without S-H and S-S bonds, Chit5-OA was used. Chemical shifts (δ, ppm) for Chit5 were observed: 4.22 (H1), 3.23 (H2), 3.79, 3.96 (H3, H4, H5, H6, H6′), and 2.11 (NH–C(=O)–CH$_3$). ^{1}H NMR spectra of Chit5-LA contain signals of chitosan indicated above, increased signals at 2.0–2.3 ppm and 1.25 ppm that were assigned to N-alkyl groups of LA, and signals of 3.64 ppm (C–H near the dithiolane fragment) and 2.3 ppm (β–H with relation to the carboxyl group) assigned to LA in polymer [45]. Upon thiol-disulfide exchange reaction, lipoic acid residues form intermolecular S-S inside the micelles (it was shown using NMR spectroscopy—Figure S2) accompanied by the particles compactization (the particle size decreases from 300–350 nm to 230–280 nm—Table 2, Figure S3), indicating increased thermodynamic stability (this is a consequence of the decline of the critical micelles concentration). The physicochemical properties of chitosan-based polymers and micelles formed from it are presented in Table 2. An increase in the zeta potential, when comparing Chit5-OA with Chit5-LA, indicates a change in the structure of the micelle and its greater homogeneity (acid residues look inward into the core). S-S crosslinking promotes the sealing of micelles and a decrease in the zeta potential.

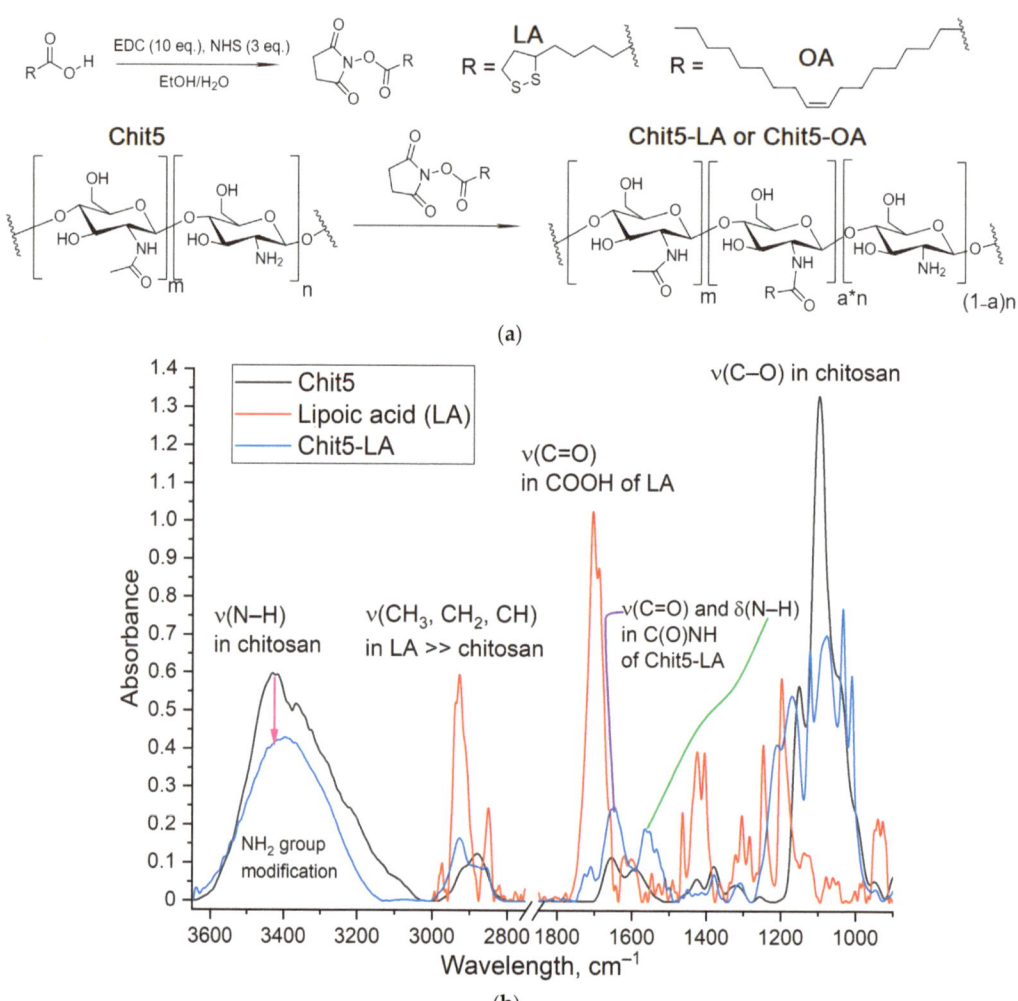

Figure 3. (a) The scheme of synthesis grafted chitosan with lipoic acid (Chit5-LA) with oleic acid (Chit5-OA). (b) FTIR spectra of Chit5, LA, and its conjugate Chit5-LA. T = 22 °C.

Table 2. Physicochemical properties of chitosan-based polymeric micelles. T = 22 °C.

Micelle *	Grafting Degree, %	M_w of One Polymeric Unit, kDa	CMC, nM	Hydrodynamic Diameter **, nm	Zeta Potential, mV
Chit5-OA	18 ± 2	6.7 ± 0.8	8 ± 2	300–450	+5 ± 1
Chit5-LA nonstitched	24 ± 3	6.4 ± 0.3	50 ± 10	300–350	+20 ± 3
Chit5-LA S-S stitched		45 ± 6 (is about 7 residues of Chit5-branches)	16 ± 2	230–280	+15 ± 2

* Chit5—chitosan 5 kDa. OA—oleic acid without SH and S-S groups. LA—lipoic acid with SH or/and S-S groups.
** Determined by nanoparticle tracking analysis (NTA).

3.4.2. Polymeric Micelles Formation and S-S Stitching Detection Using MUTMAC–R6G Probe

To show the versatility of the FRET approach to determine CMC, in addition to surfactant micelles, we studied polymeric micelles based on the chitosan (5 kDa) grafted with lipoic or oleic acid (Chit5-LA, Chit5-OA). Figure 4a shows the distance between the fluorophores pair (donor–acceptor) plotted as a function of the grafted chitosan molecules concentration. The CMC determined for Chit5-LA is 16 nM (Table 2), which is in good agreement with the pyrene-probe data described earlier for similar chitosan-based micellar systems [45].

The developed FRET-based approach is further applied to study the mechanism and the kinetics of polymeric micelles formation. Of particular interest is the aspect of the formation of the polymeric micelles (as a smart drug delivery system), where the kinetics data of formation/destruction of S-S bonds are of great importance. This can be considered as the basis for creating stimulus-sensitive drug delivery systems to tumor cells, where drug molecules will be selectively released due to the higher glutathione level in cancer cells [45,47]. The visualization of the effective penetration of R6G into A549 cancer cells in the micellar form compared with a free cytostatic is shown in Figure S4.

Therefore, based on chitosan-lipoic acid conjugates (Figure 4a–c), we studied the kinetics of the polymeric micelles formation stabilized by covalent S-S bonds (the formation of disulfide bonds was confirmed by NMR spectroscopy (Figure S2)). Such micelles compactization results in strengthening of MUTMAC->R6G FRET (Figure 4b). An appropriate analytical signal here is the I_{550}/I_{450} index (acceptor fluorescence/donor fluorescence)—the effectiveness of the FRET effect. There are no changes in the FRET status in the buffer solution. The formation of micelles was accompanied by the formation of a hydrophobic core and the compaction of $(CH_2)_n$ tails, while hydrophilic NH_2 and OH groups are exposed out in the water. During the micelles' formation, fluorescence increases, which indicates the inclusion of FRET probes in the hydrophobic areas of the micelle. On the other hand, the kinetic curve of the FRET signal (I550/I450 index) exhibits a minimum at approximately 3–5 min (Figure 4b, insert), which corresponds to an increase in FRET efficiency (in the micelles). During the formation of hydrophobic sites in the micellar particle, the drug (fluorophore) is loaded for about 5–10 min. After this point, we observe a subsequent linear increase in FRET signal up to ~1 h, which is due to the inclusion of FRET probes within the micelle cores and the continued process of micelle compactification during the formation of disulfide bonds. The crosslinking of polymeric chains in micelles (Figure 4c) causes compactization of structure; the micelle is thickened and the loading degree of FRET probes into hydrophobic core increases.

The FRET probes loading capacity was estimated to be equal to 75–80% for nonstitched polymeric micelles and 87% for S-S stitched. The crosslinking of polymer chains in micelles cause compactization of structure; the micelle is thickened and the degree of loading of FRET probes into hydrophobic venom increases. At the same time, the zeta potential of the micellar system decreases (Table 2). The dense structure of micelles is maintained at pH > 7 (typical for liquid media in body), while protonation of chitosan amino groups occurs in a weakly acidic medium, and loosening of micelles occurs with an increase in the rate of the drug release [45].

The inclusion of the studied fluorophores in polymeric micelles was demonstrated by flow cytometry (Figure S5) as a control technique, by the appearance of R6G-positive submicron micellar particles (R6G loading capacity was about 85%). The presence of fluorescent particles proves the predominant inclusion of rhodamine in the micellar system. Moreover, in the case of micelles containing covalent S-S bonds, the degree of inclusion of the fluorophore (according to the quenching of the fluorescence) is higher in comparison with nonstitched loose micelles.

Polymeric micelles with disulfide bonds formation are investigated here as a perspective stimuli-sensitive drug delivery system to tumors. Reduced glutathione (GSH) is the most important antioxidant in cells [56,57], and was found in all cell compartments in millimolar concentrations (1–10 mM). Chitosan-based polymeric micelles use this feature of cancer cells: GSH as a trigger causes accelerated release of cytostatic [44]. Evidence of the formation of S-S bonds and the possibility of their destruction by a reducing agent (glutathione excess—tumor microenvironment model) are shown in Figure 5. An increase in the concentration of glutathione is reflected in a sharp (up to 5–10 times) increase in the rate of fluorophore release from Chit5-LA micelles due to the destruction of disulfide bonds. In other words, the release of fluorophore is characterized as glutathione-dependent: with an increase in the concentration of the thiol-disulfide exchange catalyst, disulfide bonds in micelles are reduced and the micelle structure is loosened with the simultaneous release of rhodamine 6G.

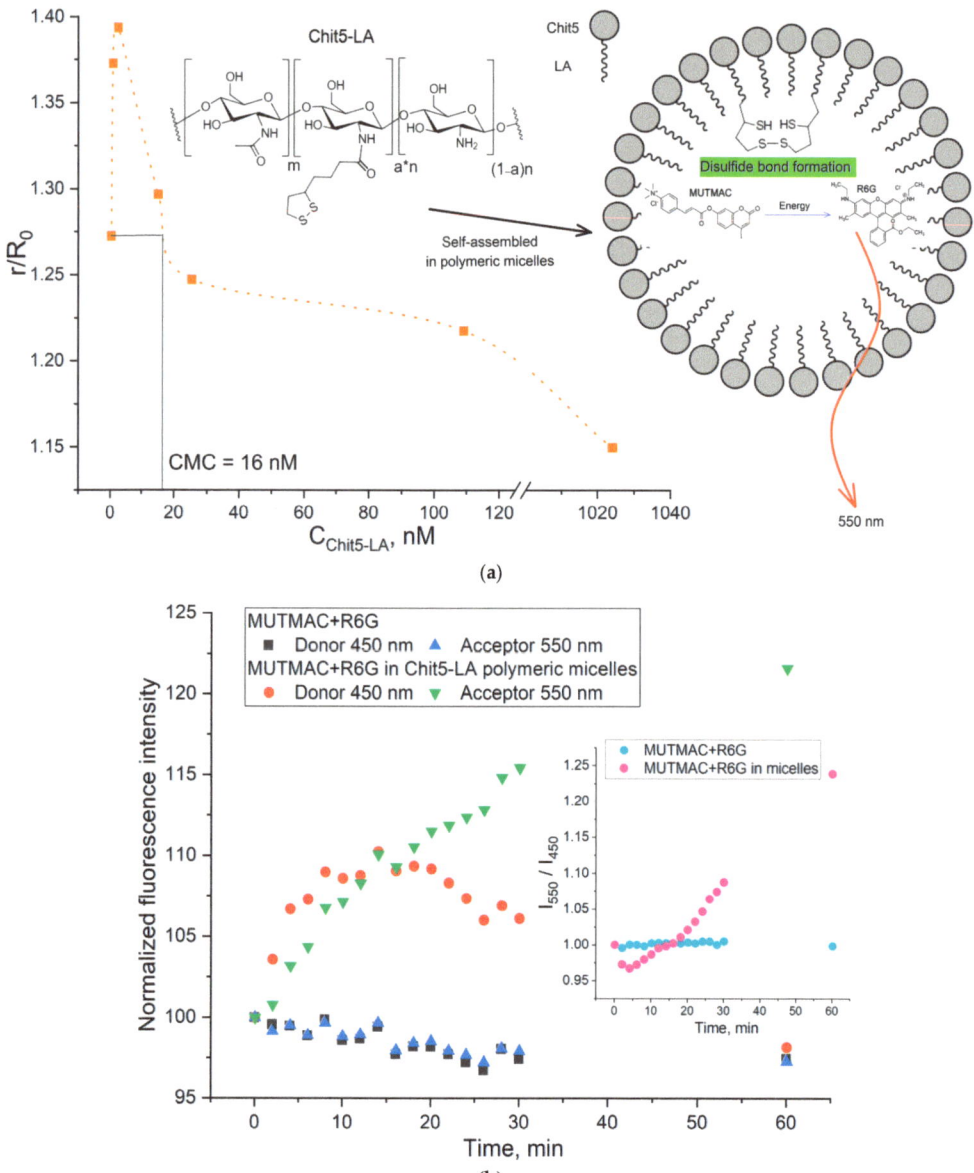

Figure 4. Cont.

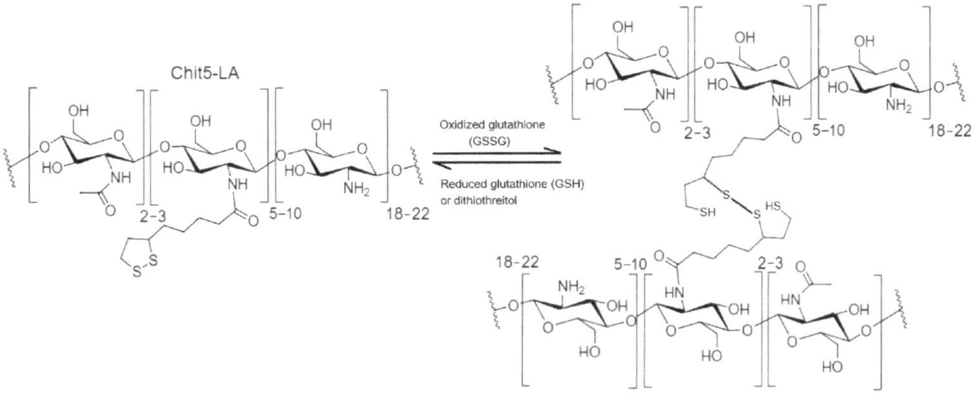

(c)

Figure 4. (a) The dependence of r/R_0 (MUTMAC–R6G, 1 µM/1 µM) on the Chit5-LA self-assembled molecules concentration. r is the distance between donor and acceptor and R_0 is Förster radius. The excitation wavelength is 360 nm. PBS buffer solution (0.01 M, pH 7.4). (b) Kinetic fluorescence curves of FRET probe components in buffer solution and during the formation of S-S bonds between crosslinking molecules of Chit5-LA. T = 37 °C. (c) The scheme for the formation and reducing of a disulfide bond.

The release of the drug from the micelles is prolonged (Figure 5). At the same time, S-S crosslinked micelles, due to their dense structure, in the absence of GSH, release no more than 20% of the loaded drug. In the presence of GSH, the release rate becomes almost constant, while the full release of the drug is achieved in 3–4 days. The plateau (1) in the case of the presence of GSH is due to the limiting stage of S-S reduction and loosening of the micelle structure (the rate of release becomes constant and is approximately equal to 20–30% per day), and (2) in the absence of GSH, the inability to release the drug from a durable micelle (release was stopped at 15–20%). The plateau can be explained by the fact that the fluorophore molecules are released from covalent micelles only from the surface layers, while the inner parts remain tightly bound for a long period of time. This means that 20% of the drug is non-firmly bound, while 80% is deeply located and firmly bound to the micelle. At the same time, the drug is almost completely released under the action of specific stimuli (the microenvironment of tumor cells), which is a key advantage of the polymeric stimulus-sensitive micelle system.

Thus, we present two pairs of FRET probes (a more suitable MUTMAC–R6G pair was used for polymer micelles) that allowed us to monitor the formation of micellar structures from amphiphilic molecules or the kinetics of polymers self-assembling in real time.

3.5. A Comparison of the Proposed FRET Probe Technique with Other Techniques Described in the Literature to Study the Properties of Micelles

Table 3 compares the informativity of different methods used to study micelles characteristics, and evaluates the expressiveness and versatility of each approach. The advantages of the FRET technique are high sensitivity and the possibility to determine the distance between the fluorophores; thus, the mechanism of the micelles formation, CMC and CPMC values, and the size of the micelles can be estimated. At the same time, the method is fast and reproducible, since the FRET signal is rather specific.

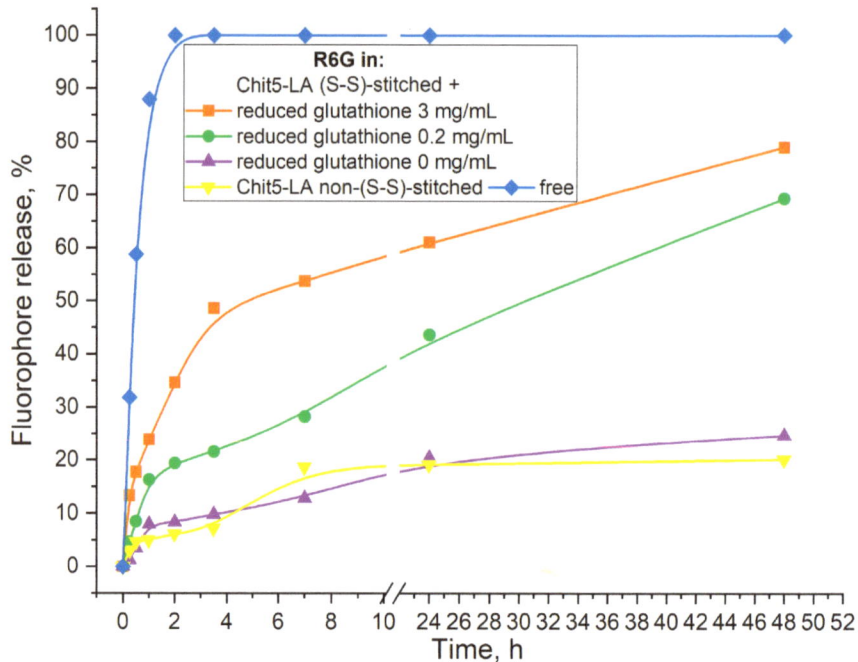

Figure 5. Release curves of rhodamine 6G (R6G) from Chit5-LA-based micelles not crosslinked with disulfide bonds and crosslinked with disulfide bonds in the presence of a reducing agent (glutathione). T = 37 °C. 0.01 M PBS (pH 7.4).

Table 3. A comparison of the methods used to investigate the micelles' properties. "+" means that the method provides relevant information about the parameter. "++" means that the method provides comprehensive information about the parameter. "±" means that the method provides indirect information about the parameter. "−" means that the method does not provide information about the parameter or the data does not follow directly.

Method	CMC Determination	CPMC Determination	Aggregation Number Determination	Size Determination	Robustness	Applicability to Different Types of Micelles	Expressiveness
Conductometry [39,58,59]	±	−	−	−	+	±	+
Surface tension [51,60,61]	+	−	−	−	±	−	+
Densitometry [59]	±	−	−	−	±	−	+
NMR spectrometry [62]	+	±	±	−	+	±	−
UV/VIS spectroscopy [63]	+	−	−	−	+	±	+
Fluorometric methods (including pyrene probe) [32,61,64–66]	++	+	±	±	+	+	+
Atomic force and electron microscopy [67]	±	−	−	++	±	+	−
FRET probes	+	+	±	±	+	++	+

4. Conclusions

In this work, we proposed the FRET probe technique (R6G with FITC or methylumbelliferone derivative, MUTMAC) as an indicator of micelle formation from surfactants or from chitosan grafted with fatty acid as promising drug carrier with stimuli-sensitivity to tumor microenvironments (pH is about 5.5–6.5 with increased concentrations of glutathione). In relation to surfactants (anionic SDS, cationic Zephirol, and nonionic Triton X-100), the FRET probe technique provides valuable information about the distance of the donor and acceptor fluorophores r/R_0 (where R_0 is the Förster distance and is about 50–60 Å), which was used to study the mechanism of micelle formation and to determine the aggregate state of the system (individual molecules/pre-micelles/micelles) and the CMC parameters. Chitosan grafted with oleic and lipoic acid was synthesized using the carbodiimide approach, followed by characterization by FTIR and NMR spectroscopy: grafting degree is about 20%, average molecular weight per one structure unit is about 6–7 kDa for Chit5-OA and non-stitched Chit5-LA, but for Chit-LA S-S stitched, molecular weight is about 45 kDa (6–8 fragments). Chitosan conjugates self-assemble into positively charged (+5–20 mV) polymeric micelles when concentration is higher than 10–20 nM. Micelles formation and functional properties, such as fluorophore loading degree, were studied using the FRET technique and were also controlled by flow cytometry and atomic force microscopy. Reductant-treated conjugate Chit5-LA, due to S-S crosslinks formation between polymer chains via lipoic acid residues, is accompanied by particles' compactization (the particle size decreases from 300–350 nm to 230–280 nm). One of the key aspects of the work is the effect of the formation and destruction of S-S bonds between polymer chains in micelles on FRET efficiency, which is important in the development of stimulus-sensitive drug delivery systems for antitumor therapy. The release of R6G (model cytostatic and fluorophore) is characterized as glutathione-dependent: with an increase in the concentration of the thiol-disulfide exchange catalyst, S-S bonds in micelles are reduced to S-H and the micelle structure is loosened with the simultaneous release of rhodamine 6G. Thus, we presented the original technique of FRET probes in relation to the study of micelle formation processes of various amphiphilic molecules, and, most importantly, demonstrated the applicability of FRET probes to study the characteristics of micellar drug delivery systems with the function of active tumor targeting.

Supplementary Materials: The following supporting information can be downloaded at: https://www.mdpi.com/article/10.3390/polym16060739/s1, Figure S1. The dependence of the fluorescence emission maximum intensity on surfactant concentration: (a) MUTMAC, (b) FITC, (c) R6G. Fluorescence emission spectra of (d) FITC, (e) R6G in free form and in micellar form. Graphs were used to determine the degree of incorporation of fluorophore into micelles. Figure S2. ^{1}H NMR of polymeric micellar systems with S-H or S-S bonds with R6G. Chit5-LA was studied as self-assembled polymer. As a control without S-H bonds, Chit5-OA was used. D_2O. T = 25 °C. Figure S3. (a) AFM image of Chit5-LA particles and (b) the corresponding section along the blue line in height, respectively. Figure S4. Fluorescence images of A549 after 60 min incubation with Rhodamine 6G 1 μg/mL free or Rhodamine 6G in micelles (S-S stitched). R6G red, DAPI blue, and FITC-labeled micelles channels and merge are shown. R6G/micelle 1:1 w/w. The scale segment is 100 μm. Figure S5. Flow cytometry assay of R6G-loaded micelles.

Author Contributions: Conceptualization, E.V.K. and I.D.Z.; methodology, I.D.Z., E.V.K. and I.V.S.; formal analysis, I.D.Z. and I.V.S.; investigation, I.D.Z. and I.V.S.; data curation, I.D.Z.; writing—original draft preparation, I.D.Z. and I.V.S.; writing—review and editing, E.V.K.; project supervision, E.V.K.; funding acquisition, E.V.K. All authors have read and agreed to the published version of the manuscript.

Funding: This research was funded by the Russian Science Foundation, grant number 24-25-00104.

Data Availability Statement: The data presented in this study are available in the main text.

Acknowledgments: The work was performed using the equipment (FTIR microscope MICRAN-3, FTIR spectrometer Bruker Tensor 27, Jasco J-815 CD Spectrometer, AFM microscope NTEGRA II) of the program for the development of Moscow State University.

Conflicts of Interest: The authors declare no conflicts of interest.

Abbreviations

Chit—chitosan; CMC—critical micelle concentration; CPMC—critical pre-micelle concentration; FITC—fluorescein isothiocyanate; FRET—Förster resonance energy transfer; LA—lipoic acid; MUmb—4-methylumbelliferone or methylumbelliferyl; MUTMAC—4-methylumbelliferyl p-trimethylammoniocinnamate chloride; R6G—rhodamine 6G; SDS—sodium dodecyl sulfate.

References

1. Algar, W.R.; Hildebrandt, N.; Vogel, S.S.; Medintz, I.L. FRET as a biomolecular research tool—Understanding its potential while avoiding pitfalls. *Nat. Methods* **2019**, *16*, 815–829. [CrossRef] [PubMed]
2. Roy, R.; Hohng, S.; Ha, T. A practical guide to single-molecule FRET. *Nat. Methods* **2008**, *5*, 507–516. [CrossRef] [PubMed]
3. Jares-Erijman, E.A.; Jovin, T.M. FRET imaging. *Nat. Biotechnol.* **2003**, *21*, 1387–1395. [CrossRef] [PubMed]
4. Wu, L.; Huang, C.; Emery, B.P.; Sedgwick, A.C.; Bull, S.D.; He, X.-P.; Tian, H.; Yoon, J.; Sessler, J.L.; James, T.D. Förster resonance energy transfer (FRET)-based small-molecule sensors and imaging agents. *Chem. Soc. Rev.* **2020**, *49*, 5110–5139. [CrossRef] [PubMed]
5. Liao, J.Y.; Song, Y.; Liu, Y. A new trend to determine biochemical parameters by quantitative FRET assays. *Acta Pharmacol. Sin.* **2015**, *36*, 1408–1415. [CrossRef] [PubMed]
6. Du, S.; Zhu, X.; Zhang, L.; Liu, M. Switchable Circularly Polarized Luminescence in Supramolecular Gels through Photomodulated FRET. *ACS Appl. Mater. Interfaces* **2021**, *13*, 15501–15508. [CrossRef]
7. Yuan, L.; Lin, W.; Zheng, K.; Zhu, S. FRET-based small-molecule fluorescent probes: Rational design and bioimaging applications. *Acc. Chem. Res.* **2013**, *46*, 1462–1473. [CrossRef] [PubMed]
8. Watrob, H.M.; Pan, C.P.; Barkley, M.D. Two-step FRET as a structural tool. *J. Am. Chem. Soc.* **2003**, *125*, 7336–7343. [CrossRef]
9. Shrestha, D.; Jenei, A.; Nagy, P.; Vereb, G.; Szöllősi, J. Understanding FRET as a research tool for cellular studies. *Int. J. Mol. Sci.* **2015**, *16*, 6718–6756. [CrossRef]
10. Panniello, A.; Trapani, M.; Cordaro, M.; Dibenedetto, C.N.; Tommasi, R.; Ingrosso, C.; Fanizza, E.; Grisorio, R.; Collini, E.; Agostiano, A.; et al. High-Efficiency FRET Processes in BODIPY-Functionalized Quantum Dot Architectures. *Chem.—A Eur. J.* **2021**, *27*, 2371–2380. [CrossRef]
11. Ngan, V.T.T.; Chiou, P.Y.; Ilhami, F.B.; Bayle, E.A.; Shieh, Y.T.; Chuang, W.T.; Chen, J.K.; Lai, J.Y.; Cheng, C.C. A CO_2-Responsive Imidazole-Functionalized Fluorescent Material Mediates Cancer Chemotherapy. *Pharmaceutics* **2023**, *15*, 354. [CrossRef] [PubMed]
12. Ilhami, F.B.; Bayle, E.A.; Cheng, C.C. Complementary nucleobase interactions drive co-assembly of drugs and nanocarriers for selective cancer chemotherapy. *Pharmaceutics* **2021**, *13*, 1929. [CrossRef]
13. Mishra, S.; Pandey, A.; Manvati, S. Coumarin: An emerging antiviral agent. *Heliyon* **2020**, *6*, e03217. [CrossRef] [PubMed]
14. Guo, C.; Yuan, H.; Yu, Y.; Gao, Z.; Zhang, Y.; Yin, T.; He, H.; Gou, J.; Tang, X. FRET-based analysis on the structural stability of polymeric micelles: Another key attribute beyond PEG coverage and particle size affecting the blood clearance. *J. Control. Release* **2023**, *360*, 734–746. [CrossRef]
15. Ghezzi, M.; Pescina, S.; Padula, C.; Santi, P.; Del Favero, E.; Cantù, L.; Nicoli, S. Polymeric micelles in drug delivery: An insight of the techniques for their characterization and assessment in biorelevant conditions. *J. Control. Release* **2021**, *332*, 312–336. [CrossRef]
16. Bhat, P.A.; Chat, O.A.; Dar, A.A. Exploiting Co-solubilization of Warfarin, Curcumin, and Rhodamine B for Modulation of Energy Transfer: A Micelle FRET On/Off Switch. *ChemPhysChem* **2016**, *17*, 2360–2372. [CrossRef] [PubMed]
17. Zhang, H.; Li, H.; Cao, Z.; Du, J.; Yan, L.; Wang, J. Investigation of the in vivo integrity of polymeric micelles via large Stokes shift fluorophore-based FRET. *J. Control. Release* **2020**, *324*, 47–54. [CrossRef]
18. Barros, C.H.N.; Hiebner, D.W.; Fulaz, S.; Vitale, S.; Quinn, L.; Casey, E. Synthesis and self-assembly of curcumin-modified amphiphilic polymeric micelles with antibacterial activity. *J. Nanobiotechnol.* **2021**, *19*, 104. [CrossRef]
19. Wang, Z.; Deng, X.; Ding, J.; Zhou, W.; Zheng, X.; Tang, G. Mechanisms of drug release in pH-sensitive micelles for tumour targeted drug delivery system: A review. *Int. J. Pharm.* **2018**, *535*, 253–260. [CrossRef]
20. Le-Vinh, B.; Le, N.M.N.; Nazir, I.; Matuszczak, B.; Bernkop-Schnürch, A. Chitosan based micelle with zeta potential changing property for effective mucosal drug delivery. *Int. J. Biol. Macromol.* **2019**, *133*, 647–655. [CrossRef]
21. Parra, A.; Jarak, I.; Santos, A.; Veiga, F.; Figueiras, A. Polymeric micelles: A promising pathway for dermal drug delivery. *Materials* **2021**, *14*, 7278. [CrossRef]
22. He, L.; Qin, X.; Fan, D.; Feng, C.; Wang, Q.; Fang, J. Dual-Stimuli Responsive Polymeric Micelles for the Effective Treatment of Rheumatoid Arthritis. *ACS Appl. Mater. Interfaces* **2021**, *13*, 21076–21086. [CrossRef]

23. Kumar, R.; Sirvi, A.; Kaur, S.; Samal, S.K.; Roy, S.; Sangamwar, A.T. Polymeric micelles based on amphiphilic oleic acid modified carboxymethyl chitosan for oral drug delivery of bcs class iv compound: Intestinal permeability and pharmacokinetic evaluation. *Eur. J. Pharm. Sci.* **2020**, *153*, 105466. [CrossRef]
24. Jiang, G.B.; Quan, D.; Liao, K.; Wang, H. Preparation of polymeric micelles based on chitosan bearing a small amount of highly hydrophobic groups. *Carbohydr. Polym.* **2006**, *66*, 514–520. [CrossRef]
25. Luo, T.; Han, J.; Zhao, F.; Pan, X.; Tian, B.; Ding, X.; Zhang, J. Redox-sensitive micelles based on retinoic acid modified chitosan conjugate for intracellular drug delivery and smart drug release in cancer therapy. *Carbohydr. Polym.* **2019**, *215*, 8–19. [CrossRef] [PubMed]
26. Almeida, A.; Araújo, M.; Novoa-Carballal, R.; Andrade, F.; Gonçalves, H.; Reis, R.L.; Lúcio, M.; Schwartz, S.; Sarmento, B. Novel amphiphilic chitosan micelles as carriers for hydrophobic anticancer drugs. *Mater. Sci. Eng. C* **2020**, *112*, 110920. [CrossRef] [PubMed]
27. Kost, B.; Brzeziński, M.; Cieślak, M.; Królewska-Golińska, K.; Makowski, T.; Socka, M.; Biela, T. Stereocomplexed micelles based on polylactides with β-cyclodextrin core as anti-cancer drug carriers. *Eur. Polym. J.* **2019**, *120*, 109271. [CrossRef]
28. Xu, W.; Wang, H.; Dong, L.; Zhang, P.; Mu, Y.; Cui, X.; Zhou, J.; Huo, M.; Yin, T. Hyaluronic acid-decorated redox-sensitive chitosan micelles for tumor-specific intracellular delivery of gambogic acid. *Int. J. Nanomed.* **2019**, *14*, 4649–4666. [CrossRef] [PubMed]
29. Lin, D.; Xiao, L.; Qin, W.; Loy, D.A.; Wu, Z.; Chen, H.; Zhang, Q. Preparation, characterization and antioxidant properties of curcumin encapsulated chitosan/lignosulfonate micelles. *Carbohydr. Polym.* **2022**, *281*, 119080. [CrossRef] [PubMed]
30. Stern, T.; Kaner, I.; Laser Zer, N.; Shoval, H.; Dror, D.; Manevitch, Z.; Chai, L.; Brill-Karniely, Y.; Benny, O. Rigidity of polymer micelles affects interactions with tumor cells. *J. Control. Release* **2017**, *257*, 40–50. [CrossRef]
31. Kim, M.P.; Kim, H.J.; Kim, B.J.; Yi, G.R. Structured nanoporous surfaces from hybrid block copolymer micelle films with metal ions. *Nanotechnology* **2015**, *26*, 095302. [CrossRef] [PubMed]
32. Zlotnikov, I.D.; Savchenko, I.V.; Kudryashova, E.V. Fluorescent Probes with Förster Resonance Energy Transfer Function for Monitoring the Gelation and Formation of Nanoparticles Based on Chitosan Copolymers. *J. Funct. Biomater.* **2023**, *14*, 401. [CrossRef]
33. Grosso, R.; De-Paz, M.V. Thiolated-polymer-based nanoparticles as an avant-garde approach for anticancer therapies—Reviewing thiomers from chitosan and hyaluronic acid. *Pharmaceutics* **2021**, *13*, 854. [CrossRef] [PubMed]
34. Liu, Z.; Jiao, Y.; Wang, Y.; Zhou, C.; Zhang, Z. Polysaccharides-based nanoparticles as drug delivery systems. *Adv. Drug Deliv. Rev.* **2008**, *60*, 1650–1662. [CrossRef] [PubMed]
35. Chaubey, P.; Mishra, B.; Mudavath, S.L.; Patel, R.R.; Chaurasia, S.; Sundar, S.; Suvarna, V.; Monteiro, M. Mannose-conjugated curcumin-chitosan nanoparticles: Efficacy and toxicity assessments against Leishmania donovani. *Int. J. Biol. Macromol.* **2018**, *111*, 109–120. [CrossRef] [PubMed]
36. Fuenzalida, J.P.; Weikert, T.; Hoffmann, S.; Vila-Sanjurjo, C.; Moerschbacher, B.M.; Goycoolea, F.M.; Kolkenbrock, S. Affinity protein-based FRET tools for cellular tracking of chitosan nanoparticles and determination of the polymer degree of acetylation. *Biomacromolecules* **2014**, *15*, 2532–2539. [CrossRef] [PubMed]
37. Bauer, T.A.; Schramm, J.; Fenaroli, F.; Siemer, S.; Seidl, C.I.; Rosenauer, C.; Bleul, R.; Stauber, R.H.; Koynov, K.; Maskos, M.; et al. Complex Structures Made Simple—Continuous Flow Production of Core Cross-Linked Polymeric Micelles for Paclitaxel Pro-Drug-Delivery. *Adv. Mater.* **2023**, *35*, 2210704. [CrossRef]
38. Mustafai, A.; Zubair, M.; Hussain, A.; Ullah, A. Recent Progress in Proteins-Based Micelles as Drug Delivery Carriers. *Polymers* **2023**, *15*, 836. [CrossRef]
39. Negut, I.; Bita, B. Polymeric Micellar Systems—A Special Emphasis on "Smart" Drug Delivery. *Pharmaceutics* **2023**, *15*, 976. [CrossRef]
40. Gong, J.; Chen, M.; Zheng, Y.; Wang, S.; Wang, Y. Polymeric micelles drug delivery system in oncology. *J. Control. Release* **2012**, *159*, 312–323. [CrossRef]
41. Gaucher, G.; Satturwar, P.; Jones, M.C.; Furtos, A.; Leroux, J.C. Polymeric micelles for oral drug delivery. *Eur. J. Pharm. Biopharm.* **2010**, *76*, 147–158. [CrossRef] [PubMed]
42. Wakaskar, R.R. Polymeric Micelles for Drug Delivery. *Int. J. Drug Dev.* **2017**, *9*, 12–13.
43. Zlotnikov, I.D.; Davydova, M.P.; Danilov, M.R.; Krylov, S.S.; Belogurova, N.G.; Kudryashova, E.V. Covalent Conjugates of Allylbenzenes and Terpenoids as Antibiotics Enhancers with the Function of Prolonged Action. *Pharmaceuticals* **2023**, *16*, 1102. [CrossRef]
44. Zlotnikov, I.D.; Ezhov, A.A.; Dobryakova, N.V.; Kudryashova, E.V. Disulfide Cross-Linked Polymeric Redox-Responsive Nanocarrier Based on Heparin, Chitosan and Lipoic Acid Improved Drug Accumulation, Increased Cytotoxicity and Selectivity to Leukemia Cells by Tumor Targeting via "Aikido" Principle. *Gels* **2024**, *10*, 157. [CrossRef]
45. Zlotnikov, I.D.; Streltsov, D.A.; Ezhov, A.A.; Kudryashova, E.V. Smart pH- and Temperature-Sensitive Micelles Based on Chitosan Grafted with Fatty Acids to Increase the Efficiency and Selectivity of Doxorubicin and Its Adjuvant Regarding the Tumor Cells. *Pharmaceutics* **2023**, *15*, 1135. [CrossRef] [PubMed]
46. Zlotnikov, I.D.; Streltsov, D.A.; Belogurova, N.G.; Kudryashova, E.V. Chitosan or Cyclodextrin Grafted with Oleic Acid Self-Assemble into Stabilized Polymeric Micelles with Potential of Drug Carriers. *Life* **2023**, *13*, 446. [CrossRef] [PubMed]

47. Zlotnikov, I.D.; Ezhov, A.A.; Ferberg, A.S.; Krylov, S.S.; Semenova, M.N.; Semenov, V.V.; Kudryashova, E.V. Polymeric Micelles Formulation of Combretastatin Derivatives with Enhanced Solubility, Cytostatic Activity and Selectivity against Cancer Cells. *Pharmaceutics* **2023**, *15*, 1613. [CrossRef] [PubMed]
48. Zlotnikov, I.D.; Vigovskiy, M.A.; Davydova, M.P.; Danilov, M.R.; Dyachkova, U.D.; Grigorieva, O.A.; Kudryashova, E.V. Mannosylated Systems for Targeted Delivery of Antibacterial Drugs to Activated Macrophages. *Int. J. Mol. Sci.* **2022**, *23*, 16144. [CrossRef]
49. Bales, B.L.; Messina, L.; Vidal, A.; Peric, M.; Nascimento, O.R. Precision relative aggregation number determinations of SDS micelles using a spin probe. A model of micelle surface hydration. *J. Phys. Chem. B* **1998**, *102*, 10347–10358. [CrossRef]
50. Robson, R.J.; Dennis, E.A. The size, shape, and hydration of nonionic surfactant micelles. Triton X-100. *J. Phys. Chem.* **1977**, *81*, 1075–1078. [CrossRef]
51. Wu, S.; Liang, F.; Hu, D.; Li, H.; Yang, W.; Zhu, Q. Determining the Critical Micelle Concentration of Surfactants by a Simple and Fast Titration Method. *Anal. Chem.* **2020**, *92*, 4259–4265. [CrossRef] [PubMed]
52. Deutschle, T.; Porkert, U.; Reiter, R.; Keck, T.; Riechelmann, H. In vitro genotoxicity and cytotoxicity of benzalkonium chloride. *Toxicol. Vitr.* **2006**, *20*, 1472–1477. [CrossRef] [PubMed]
53. Mutlu-Agardan, N.B.; Sarisozen, C.; Torchilin, V.P. Cytotoxicity of Novel Redox Sensitive PEG2000-S-S-PTX Micelles against Drug-Resistant Ovarian and Breast Cancer Cells. *Pharm. Res.* **2020**, *37*, 65. [CrossRef] [PubMed]
54. Ensor, C.M.; Holtsberg, F.W.; Bomalaski, J.S.; Clark, M.A. Pegylated arginine deiminase (ADI-SS PEG20,000 mw) inhibits human melanomas and hepatocellular carcinomas in vitro and in vivo. *Cancer Res.* **2002**, *62*, 5443–5450. [CrossRef]
55. Zhong, P.; Zhang, J.; Deng, C.; Cheng, R.; Meng, F.; Zhong, Z. Glutathione-Sensitive Hyaluronic Acid-SS-Mertansine Prodrug with a High Drug Content: Facile Synthesis and Targeted Breast Tumor Therapy. *Biomacromolecules* **2016**, *17*, 3602–3608. [CrossRef] [PubMed]
56. Balendiran, G.K.; Dabur, R.; Fraser, D. The role of glutathione in cancer. *Cell Biochem. Funct.* **2004**, *22*, 343–352. [CrossRef] [PubMed]
57. Kennedy, L.; Sandhu, J.K.; Harper, M.E.; Cuperlovic-culf, M. Role of glutathione in cancer: From mechanisms to therapies. *Biomolecules* **2020**, *10*, 1429. [CrossRef] [PubMed]
58. Sharma, S.; Kumar, K.; Chauhan, S.; Chauhan, M.S. Conductometric and spectrophotometric studies of self-aggregation behavior of streptomycin sulphate in aqueous solution: Effect of electrolytes. *J. Mol. Liq.* **2020**, *297*, 111782. [CrossRef]
59. Perinelli, D.R.; Cespi, M.; Lorusso, N.; Palmieri, G.F.; Bonacucina, G.; Blasi, P. Surfactant Self-Assembling and Critical Micelle Concentration: One Approach Fits All? *Langmuir* **2020**, *36*, 5745–5753. [CrossRef]
60. Vargas, M.; Albors, A.; Chiralt, A.; González-Martínez, C. Characterization of chitosan-oleic acid composite films. *Food Hydrocoll.* **2009**, *23*, 536–547. [CrossRef]
61. Fluksman, A.; Benny, O. A robust method for critical micelle concentration determination using coumarin-6 as a fluorescent probe. *Anal. Methods* **2019**, *11*, 3810–3818. [CrossRef]
62. Lesemann, M.; Thirumoorthy, K.; Kim, Y.J.; Jonas, J.; Paulaitis, M.E. Pressure dependence of the critical micelle concentration of a nonionic surfactant in water studied by 1H-NMR. *Langmuir* **1998**, *14*, 5339–5341. [CrossRef]
63. Mabrouk, M.M.; Hamed, N.A.; Mansour, F.R. Spectroscopic methods for determination of critical micelle concentrations of surfactants; a comprehensive review. *Appl. Spectrosc. Rev.* **2023**, *58*, 206–234. [CrossRef]
64. Ollmann, M.; Galla, H.J.; Schwarzmann, G.; Sandhoff, K. Pyrene-Labeled Gangliosides: Micelle Formation in Aqueous Solution, Lateral Diffusion, and Thermotropic Behavior in Phosphatidylcholine Bilayers. *Biochemistry* **1987**, *26*, 5943–5952. [CrossRef] [PubMed]
65. Mohr, A.; Talbiersky, P.; Korth, H.G.; Sustmann, R.; Boese, R.; Bläser, D.; Rehage, H. A new pyrene-based fluorescent probe for the determination of critical micelle concentrations. *J. Phys. Chem. B* **2007**, *111*, 12985–12992. [CrossRef]
66. Berghmans, M.; Govaers, S.; Berghmans, H.; De Schryver, F.C. Study of polymer gelation by fluorescence spectroscopy. *Polym. Eng. Sci.* **1992**, *32*, 1466–1470. [CrossRef]
67. Kotta, S.; Aldawsari, H.M.; Badr-Eldin, S.M.; Nair, A.B.; YT, K. Progress in Polymeric Micelles for Drug Delivery Applications. *Pharmaceutics* **2022**, *14*, 1636. [CrossRef]

Disclaimer/Publisher's Note: The statements, opinions and data contained in all publications are solely those of the individual author(s) and contributor(s) and not of MDPI and/or the editor(s). MDPI and/or the editor(s) disclaim responsibility for any injury to people or property resulting from any ideas, methods, instructions or products referred to in the content.

Article

Activity of a Recombinant Chitinase of the *Atta sexdens* Ant on Different Forms of Chitin and Its Fungicidal Effect against *Lasiodiplodia theobromae*

Katia Celina Santos Correa [1], William Marcondes Facchinatto [2], Filipe Biagioni Habitzreuter [3], Gabriel Henrique Ribeiro [4], Lucas Gomes Rodrigues [1], Kelli Cristina Micocci [1], Sérgio Paulo Campana-Filho [3], Luiz Alberto Colnago [4] and Dulce Helena Ferreira Souza [1,*]

1. Department of Chemistry, Federal University of Sao Carlos, 13565-905 Sao Carlos, Brazil; katiacorrea12@gmail.com (K.C.S.C.); lucasgomes@estudante.ufscar.br (L.G.R.); kelli.micocci@gmail.com (K.C.M.)
2. Aveiro Institute of Materials, CICECO, Department of Chemistry, University of Aveiro, St. Santiago, 3810-193 Aveiro, Portugal; williamfacchinatto@alumni.usp.br
3. Sao Carlos Institute of Chemistry, University of Sao Paulo, Ave. Trabalhador Sao-carlense 400, 13560-590 Sao Carlos, Brazil; filipeh@usp.br (F.B.H.); scampana@iqsc.usp.br (S.P.C.-F.)
4. Brazilian Corporation for Agricultural Research, Embrapa Instrumentation, St. XV de Novembro 1452, 13560-970 Sao Carlos, Brazil; gabrielhenri10@hotmail.com (G.H.R.); luiz.colnago@embrapa.br (L.A.C.)
* Correspondence: dulce@ufscar.br

Abstract: This study evaluates the activity of a recombinant chitinase from the leaf-cutting ant *Atta sexdens* (AsChtII-C4B1) against colloidal and solid α- and β-chitin substrates. ^{1}H NMR analyses of the reaction media showed the formation of N-acetylglucosamine (GlcNAc) as the hydrolysis product. Viscometry analyses revealed a reduction in the viscosity of chitin solutions, indicating that the enzyme decreases their molecular masses. Both solid state ^{13}C NMR and XRD analyses showed minor differences in chitin crystallinity pre- and post-reaction, indicative of partial hydrolysis under the studied conditions, resulting in the formation of GlcNAc and a reduction in molecular mass. However, the enzyme was unable to completely degrade the chitin samples, as they retained most of their solid-state structure. It was also observed that the enzyme acts progressively and with a greater activity on α-chitin than on β-chitin. AsChtII-C4B1 significantly changed the hyphae of the phytopathogenic fungus *Lasiodiplodia theobromae*, hindering its growth in both solid and liquid media and reducing its dry biomass by approximately 61%. The results demonstrate that AsChtII-C4B1 could be applied as an agent for the bioproduction of chitin derivatives and as a potential antifungal agent.

Keywords: insect chitinase; chitin; fungicide

1. Introduction

Chitin is a polysaccharide that occurs abundantly in nature and is a structural component of many organisms, such as mollusks, fungi, and arthropods [1]. Composed of a linear chain of N-acetyl-D-glucosamine (GlcNAc) monomers, linked by β-(1-4) glycosidic bonds [2], chitin occurs in nature as three polymorphs called α-, β-, and γ-chitin [3]. These polymorphs have different arrangements of their polymer chains in the crystalline domains, which results in marked differences in their physicochemical properties, such as their crystallinity and swelling capacity. The polymorphs also differ in their degree of hydration, the size of the unit cell, and the number of chitin chains per unit cell [4]. The molecular organization of chitin involves macromolecules that interact with other elements either covalently or supramolecularly, which defines many of the functions in the organism, ranging from growth and mechanical resistance to defense against microorganisms and diseases [5]. The microfibrils combine with sugars, proteins, glycoproteins, and proteoglycans to form cell walls in fungi, as well as the arthropod cuticles and peritrophic matrices present in crustaceans and insects, respectively [4,6].

Due to its abundance and also to the numerous possibilities for carrying out chemical modifications, which can result in chitin whiskers via acidolysis, chitosan via N-deacetylation, oligomers and GlcNAc units via hydrolysis, and chitosan, chitin derivatives have enormous potential for use in the pharmaceutical, cosmetics, and nutritional supplement industries [7,8] and have also aroused the interest of the scientific community [9]. The characteristics of this material include its biodegradability, biocompatibility, and antioxidant [10] and antibacterial activities [11], and thus it has been used in the food and health industries [12].

Chemical, enzymatic, and physicochemical treatments can be used to make chemical changes to chitin's structure, including reducing the average molecular mass via depolymerization [13]. Depolymerization via chemical treatment is the most traditional method and is mainly performed via deacetylation/degradation through alkaline/acid treatment and the introduction of new chemical groups using specific solvents. The modification effect depends on the established conditions, including the type of reagent, time, temperature, and pH [14,15]. Chemical methods are efficient, but it is necessary to rethink their use because of the formation of polluting residues [16]. Physical methods are employed to reduce the use of chemical reagents, such as the use of ultrasound, which has been applied to provoke modifications in the structure of chitin, providing porosity and reducing the size of fibers or particles. However, the use of these methods on an industrial scale presents complications such as a high energy consumption and poor control of the products formed [17].

Thus, enzymatic hydrolysis, which has a low energy consumption, reduces the generation of polluting residues, and enables greater control of the products formed, is a very promising methodology and is being increasingly employed [18]. Enzymatic hydrolysis processes are generally conducted between 30 and 60 °C, with a pH ranging from 4 to 12 and a duration of a few hours [17–20]. The enzymes that catalyze hydrolysis reactions of chitin, called chitinolytic enzymes, belong to the family of glycosyl hydrolases (GHs) [21]. According to the carbohydrate-active enzyme database (CAZy), chitinolytic enzymes are classified into the families GH18, GH19, GH23, and GH48, with the difference between the GHs being the composition of their amino acids and catalytic properties [22]. Chitinases of the GH18 family are present in almost all organisms, including plants and mammals, and are classified according to the type of cleavage they promote. Endo-chitinases catalyze the internal hydrolysis of chitin chains at random positions, whereas exo-chitinases hydrolyze the chitin chain at its terminal, whether at the reducing or non-reducing end [13].

In organisms, chitinases are involved in tissue degradation, developmental regulation, pathogenicity, and immunological defense. Although the most extensive function of GH18 chitinases is to degrade endogenous chitin, many microorganisms produce them to utilize chitin as a nutritional source [23]. Insect chitinolytic enzymes, for example, have been identified as potential biopesticides against organisms that contain chitin in vital structures, such as the peritrophic membrane or cuticle of insects; eggshells; nematode sheaths; and the cell walls of pathogenic fungi [24]. For instance, a chitinase from *Bombyx mori* was evaluated for its potential use as a biopesticide against the *Monochamus alternatus* beetle. The oral ingestion of chitinase induced modifications in the beetle's peritrophic membrane chitin, leading to a reduced body weight and mortality [25]. A chitinase from *Ostrinia furnacalis* showed activity against phytopathogens such as *Fusarium graminearum*, *Botrytis cinerea*, *Rhizoctonia solani*, *Phytophthora capsici*, and *Colletotrichum gloeosporiodes* [26].

Recently, our research group expressed in *Pichia pastoris* a chitinase from the leaf-cutting ant *Atta sexdens* (AsChtII-C4B1), which consists of a catalytic domain and a chitin-binding domain (CBM) belonging to the GH18 family of GHs [27]. AsChtII-C4B1 exhibited larvicidal activity against the tested model *Spodoptera frugiperda* and fungicidal activity against human pathogenic fungi *Candida albicans* and *Aspergillus fumigatus*. In this present study, we investigate the hydrolytic action of AsChtII-C4B1 on different forms of chitin (α- and β-chitin) and evaluate the activity of the recombinant chitinase against the phytopathogenic fungus *Lasiodiplodia theobromae* a cause of severe losses in agricultural production.

2. Materials and Methods

2.1. Materials

The fungus *Lasiodiplodia theobromae* (CDA 1169), isolated from the stem of soursop (*Annona muricata*), Jaiba, MG, Brazil, was acquired from the Mycological Collection of the Federal University of Pernambuco, Brazil, and maintained in sterile water according to the CASTELLANI method [28] at room temperature until use. Subsequently, the fungus was inoculated at 28 °C in a plate containing potato dextrose agar (PDA).

The recombinant chitinase AsChtII-C4B1 was obtained as reported in the literature, being expressed into the extracellular medium in a *Picchia pastoris* system and purified on a nickel resin affinity column [27].

2.2. Methods

2.2.1. Evaluation of Enzyme Activity against Various Forms of Chitin

α- and β-Chitin Substrates

β-chitin was extracted from the gladii of squids (*Doryteuthis* spp.) following the methodology described in the literature [29], ground in a knife mill equipped with a 1 mm sieve, and then separated into different powder fractions with average diameters (d) varying from <0.125 to >0.425 mm [30]. α-chitin, isolated from crab shells, was purchased from SIGMA ALDRICH (St. Louis, MI, USA). α-chitinase and colloidal β-chitin were prepared according to a method described in the literature [31] and used at a concentration of 5% (*w/v*) in citrate-phosphate buffer, pH 5.0 [32].

Enzymatic Activity against Various TYPES of Chitin

For the reaction medium, 1.0 mL of purified protein (0.713 mg) was added to individual tubes containing the solid substrates βQ125SE (β-chitin d < 125 mm) and βQ425SE (β-chitin d > 0.425 mm), both at a concentration of 5% (*w/v*) in citrate-phosphate buffer, pH 5.0, totaling a final volume of 2 mL. The colloidal β-chitin (βQ125CE and βQ425CE) and α-chitin (αCenz) substrates underwent the same enzymatic treatment. The reaction medium contained in the tubes was stirred at 250 rpm at 55 °C for 48 h. The solution was centrifuged at 10,000× *g* at 4 °C for 40 min and the pellets were homogenized in 5 mL of MilliQ H_2O (St. Louis, MI, USA). After this step, dialysis of the reaction medium was performed using the dialysis membrane (3.5K MWCO) against 600 mL of ultrapure H_2O at 4 °C overnight.

The same procedure previously described, but in the absence of the enzyme, was conducted using colloidal α-chitin at 55 °C (here called CT55) and at 25 °C (here called chitin CT25).

2.2.2. Evaluation of GlcNAc Production via ^1H Nuclear Magnetic Resonance (NMR) Spectroscopy Post Enzymatic Reaction

Aliquots of 350 µL from the samples (supernatant) of the enzymatic reaction against colloidal α- and β-chitin substrates were diluted in 250 µL of an internal standard solution of sodium 3-trimethylsilylpropionate-d4 (0.50 mM, TMSP in D_2O), used as an internal standard at 0.00 ppm. The samples were transferred to a standard 5 mm NMR tube for analysis.

The 1D and 2D NMR experiments of the samples were performed at 25 °C using a Bruker 14.1 Tesla instrument, AVANCE III (Billerica, MA, USA), equipped with a 5 mm PABBO (Broad Band Observe) direct detection probe with ATMA® (Automatic Tuning Matching Adjustment) and a BCU-I variable temperature unit. The ^1H NMR spectra were acquired using a pre-saturation solvent suppression pulse sequence of the water signal (here called noesypr1d), Bruker TopSpin, with a field gradient and with water signal suppression by irradiation at the frequency of 2822.04 Hz (O1). The conditions were as follows: 64 averages (ns), 4 dummy scans (ds), 65,536 data points during acquisition (td), a spectral window (sw) of 20.03 ppm, a receiver gain (rg) of 80.6, a 90° pulse of 11.850 µs, an acquisition time between each acquisition (aq) of 2.73 s, and a 5 ms mixing time (d8). The ^{1}H NMR spectra were referenced through the TMSP-d4 signal at 0.0 ppm. To support the assignment of the compound of interest, 2D NMR experiments were conducted, such as ^{1}H-^{1}H COSY.

The ^{1}H NMR spectra were processed using TopSpinTM 3.6.1 software (Bruker, Biospin, Ettlingen, Germany). To determine the product of interest, N-acetylglucosamine (GlcNAc), a database query was carried out (human metabolome; HMDB—Human Metabolome Database; N-acetylglucosamine (HMDB0000215)). The GlcNAc assignments were confirmed by an analysis of 2D NMR correlation maps. The compound was quantified in the ^{1}H NMR spectra using Chenomx NMR Suite 8.4 software (Chenomix Inc, Edmonton, Edmonton, AB, Canada). ^{1}H-^{13}C HMBC, Jres, was performed on a selected sample.

2.2.3. Analysis of the Morphology of Various Types of Chitin

Solid-State ^{13}C Nuclear Magnetic Resonance (NMR) Spectroscopy

The solid-state NMR ^{13}C spectra of the chitin samples (βQ125SE, βQ425SE, βQ125CE, βQ425CE, CEnz, αQCT25, and αQCT55 °C) were obtained at 25 ± 1 °C using an Advance 400 spectrometer (Bruker) coupled to a 4 mm dual-resonance probe with magic-angle spinning (MAS), operating at 100.5 MHz for the carbon nucleus and 400 MHz for the hydrogen nucleus. The average degree of acetylation (GA) was calculated and short-range molecular ordering was assessed. The short-range crystallinity index (CrI$_{SR}$) was calculated from the fitting of signals from carbon C4 and C6 following the deconvolution method proposed in the literature [33].

X-ray Diffraction (XRD)

XRD patterns of the chitin samples (βQ125SE, βQ425SE, βQ125CE, βQ425CE, Cenz, CT25, and C55 °C) were acquired using an AXS D8 Advance diffractometer (Bruker; Billerica, Billerica, MA, USA) with a Cu Kχ radiation source (λ = 0.1548 nm). Measurements were performed in the range 5° < 2θ < 50° at a scan speed of 5° min. The crystallinity indices (ICr) were estimated by subtracting the contribution of the amorphous region (A$_{am}$) by fitting a cubic spline curve from the total diffraction pattern area (A$_{tot}$). Data treatment was conducted using Microcal Origin 2020 software [34]. In addition, the apparent crystalline dimensions (L$_{hkl}$) referring to (020)$_h$ and (200)$_h$ reflections were calculated by applying the Scherrer relation through the width at half-heights (FWHM) of diffraction peaks at 2θ ≈ 8.3° and ≈ 19.7°, respectively. Data treatment was conducted by fitting Lorentzian curves, as described in previous studies [30,33].

Capillary Viscometry in Dilute Regime

The viscosity average molecular masses (M_v) of the chitin samples were determined from their intrinsic viscosities (η) and average degrees of acetylation (GA). To determine η, pre- and post-enzymatic treatment chitins were dissolved in N,N-dimethylacetamide (DMAc) containing 5% LiCl (w/v). The samples were dissolved in DMAc/LiCl at room temperature for 24 h and then filtered under positive pressure (0.45 um). The flow times in a glass capillary (ϕ = 0.84 mm) were determine at 25.00 ± 0.01 ^{0}C using an AVS-360 viscometer connected to an AVS-20 automatic burette, both from Schott-Geräte (Mainz, Germany). From the extrapolation of the line obtained at infinite dilution, it is possible to determine the η value of the polymer, which is necessary to calculate M_v, as previously established in the literature [35,36]. The values of M_v, GA, and the molecular masses (g/mol) of GlcN and GlcNAc units were used to calculate the average degree of polymerization (GP_v), as indicated in previous studies [31,34]. The assays were conducted in triplicate.

2.3. Evaluation of the AsChtII-C4B1 Enzyme Activity on the Growth of the Fungus L. theobromae

2.3.1. Assays of Activity and Thermostability of AsChtII-C4B1

The chitinase activity in biological assays was determined by the 3,5-dinitrosalicylic acid (DNS) method (Sigma-Aldrich; St. Louis, MI, USA), according to the literature [37,38], using colloidal α-chitin as a substrate. Briefly, 200 µL (0.138 mg) of the purified enzyme was homogenized with 200 µL of 5% (w/v) colloidal α-chitin in citrate-phosphate buffer, pH 5.0. The solution was incubated at 55 °C with shaking at 250 rpm for 1 h. After that, 400 µL of DNS was added to the reaction. The reaction mixture was heated at 100 °C for 10 min and then cooled to −20 °C for 5 min. The solution was centrifuged at 10,000× g for 5 min and the supernatant was subjected to absorbance measurements at 540 nm in a spectrophotometer (BIOMATE 160; Mettler Toledo; Langacher; Greifensee, Switzerland). The blank control mixture (without the enzyme) underwent the same treatment and was used to zero the equipment. The results obtained were compared with a standard curve of (GlcNAc) ranging from 0.1 to 1 mg mL^{-1}.

The enzymatic activity was also assessed on both colloidal and solid α- and β-chitin substrates by varying the incubation time (1–72 h), measuring the enzyme concentration in the reaction solution via the Bradford method [39], using BSA as a standard. The percentage of enzyme in the solution was calculated from the difference between the total enzyme concentration and the remaining enzyme concentration in the supernatant during the assay period. All experiments were conducted in triplicate.

2.3.2. Antifungal Assays

From the fungus cultivated on a Petri dish (90 cm × 15 mm) in PDA medium (Potato Dextrose Agar M096, HIMEDIA), an 8 mm halo was removed and placed in 10 mL of liquid medium (5.6 g L^{-1} (NH4)$_2$SO$_4$, 4 g L^{-1} KH$_2$PO$_4$, 0.6 g L^{-1} MgSO$_4$.7H$_2$O, 1.8 g L^{-1} peptone, 0.5 g L^{-1} yeast extract, 0.02 g L^{-1} MnSO$_4$.H$_2$O, 0.002 g L^{-1} ZnSO$_4$.7H$_2$O, 0.04 g L^{-1} CoCl$_2$.6H$_2$O) [40]. The culture medium was incubated at 28 °C with constant agitation at 150 rpm for 72 h. Subsequently, purified chitinase (0.713 mg) was added to the medium and the culture was maintained for another 72 h under the same conditions to verify the fungal mycelial growth.

The fungus's dry mass was calculated after cultivation under the previously described conditions at three different times: 24, 48, and 72 h. To this end, after the desired time, the culture medium was centrifuged, rinsed with Milli-Q water, dried at 60 °C, and weighed.

In another experiment, an 8 mm mycelial sample was removed from the Petri dish cultivated with the fungus, plated on a new Petri dish, and a solution containing 0.713 mg of AsChtII-C4B1 was dripped onto the halo. The same procedure was conducted in the absence of the enzyme (positive growth control) and the presence of 2 µg mL^{-1} of the commercial fungicide Amphotericin-B (negative growth control). The results are expressed as the mean of three replicates of three independent experiments with the indicated standard deviations.

2.3.3. Analysis of Biological Samples via Scanning Electron Microscopy (SEM)

The mycelial samples from the liquid medium experiment mentioned in Section 2.3.2. were incubated in a Karnovsky solution (4% paraformaldehyde, 5% glutaraldehyde, 0.05% CaCl$_2$) [41] at room temperature for 24 h to fix and preserve the biological material. Subsequently, the Karnovsky solution was discarded and the samples were dehydrated by varying the percentage of acetone every 10 min (30, 50, 70, 90, and 100%), with the 100% step performed 3 times every 10 min. The mycelia were lyophilized and coated with a gold layer for SEM analysis to observe possible morphological changes in the fungal hyphae. Micrographs were obtained using JEOL equipment, model JSM 6510 (Tokio; Tokyo, Japan), with an electron acceleration voltage of 5 kV and a working distance (WD) of 10 mm.

3. Results and Discussion

3.1. Activity of Recombinant Chitinase on Various Substrates

The recombinant chitinase AsChtII-C4B1 contains a catalytic domain and a carbohydrate-binding module (CBM), which has been reported to assist in anchoring the enzyme to the insoluble substrate through the interaction of conserved aromatic residues, breaking the crystalline structure of the substrate, resulting in the formation of free chain ends [23,42]. This enzyme shows activity against colloidal α-chitin and has been shown to have fungicidal and larvicidal activity [27]; it was inferred that it might also act on solid-state chitin. Thus, this study assessed the activity of the enzyme against solid α-chitin and colloidal and solid β-chitin.

Considering the modes of enzymatic activity, GH18 chitinases can be divided into processive and non-processive chitinases. Processive chitinases can slide along the substrate chain and continue hydrolysis without the enzyme, detaching from the chitin chain after each catalytic event, thus producing soluble reducing ends in contrast to non-processive chitinases. In general, exo-chitinases are processive enzymes, whereas endo-chitinases are non-processive [43].

To infer the mode of action of the AsChtII-C4B1 enzyme, the protein concentration in solution was quantified post-reaction (monitored up to 72 h) with α- and β-chitin substrates in both solid and colloidal states (Figure 1). In all experiments, the protein concentration in solution decreased over the reaction duration, suggesting that AsChtII-C4B1 is a processive enzyme, binding to the substrate and catalyzing the cleavage of consecutive bonds without dissociating from it.

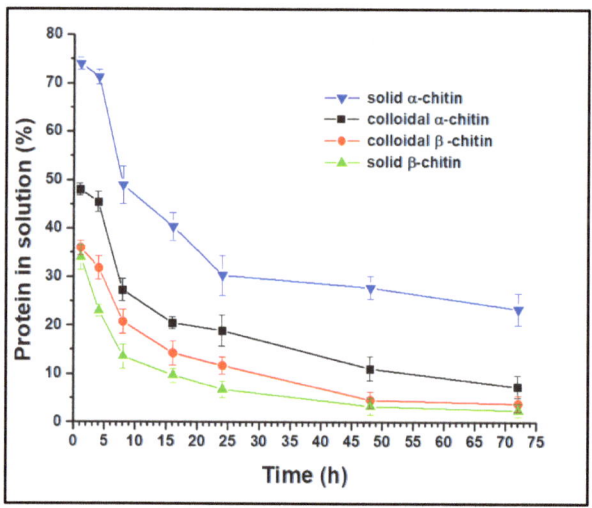

Figure 1. Analysis of free protein concentration in solution in reactions with α- and β-chitin substrates in both solid and colloidal forms. The experiments were performed in triplicate.

Analysis of the free protein concentration demonstrated that solid β-chitin was the most accessible substrate for AsChtII-C4B1 binding, as it showed the lowest concentration in solution at all evaluated reaction times. One hour post-reaction with this chitin, only 34% of the enzyme was free, while in the reaction with solid α-chitin, 74% of the enzyme was free. For colloidal substrates, β-chitin also proved to be more accessible to enzyme binding than α-chitin. The chains of chitin molecules are organized in sheets, strongly held together by hydrogen bonds, and the structure of β-chitin has fewer hydrogen bonds between neighboring chains compared with that of α-chitin, forming less dense fibrils that are more susceptible to swelling and hydrolysis reactions [44]. Thus, the results obtained in this experiment, showing that β-chitin is more susceptible to enzyme binding, can be explained by the structural nature of the substrates.

Studies have shown that the major impediment to enzymatic hydrolysis is the crystallinity of the substrate [7,45]. By synthesizing colloidal chitins, molecules with a lower crystallinity are obtained, increasing the amorphous regions, which are more accessible to enzymatic action. Experiments with α-chitin revealed that the colloidal molecule has a higher concentration of protein bound to the substrate than in the solid state, as expected. However, this behavior was not observed with β-chitin, as the enzyme bound at greater concentrations in the solid form than in the colloidal form. Differences in substrate preferences by chitinases have been associated with the presence/absence of a CBM, with chitinases having a CBM reported as being more efficient at degrading crystalline chitin, while those without a CBM act on less crystalline chitin [22]. AsChtII-C4B1 has a CBM, and this enzyme is capable of catalyzing the cleavage reactions of the substrate bonds in both solid and colloidal forms, other factors, besides CBM, might be involved in the enzyme's binding to the substrates.

3.2. H NMR Spectroscopy of Enzymatic Hydrolysis on α- and β-Chitin Colloidal Substrates

^{1}H NMR spectroscopy is a highly specific and sensitive method for quantifying and determining the products resulting from chitin hydrolysis. Since enzymatic hydrolysis is conducted under mild conditions and shows high selectivity in producing, for example, GlcNAc and (GlcNAc)$_2$, this technique is useful in assessing chitinase activity, exhibiting an excellent correlation between product concentration and peak integrals [46].

Therefore, the product generated in the reactions of AsChtII-C4B1 with the colloidal α- and β-chitin substrates was identified and quantified via ^{1}H NMR spectroscopy (Figure 2). Signals corresponding to the hydrogens of GlcNAc were identified in the spectrum. Their chemical structures and assignments are presented in the Supplementary Materials (Figure S1 and Table S1). To support the assignment of the produced compound, 2D NMR experiments, such as ^{1}H-^{1}H COSY (Figure S2), were performed. For the quantification of GlcNAc, signals from the hydrogens of the N-acetyl group (5.19 ppm) and the anomeric hydrogens of GlcNAc (2.03 ppm) were selected. The quantities of GlcNAc found were 2.64 and 2.16 mmol L^{-1} for the α- and β-chitin substrates, respectively, showing that the AsChtII-C4B1 enzyme is active on both substrates.

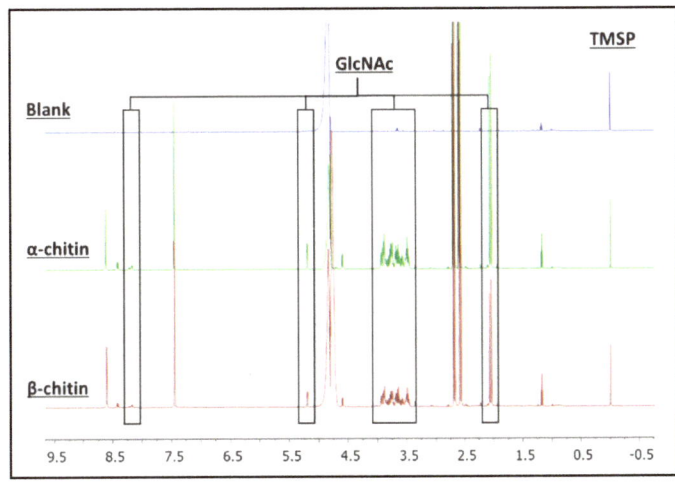

Figure 2. ^{1}H NMR spectra of the enzymatic reaction samples on colloidal α- and β-chitin. Signals of the hydrolyzed product, N-acetyl-d-glucosamine (GlcNAc), are highlighted. The concentration of the compound (GlcNAc) was determined from the signal of the internal standard (TMSP).

3.3. Activity of AsChtII-C4B1 on Different Types of Chitin

The enzyme activity on various types of solid and colloidal chitins was evaluated pre- and post-enzymatic action using the solid-state ^{13}C NMR technique. Before treatment, the spectra of βQ425 chitin (β-chitin with particle d > 425 nm) and βQ125 chitin (β-chitin with particle d < 125 nm) showed only one signal around 77 ppm, which is due to carbons C3 and C5. After enzymatic treatment of the solid chitin samples (βQ425SE and βQ125SE) (Figure 3a,b), the signal at 77 ppm showed a shoulder, and the signal at 63 ppm became slightly wider.

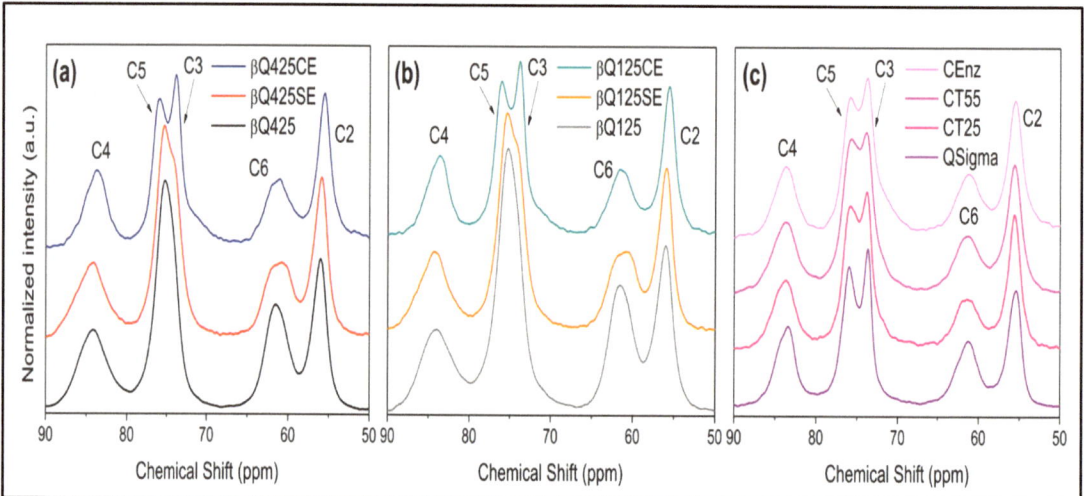

Figure 3. Structural characterization of chitin samples from the solid-state ^{13}C NMR spectra: (**a**) βQ425: solid β-chitin, βQ425SE: solid β-chitin treated with the enzyme, and βQ425CE: colloidal β-chitin treated with the enzyme; (**b**) βQ125: solid β-chitin, βQ125SE: solid β-chitin treated with the enzyme, and βQ125CE: colloidal β-chitin treated with the enzyme; (**c**) Qsigma: solid α-chitin, QCT25, QCT55: colloidal α-chitin at varying temperatures of 25 and 55 °C, and QEnz: colloidal α-chitin treated with the enzyme.

In the case of the reaction with the 425 and 125 colloidal β-chitins (βQ425CE and βQ125CE) (Figure 3a,b), the signal around 77 ppm was split into two signals at 73 and 78 ppm, corresponding to carbon C3 and C5 (references). Typically, for β-chitin samples, these signals overlap, while for α-chitin (Figure 3c), they are in slightly different chemical environments and do not overlap [47,48]. The separation of the signals of carbon C5 and C3 indicates a greater structural homogeneity, resulting from the antiparallel arrangement of the α-chitin chains [45]. Thus, it seems plausible to consider that, in the βQ425CE and βQ125CE substrates, new ordered arrangements are formed, resembling those of α-chitin, suggesting an increase in the substrate's relative crystallinity [7,49]. Conversion from the β-chitin allomorph to α-chitin has been reported in previous studies [50,51]. According to these studies, the process of recrystallization into a more thermodynamically stable morphology can be achieved under conditions of heating/cooling the sample, as well as solubilization in strongly acidic systems. In both cases, the breakdown of microfibrils is ensured, forcing the system to adopt an arrangement that provides greater packing of the chains. In the present case, the separation of carbon C5 and C3 signals appears as the first evidence that recrystallization can also be achieved under enzymatic conditions.

It is worth noting that the profiles of the other spectral signals in Figure 3 do not indicate a significant alteration in the structure. In particular, slight changes quantified from the signals of carbons C4 and C6 indicate a small increase in the short-range crystallinity index (CrI$_{SR}$) in the colloidal substrates compared with the solid substrates treated with the enzyme, but both are lower than the values of the starting substrates. Indeed, the enzyme's action on any polymorph proves to be efficient in breaking the crystalline domains. It is also worth highlighting that the effect of different processing temperatures on the commercial reference substrate led to an equivalent decrease in CrI$_{SR}$ values compared with those achieved via enzymatic treatment (Figure 3c).

The XRD patterns of the α- and β-chitin samples are illustrated in Figure 4. Even with enzymatic treatment, the solid substrates (βQ425SE and βQ125SE) exhibit profiles similar to those of the original substrates (βQ425 and βQ125), as reflected in the long-range crystallinity index (CrI$_{SR}$) (Table 1). A notable change is observed in the XRD profiles of the colloidal substrates (βQ425CE and βQ125CE). Despite a considerable decrease in CrI$_{XRD}$ values, there was a significant increase in the dimensions of the crystallites L$_{020}$ and L$_{200}$, indicating a relative increase in the crystalline domains dispersed throughout the matrix. Indeed, the profiles of βQ425CE and βQ125CE highlight diffraction peaks that are otherwise less evident in other β-chitin substrates. Furthermore, the average profile achieved by βQ425CE and βQ125CE resembles those of the substrates treated from α-chitins, as shown in (Figure 4c), especially in the emergence of peaks centered at 12.8^0, 22.8^0, and 26.5^0, and the peak's shift from 8.6^0, to 9.4^0, similar to what was observed for the spectral profile resulting from carbons C5 and C3 (Figure 3). The convergence of these trends suggests that the enzyme acts in favor of increasing the regularity and orderly packing of the molecular chains, although this trend cannot be deduced from the colloidal samples. In this case, αQCT25 and αQCT55 already consist of the structure of the more stable alpha polymorph, and there is not enough sensitivity in the diffractograms to indicate the influence of the enzymatic treatment.

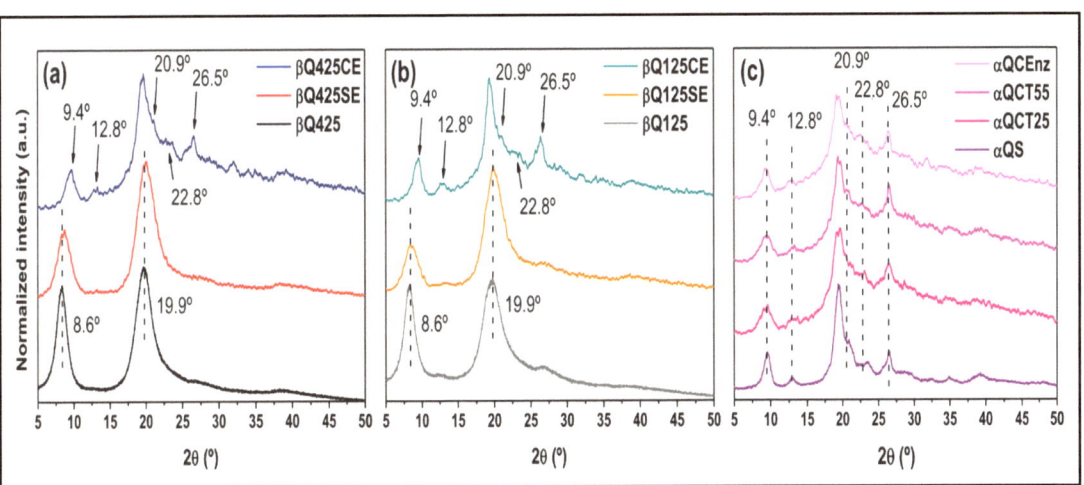

Figure 4. XRD patterns of β-chitin 125 (**a**), β-chitin 425 (**b**), and α-chitin (**c**) substrates.

Table 1. Short-range (CrI_{SR}) and long-range (CrI_{XRD}) crystallinity indexes calculated via NMR and XRD, respectively; the dimensions of the crystallites correspond to the reflection planes $(020)_h$ and $(200)_h$.

Sample	CrI_{SR} (%)	CrI_{XRD} (%)	L_{020} (nm)	L_{200} (nm)
βQ425	83.8 ± 1.7	71.2	4.70	3.43
βQ425SE	75.7 ± 9.0	66.4	4.30	3.80
βQ425CE	79.0 ± 9.4	37.8	7.13	5.32
βQ125	80.7 ± 1.2	61.1	4.67	3.19
βQ125SE	76.3 ± 9.1	61.0	5.92	3.79
βQ125CE	79.5 ± 9.9	46.5	7.00	5.18
QSigma	85.7 ± 0.2	58.8	8.02	6.03
QCenz	75.1 ± 2.7	42.7	7.20	5.11

3.4. Capillary Viscometry in Dilute Regime

To study the enzymatic activity on solid substrates and assess the enzyme's behavior towards various substrates, reactions were conducted with solid substrates and the reaction medium was studied using the capillary viscometry technique. Table 2 shows that depolymerization of solid substrates occurred after enzymatic reaction (βQ425SE, βQ125SE, and αQSE), as the results of the intrinsic viscosity (η) and viscosity average molecular mass (M_v) were lower than those for the solid substrates not treated with AsChtII-C4B1 (βQ425, βQ125, and αQ). Moreover, these results indicate that the enzyme's action differs for each sample; there is about 4-fold and 2-fold reductions in M_v for βQ125SE βQ425SE, respectively, after enzymatic treatment. This difference in M_v possibly occurred because the βQ125 sample comprises smaller particles (i.e., a larger available surface area), facilitating the interaction between its surface and the enzyme. In contrast, the α-chitin sample treated with the enzyme (αQSE) exhibited a 12-fold reduction in its M_v, indicating a greater enzyme activity in this polymorphic form of chitin, possibly due to its more ordered structure.

Table 2. Capillary viscometry results for the enzyme reaction medium with chitins: intrinsic viscosity (n), viscosity average molecular mass (M_v) average degree of polymerization (GP_v), and average degree of acetylation (GA).

Sample	η (mL·mg^{-1})	$M_v \times 10^6$ (g mol^{-1})	GP_v	GA (%)
BQ 125	3.30 ± 0.22	996 ± 92	6187 ± 581	94.3 ± 4.35
BQ125SE	1.28 ± 0.02	253 ± 5.0	1465 ± 198	>95
BQ425	5.06 ± 0.74	1860 ± 39	11,546 ± 2400	90.3 ± 3.74
BQ425SE	3.03 ± 0.07	876 ± 30	4319 ± 148	>95
AQ	3.18 ± 0.18	943 ± 76.5	5858 ± 474	95.4 ± 5.28
AQ SE	0.57 ± 0.04	79.3 ± 7.8	492 ± 48	>95

The average degree of acetylation (GA) of the samples pre- and post-enzymatic treatment was calculated based on the results of the ^{13}C NMR analysis. According to the results for solid substrates assessed without enzyme treatment, the results for solid substrates treated with the enzyme demonstrated that there was no significant change in GA.

3.5. Analysis of Enzyme Activity on the Growth of Fungus L. theobromae

Chitinases catalyze the degradation of α-chitin present in the shells of crabs and shrimps and in the cell walls of fungi [52], as well as catalyzing the degradation of α and β forms of chitin found in insects [53].

A previous study reported that the recombinant chitinase AsChtII-C4B1 demonstrated fungicidal activity against the filamentous fungus *Aspergillus fumigatus* [27]. In the present study, the enzyme was purified via ammonium sulfate precipitation and affinity chromatography and evaluated against the phytopathogenic fungus *L. theobromae*, an ascomycete belonging to class *Dothideomycetes*, order *Botryospheriales*, family *Botryosphaeriaceae* [54]. This phytopathogen attacks more than 500 species of plants, mainly in tropical and subtropical regions [55]. It causes fungal gummosis of peaches—a disease that severely restricts the the growth and production of this fruit in orchards in southern China, the United States, and Japan [56]. It causes leaf blight, stem cancer, and fruit rot in *Theobroma* cacao in Malaysia [57]. In Brazil, *L. theobromae* is a serious threat to cashew cultivation areas, causing resinosis and black rot of the stem [58]. In humans, this fungus has been associated with clinical manifestations such as corneal ulcers, rhinosinusitis, and mycosis in immunodeficient patients [59,60].

Given that *L. theobromae* grows optimally at 28 °C, the enzyme activity was tested at this temperature using colloidal α-chitin as a substrate, varying the reaction time. It was observed that the enzyme's activity increased with reaction time, reaching a plateau at 48 h (Figure S3, Supplementary Materials). Analysis of the enzyme's thermostability showed that it retains 55% of its activity even after 72 h (data not shown), allowing for an evaluation of AsChtII-C4B1's interference in the growth of this fungus in both solid and liquid media. Figure 5a–c show the growth of *L. theobromae* in a solid medium (Figure 5a) in the presence of a commercial fungicide (Figure 5b) and in the presence of the enzyme (Figure 5c). It was observed that the commercial fungicide *Amphotericin-B* completely inhibited mycelial growth, and the presence of the enzyme substantially inhibited this growth.

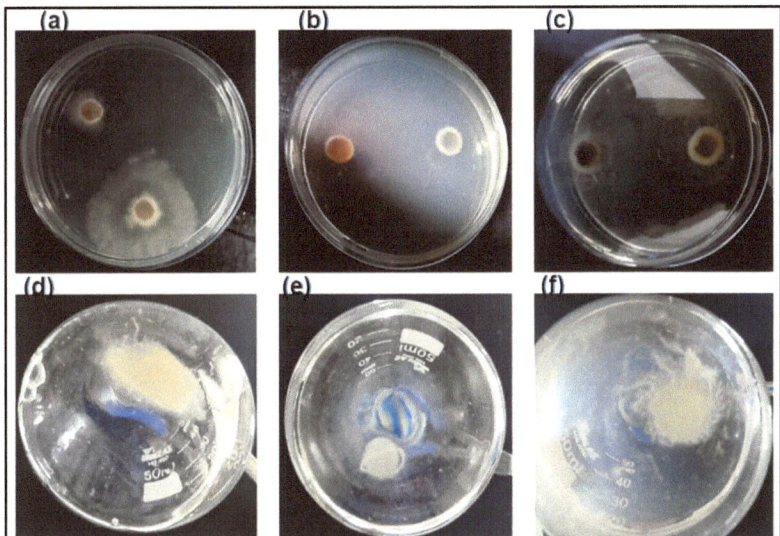

Figure 5. Chitinase activity on the growth of fungus *L. theobromae* in solid and liquid media: fungal growth (**a**) in the absence of the enzyme (positive control), (**b**) in the presence of commercial fungicide *Amphotericin-B* (negative control), and (**c**) in the presence of chitinase, conducted in Petri dishes containing solid medium (PDA); fungal growth (**d**) in the absence of the enzyme (positive control), (**e**) in the presence of commercial fungicide *Amphotericin-B* (negative control), and (**f**) in the presence of chitinase, conducted in liquid medium.

In liquid medium, the enzyme's presence caused hyphae dispersion, suggesting that they were digested (Figure 5f). The fungus hyphae, both in the presence and the absence of the enzyme, were dried and weighed, and analyses showed a loss of mycelial mass after treatment with chitinase, as illustrated in Figure 6. Reductions in the mycelial mass of 20, 48.2, and 61.3% were observed after 24, 48, and 72 h, respectively, compared with the dry biomass of the fungus not subjected to the enzymatic process.

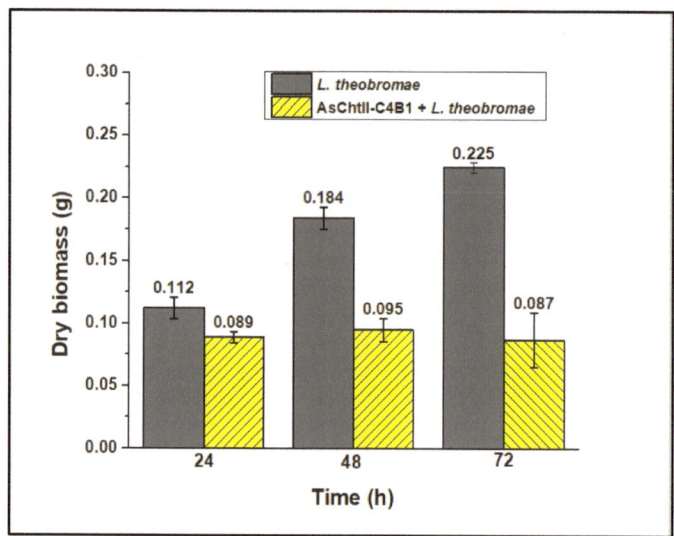

Figure 6. Evaluation of the dry mass of fungal mycelia before and after the action of chitinase. The assessment was conducted in three individual experiments varying the incubation time (24, 48, and 72 h). The experiments were performed in triplicate.

Although insects produce more chitinases than other organisms [61], most chitinases described in the literature with fungicidal activity are those from bacteria, fungi, and plants [62]. For example, the chitinase from *Trichoderma asperellum*, a fungus from the Hypocreaceae family, exhibited activity against the mycelium of fungus *Aspergillus nigers* [63], and the recombinant chitinase from cowpeas (*Vigna unguiculata*) inhibited the germination of spores and mycelial growth of the fungus *Penicillium herquei* [64].

The antifungal potential of chitinases depends on the morphology of the fungal cell walls, as the chitin contained in these microorganisms' cell walls is associated with β-glucans 1/3 and 1/6 or other polysaccharides whose composition can vary according to the species [62,65]. Most fungal pathogens contain chitin, as a dominant component in the cell wall, varying in the range of 5 to 27% in dry mass. They are found mainly in the mitosis cycle of fungal cells and at the growing hyphal tip [5]. Thus, the fungicidal activity of a chitinase may present a specificity not only associated with the microstructure of the surface, but also with the proportion of chitin in the fungal cell wall.

To investigate the effect of AsChtII-C4B1 on the morphology of the chitin contained in the fungal cell wall, mycelium samples from experiments in a liquid medium were analyzed via microscopy. SEM images of fungus mycelium treated with recombinant chitinase show morphological changes in the hyphae (Figure 7). The fungus hyphae, in the absence of the enzyme, presented a dense network of long tubular structures with smooth surfaces (Figure 7a,b). The fungus hyphae treated with recombinant chitinase exhibited a disordered network and brittle, thinner tubular structures (Figure 7c). Other modifications observed in the hyphae included crushing of the tubular parts, long translucent hyphal tips, and holes on the surface (Figure 7d–l). It was also possible to observe the outer layers of the hyphae detaching from the fungal cell wall and to observe protruding, broken, and translucent tubular structures. Additionally, hyphae with wrinkled surfaces were observed in SEM images. Some cut hyphae, with rounded and brittle ends, were also observed.

Therefore, these results demonstrate that the chitinase AsChtII-C4B1 inhibited the growth of phytopathogenic fungus *L. theobromae*, acting on the morphology of the hyphae both in the cell walls and in the degradation of nascent chitin corresponding to the hyphal tips, as demonstrated in SEM images.

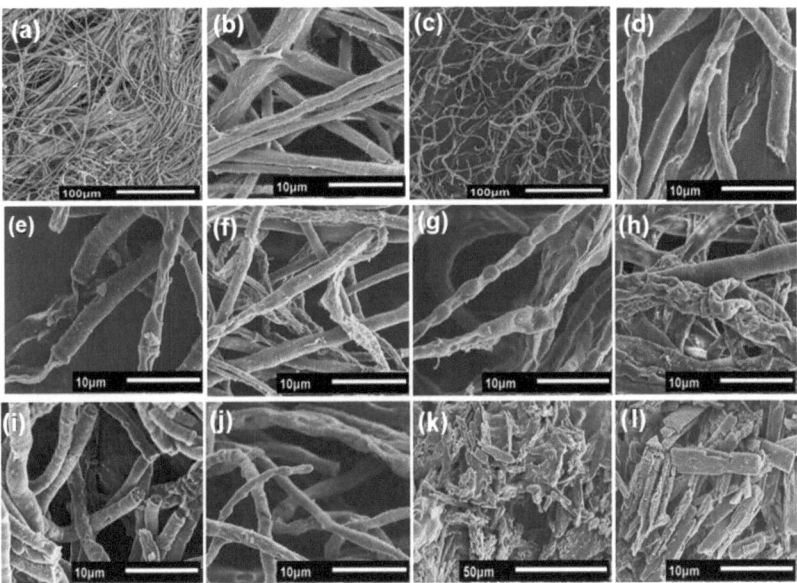

Figure 7. Morphological changes in the hyphae of fungus *L. theobromae* under the activity of the recombinant chitinase AsChtII-C4B1. (**a,b**): hyphae of the fungus in the absence of the enzyme. (**c**): hyphae of the fungus under the action of chitinase. (**d–l**): morphological alterations of the hyphae, such as crushing of the tubular parts, long translucent tips with holes on the surface, swollen tubes, rounded ends, and brittleness.

Observations similar to those of this study were reported in pathogenic fungi affected by a recombinant bacterial chitinase (ChiKJ406136) [66]. Studies indicated that the mycelial cells of the fungus were ruptured, broken, and distorted. The antifungal activity of a recombinant plant chitinase (GlxChiB) also influenced the growth of the fungus *Trichoderma viride*, causing damage not only to the hyphal tips but also to the lateral cell walls [67].

Biological control using microorganisms that produce chitinases has favorable effects against many post-harvest fungal pathogens [68]. The application of chitinases in food preservation is promising, as they can degrade the cell wall of contagious fungi—one of the main problems in terms of food deterioration—and prevent the germination process of fungal spores, thus helping to reduce food decomposition and degradation [69].

4. Conclusions

Chitinases are enzymes of biotechnological interest because they play a key role in the degradation of polysaccharide chitin. The results presented here demonstrate that it is possible to produce GlcNAc from colloidal α- and β-chitin using a recombinant insect chitinase. AsChtII-C4B1 was capable of degrading various types of chitin substrates—an important characteristic for applications in biocatalytic processes such as the production of chitin derivatives. This study also investigated the fungicidal activity of the enzyme on the phytopathogenic fungus *L. theobromae*, which affects various economically important crops, causing a reduction in the quality and quantity of agricultural commodities. Chitinase reduced the biomass of the fungus and modified the structures of its hyphae, confirming the efficiency of chitinase in reducing fungal mycelial growth. The results shown in this work indicate that the chitinase AsChtII-C4B1 presents important characteristics for potential biotechnological applications and future studies need to be developed to optimize its properties so that it can effectively be used as a fungicide.

Supplementary Materials: The following supporting information can be downloaded at: https://www.mdpi.com/article/10.3390/polym16040529/s1, Figure S1: GlcNAc chemical structure; Figure S2: COZY ^{1}H-^{1}H NMR spectrum for GlcNAc; Figure S3: Enzymatic activity in the presence of colloidal α-chitin substrate at 28 °C; Table S1: Assignments of the ^{1}H NMR spectra in D$_2$O for the N-acetyl-d-glucosamine compound.

Author Contributions: Conceptualization, K.C.S.C., W.M.F., D.H.F.S. and L.A.C.; methodology, K.C.S.C., W.M.F., F.B.H., G.H.R., L.G.R. and K.C.M.; validation, K.C.S.C., W.M.F., F.B.H. and G.H.R.; formal analysis, K.C.S.C., W.M.F., F.B.H. and G.H.R.; investigation, K.C.S.C., W.M.F., F.B.H. and G.H.R.; data curation, K.C.S.C., W.M.F., F.B.H. and G.H.R.; writing—original draft preparation, K.C.S.C., W.M.F., F.B.H. and G.H.R.; writing—review and editing, D.H.F.S., L.A.C. and S.P.C.-F.; supervision, D.H.F.S.; project administration, D.H.F.S.; funding acquisition, D.H.F.S. and L.A.C. All authors have read and agreed to the published version of the manuscript.

Funding: This research was funded by the São Paulo Research Foundation (FAPESP), grant no. 2018/06297-9 to D.H.F.S. 2021/12694-3 (LAC), and the Brazilian National Council for Scientific and Technological Development (CNPq), grant no. 141424/2020-6 to K.C.S.C., 116875/2023-2 to K.C.M. and 404690/2023-8 and 307635/2021-0 (LAC). National Council for Scientific and Technological Development: 311464/2022-2 to S.P.Campana-Filho.

Data Availability Statement: Data are contained within the article and Supplementary Materials.

Conflicts of Interest: Author Gabriel Henrique Ribeiro and Luiz Alberto Colnago was employed by the company Bra-zilian Corporation for Agricultural Research. The remaining authors declare that the research was conducted in the absence of any commercial or financial relationships that could be construed as a potential conflict of interest.

References

1. Patel, S.; Goyal, A. Chitin and chitinase: Role in pathogenicity, allergenicity and health. *Int. J. Biol. Macromol.* **2017**, *97*, 331–338. [CrossRef]
2. Zuhairah Zainuddin, S.; Abdul Hamid, K. *Chitosan-Based Oral Drug Delivery System for Peptide, Protein and Vaccine Delivery. Chitin and Chitosan-Physicochemical Properties and Industrial Applications*; BoD–Books on Demand: Norderstedt, Germany, 2021. [CrossRef]
3. Kaya, M.; Mujtaba, M.; Ehrlich, H.; Salaberria, A.M.; Baran, T.; Amemiya, C.T.; Galli, R.; Akyuz, L.; Sargin, I.; Labidi, J. On chemistry of γ-chitin. *Carbohydr. Polym.* **2017**, *176*, 177–186. [CrossRef] [PubMed]
4. Merzendorfer, H. Insect chitin synthases: A review. *J. Comp. Physiol. B* **2006**, *176*, 1–15. [CrossRef] [PubMed]
5. Bai, L.; Liu, L.; Esquivel, M.; Tardy, B.L.; Huan, S.; Niu, X.; Liu, S.; Yang, G.; Fan, Y.; Rojas, O.J. Nanochitin: Chemistry, Structure, Assembly, and Applications. *Chem. Rev.* **2022**, *122*, 11604–11674. [CrossRef] [PubMed]
6. Zhang, X.; Yuan, J.; Li, F.; Xiang, J. Chitin Synthesis and Degradation in Crustaceans: A Genomic View and Application. *Mar. Drugs* **2021**, *19*, 153. [CrossRef] [PubMed]
7. Chen, J.K.; Shen, C.R.; Liu, C.L. N-acetylglucosamine: Production and applications. *Mar. Drugs* **2010**, *8*, 2493–2516. [CrossRef] [PubMed]
8. Liu, L.; Liu, Y.; Shin, H.D.; Chen, R.; Li, J.; Du, G.; Chen, J. Microbial production of glucosamine and N-acetylglucosamine: Advances and perspectives. *Appl. Microbiol. Biotechnol.* **2013**, *97*, 6149–6158. [CrossRef] [PubMed]
9. Ahmad, S.I.; Ahmad, R.; Khan, M.S.; Kant, R.; Shahid, S.; Gautam, L.; Hasan, G.M.; Hassan, M.I. Chitin and its derivatives: Structural properties and biomedical applications. *Int. J. Biol. Macromol.* **2020**, *164*, 526–539. [CrossRef]
10. Kidibule, P.E.; Santos-Moriano, P.; Plou, F.J.; Fernández-Lobato, M. Endo-chitinase Chit33 specificity on different chitinolytic materials allows the production of unexplored chitooligosaccharides with antioxidant activity. *Biotechnol. Rep.* **2020**, *27*, e00500. [CrossRef]
11. Sánchez, Á.; Mengíbar, M.; Rivera-Rodríguez, G.; Moerchbacher, B.; Acosta, N.; Heras, A. The effect of preparation processes on the physicochemical characteristics and antibacterial activity of chitooligosaccharides. *Carbohydr. Polym.* **2017**, *157*, 251–257. [CrossRef]
12. Wang, Y.-T.; Wu, P.-L. Gene Cloning, Characterization, and Molecular Simulations of a Novel Recombinant Chitinase from Chitinibacter Tainanensis CT01 Appropriate for Chitin Enzymatic Hydrolysis. *Polymers* **2020**, *12*, 1648. [CrossRef]
13. Chen, W.; Jiang, X.; Yang, Q. Glycoside hydrolase family 18 chitinases: The known and the unknown. *Biotechnol. Adv.* **2020**, *43*, 107553. [CrossRef]

14. Li, F.; You, X.; Li, Q.; Qin, D.; Wang, M.; Yuan, S.; Chen, X.; Bi, S. Homogeneous deacetylation and degradation of chitin in NaOH/urea dissolution system. *Int. J. Biol. Macromol.* **2021**, *189*, 391–397. [CrossRef]
15. Allison, C.L.; Lutzke, A.; Reynolds, M.M. Identification of low molecular weight degradation products from chitin and chitosan by electrospray ionization time-of-flight mass spectrometry. *Carbohydr. Res.* **2020**, *493*, 108046. [CrossRef] [PubMed]
16. Yu, X.; Jiang, Z.; Xu, X.; Huang, C.; Yao, Z.; Yang, X.; Zhang, Y.; Wang, D.; Wei, C.; Zhuang, X. Mechano-Enzymatic Degradation of the Chitin from Crustacea Shells for Efficient Production of N-acetylglucosamine (GlcNAc). *Molecules* **2022**, *27*, 4720. [CrossRef] [PubMed]
17. Hou, F.; Gong, Z.; Jia, F.; Cui, W.; Song, S.; Zhang, J.; Wang, Y.; Wang, W. Insights into the relationships of modifying methods, structure, functional properties and applications of chitin: A review. *Food Chem.* **2023**, *409*, 135336. [CrossRef] [PubMed]
18. Kaczmarek, M.B.; Struszczyk-Swita, K.; Li, X.; Szczęsna-Antczak, M.; Daroch, M. Enzymatic Modifications of Chitin, Chitosan, and Chitooligosaccharides. *Front. Bioeng. Biotechnol.* **2019**, *7*, 243. [CrossRef] [PubMed]
19. Arnold, N.D.; Brück, W.M.; Garbe, D.; Brück, T.B. Enzymatic Modification of Native Chitin and Conversion to Specialty Chemical Products. *Mar. Drugs* **2020**, *18*, 93. [CrossRef] [PubMed]
20. Krolicka, M.; Hinz, S.W.A.; Koetsier, M.J.; Joosten, R.; Eggink, G.; Van Den Broek, L.A.M.; Boeriu, C.G. Chitinase Chi1 from Myceliophthora thermophila C1, a Thermostable Enzyme for Chitin and Chitosan Depolymerization. *J. Agric. Food Chem.* **2018**, *66*, 1658–1669. [CrossRef]
21. Poria, V.; Rana, A.; Kumari, A.; Grewal, J.; Pranaw, K.; Singh, S. Current perspectives on chitinolytic enzymes and their agro-industrial applications. *Biology* **2021**, *10*, 1319. [CrossRef]
22. Oyeleye, A.; Normi, Y.M. Chitinase: Diversity, limitations, and trends in Engineering for suitable applications. *Biosci. Rep.* **2018**, *38*, 1–21. [CrossRef] [PubMed]
23. Huang, Q.S.; Xie, X.L.; Liang, G.; Gong, F.; Wang, Y.; Wei, X.Q.; Wang, Q.; Ji, Z.L.; Chen, Q.X. The GH18 family of chitinases: Their domain architectures, functions and evolutions. *Glycobiology* **2012**, *22*, 23–34. [CrossRef]
24. Kramer, K.J.; Muthukrishnan, S. Insect chitinases: Molecular biology and potential use as biopesticides. *Insect Biochem. Mol. Biol.* **1997**, *27*, 887–900. [CrossRef]
25. Kabir, K.E.; Sugimoto, H.; Tado, H.; Endo, K.; Yamanaka, A.; Tanaka, S.; Koga, D. Effect of Bombyx mori Chitinase against Japanese Pine Sawyer (*Monochamus alternatus*) Adults as a Biopesticide. *Biosci. Biotechnol. Biochem.* **2006**, *70*, 219–229. [CrossRef]
26. Liu, T.; Guo, X.; Bu, Y.; Zhou, Y.; Duan, Y.; Yang, Q. Structural and biochemical insights into an insect gut-specific chitinase with antifungal activity. *Insect Biochem. Mol. Biol.* **2020**, *119*, 103326. [CrossRef] [PubMed]
27. Micocci, K.C.; Moreira, A.C.; Sanchez, A.D.; Pettinatti, J.L.; Rocha, M.C.; Dionizio, B.S.; Correa, K.C.S.; Malavazi, I.; Wouters, F.C.; Bueno, O.C.; et al. Identification, cloning, and characterization of a novel chitinase from leaf-cutting ant Atta sexdens: An enzyme with antifungal and insecticidal activity. *Biochim. Biophys. Acta-Gen. Subj.* **2023**, *1867*, 130249. [CrossRef] [PubMed]
28. Castellani, A. Further researches on the long viability and growth of many pathogenic fungi and some bacteria in sterile distilled water. *Mycopathol. Et Mycol. Appl.* **1963**, *20*, 1–6. [CrossRef]
29. Campana-filho, P.; Lavall, R.L.; Assis, O.B.G. b -Chitin from the pens of Loligo sp.: Extraction and characterization. *Bioresour. Technol.* **2007**, *98*, 2465–2472. [CrossRef]
30. Facchinatto, W.M.; dos Santos, D.M.; Bukzem, A.d.L.; Moraes, T.B.; Habitzreuter, F.; de Azevedo, E.R.; Colnago, L.A.; Campana-Filho, S.P. Insight into morphological, physicochemical and spectroscopic properties of β-chitin nanocrystalline structures. *Carbohydr. Polym.* **2021**, *273*, 118563. [CrossRef]
31. Souza, C.P.; Burbano-Rosero, E.M.; Almeida, B.C.; Martins, G.G.; Albertini, L.S.; Rivera, I.N.G. Culture medium for isolating chitinolytic bacteria from seawater and plankton. *World J. Microbiol. Biotechnol.* **2009**, *25*, 2079–2082. [CrossRef]
32. McIlvaine, T.C. A Buffer Solution for Colorimetric Comparison. *J. Biol. Chem.* **1921**, *49*, 183–186. [CrossRef]
33. Facchinatto, W.M.; dos Santos, D.M.; Fiamingo, A.; Bernardes-Filho, R.; Campana-Filho, S.P.; de Azevedo, E.R.; Colnago, L.A. Evaluation of chitosan crystallinity: A high-resolution solid-state NMR spectroscopy approach. *Carbohydr. Polym.* **2020**, *250*, 116891. [CrossRef]
34. Osorio-Madrazo, A.; David, L.; Trombotto, S.; Lucas, J.M.; Peniche-Covas, C.; Domard, A. Kinetics study of the solid-state acid hydrolysis of chitosan: Evolution of the crystallinity and macromolecular structure. *Biomacromolecules* **2010**, *11*, 1376–1386. [CrossRef] [PubMed]
35. Huggins, M.L. The Viscosity of Dilute Solutions of Long-Chain Molecules. IV. Dependence on Concentration. *J. Am. Chem. Soc.* **1942**, *64*, 2716–2718. [CrossRef]
36. Rinaudo, M.; Milas, M.; Dung, P. Le Characterization of chitosan. Influence of ionic strength and degree of acetylation on chain expansion. *Int. J. Biol. Macromol.* **1993**, *15*, 281–285. [CrossRef] [PubMed]
37. Miller, G.L. Use of Dinitrosalicylic Acid Reagent for Determination of Reducing Sugar. *Anal. Chem.* **1959**, *31*, 426–428. [CrossRef]
38. Panwar, P.; Cui, H.; O'Donoghue, A.J.; Craik, C.S.; Sharma, V.; Brömme, D.; Guido, R.V.C. Structural requirements for the collagenase and elastase activity of cathepsin K and its selective inhibition by an exosite inhibitor. *Biochem. J.* **2014**, *465*, 163–173. [CrossRef]
39. Bradford, M.M. A dye binding assay for protein. *Anal. Biochem.* **1976**, *72*, 248–254. [CrossRef]
40. Mandels, M.; Weber, J. The Production of Cellulases. *Gend. Educ.* **1991**, *7*, 143–155.
41. Karnovsky, M.J. The ultrastructural basis of capillary permeability studied with peroxidase as a tracer. *J. Cell Biol.* **1967**, *35*, 213–236. [CrossRef] [PubMed]

42. Adrangi, S.; Faramarzi, M.A. From bacteria to human: A journey into the world of chitinases. *Biotechnol. Adv.* **2013**, *31*, 1786–1795. [CrossRef]
43. Beckham, G.T.; Ståhlberg, J.; Knott, B.C.; Himmel, M.E.; Crowley, M.F.; Sandgren, M.; Sørlie, M.; Payne, C.M. Towards a molecular-level theory of carbohydrate processivity in glycoside hydrolases. *Curr. Opin. Biotechnol.* **2014**, *27*, 96–106. [CrossRef]
44. Cardozo, F.A.; Facchinatto, W.M.; Colnago, L.A.; Campana-Filho, S.P.; Pessoa, A. Bioproduction of N-acetyl-glucosamine from colloidal α-chitin using an enzyme cocktail produced by Aeromonas caviae CHZ306. *World J. Microbiol. Biotechnol.* **2019**, *35*, 114. [CrossRef]
45. Zeng, J.B.; He, Y.S.; Li, S.L.; Wang, Y.Z. Chitin whiskers: An overview. *Biomacromolecules* **2012**, *13*, 1–11. [CrossRef] [PubMed]
46. Liu, F.C.; Su, C.R.; Wu, T.Y.; Su, S.G.; Yang, H.L.; Lin, J.H.Y.; Wu, T.S. Efficient 1H-NMR quantitation and investigation of N-Acetyl-D-glucosamine (GlcNAc) and N,N′-diacetylchitobiose (GlcNAc)2 from chitin. *Int. J. Mol. Sci.* **2011**, *12*, 5828–5843. [CrossRef] [PubMed]
47. Abdelmalek, B.E.; Sila, A.; Haddar, A.; Bougatef, A.; Ayadi, M.A. β-Chitin and chitosan from squid gladius: Biological activities of chitosan and its application as clarifying agent for apple juice. *Int. J. Biol. Macromol.* **2017**, *104*, 953–962. [CrossRef] [PubMed]
48. Focher, B.; Naggi, A.; Torri, G.; Cosani, A.; Terbojevich, M. Structural differences between chitin polymorphs and their precipitates from solutions-evidence from CP-MAS 13C-NMR, FT-IR and FT-Raman spectroscopy. *Carbohydr. Polym.* **1992**, *17*, 97–102. [CrossRef]
49. Rinaudo, M. Chitin and chitosan: Properties and applications. *Prog. Polym. Sci.* **2006**, *31*, 603–632. [CrossRef]
50. Focher, B.; Beltrame, P.L.; Naggi, A.; Torri, G. Alkaline N-deacetylation of chitin enhanced by flash treatments. Reaction kinetics and structure modifications. *Carbohydr. Polym.* **1990**, *12*, 405–418. [CrossRef]
51. Saito, Y.; Putaux, J.L.; Okano, T.; Gaill, F.; Chanzy, H. Structural aspects of the swelling of β chitin in HCl and its conversion into α chitin. *Macromolecules* **1997**, *30*, 3867–3873. [CrossRef]
52. Beier, S.; Bertilsson, S. Bacterial chitin degradation-mechanisms and ecophysiological strategies. *Front. Microbiol.* **2013**, *4*, 149. [CrossRef]
53. Merzendorfer, H. Chitin metabolism in insects: Structure, function and regulation of chitin synthases and chitinases. *J. Exp. Biol.* **2003**, *206*, 4393–4412. [CrossRef]
54. Phillips, A.J.L.; Alves, A.; Abdollahzadeh, J.; Slippers, B.; Wingfield, M.J.; Groenewald, J.Z.; Crous, P.W. The Botryosphaeriaceae: Genera and species known from culture. *Stud. Mycol.* **2013**, *76*, 51–167. [CrossRef]
55. Gunamalai, L.; Duanis-Assaf, D.; Sharir, T.; Maurer, D.; Feygenberg, O.; Sela, N.; Alkan, N. Comparative characterization of virulent and less virulent. *Mol. Plant Microbe Interact.* **2023**, *36*, 502–515. [CrossRef]
56. Meng, J.; Zhang, D.; Pan, J.; Wang, X.; Zeng, C.; Zhu, K.; Wang, F.; Liu, J.; Li, G. High-Quality Genome Sequence Resource of Lasiodiplodia theobromae JMB122, a Fungal Pathogen Causing Peach Gummosis. *Mol. Plant-Microbe Interact.* **2022**, *35*, 938–940. [CrossRef] [PubMed]
57. Huda-Shakirah, A.R.; Mohamed Nor, N.M.I.; Zakaria, L.; Leong, Y.H.; Mohd, M.H. Lasiodiplodia theobromae as a causal pathogen of leaf blight, stem canker, and pod rot of Theobroma cacao in Malaysia. *Sci. Rep.* **2022**, *12*, 1–14. [CrossRef] [PubMed]
58. Cardoso, J.E.; Bezerra, M.A.; Viana, F.M.P.; de Sousa, T.R.M.; Cysne, A.Q.; Farias, F.C. Ocorrência endofítica de Lasiodiplodia theobromae em tecidos de cajueiro e sua transmissão por propágulos. *Summa Phytopathol.* **2009**, *35*, 262–266. [CrossRef]
59. Saha, S.; Sengupta, J.; Banerjee, D.; Khetan, A. Lasiodiplodia theobromae Keratitis: A Case Report and Review of Literature. *Mycopathologia* **2012**, *174*, 335–339. [CrossRef]
60. Maurya, A.K.; Kumari, S.; Behera, G.; Bhadade, A.; Tadepalli, K. Rhino sinusitis caused by Lasiodiplodia theobromae in a diabetic patient. *Med. Mycol. Case Rep.* **2023**, *40*, 22–24. [CrossRef] [PubMed]
61. Arakane, Y.; Muthukrishnan, S. Insect chitinase and chitinase-like proteins. *Cell. Mol. Life Sci.* **2010**, *67*, 201–216. [CrossRef] [PubMed]
62. Singh, G.; Kumar, S. Biocatalysis and Agricultural Biotechnology Antifungal and insecticidal potential of chitinases: A credible choice for the eco-friendly farming. *Biocatal. Agric. Biotechnol.* **2019**, *20*, 101289. [CrossRef]
63. Luong, N.N.; Tien, N.Q.D.; Huy, N.X.; Tue, N.H.; Man, L.Q.; Sinh, D.D.H.; Van Thanh, D.; Chi, D.T.K.; Hoa, P.T.B.; Loc, N.H. Expression of 42 kDa chitinase of *Trichoderma asperellum* (Ta-CHI42) from a synthetic gene in *Escherichia coli*. *FEMS Microbiol. Lett.* **2021**, *368*, fnab110. [CrossRef]
64. Landim, P.G.C.; Correia, T.O.; Silva, F.D.A.; Nepomuceno, D.R.; Costa, H.P.S.; Pereira, H.M.; Lobo, M.D.P.; Moreno, F.B.M.B.; Brandão-Neto, J.; Medeiros, S.C.; et al. Production in *Pichia pastoris*, antifungal activity and crystal structure of a class I chitinase from cowpea (*Vigna unguiculata*): Insights into sugar binding mode and hydrolytic action. *Biochimie* **2017**, *135*, 89–103. [CrossRef]
65. Gow, N.A.R.; Latge, J.; Munro, C.A.; Group, A.F.; Kingdom, U.; Group, A.F.; Kingdom, U. The Fungal Cell Wall: Structure, Biosynthesis, and Function. *Microbiol. Spectr.* **2017**, *5*. [CrossRef] [PubMed]
66. Li, S.; Zhang, B.; Zhu, H.; Zhu, T. Cloning and expression of the Chitinase Encoded by ChiKJ406136 from *Streptomyces Sampsonii* (Millard & Burr) Waksman KJ40 and its antifungal effect. *Forests* **2018**, *9*, 699. [CrossRef]
67. Takashima, T.; Henna, H.; Kozome, D.; Kitajima, S.; Uechi, K.; Taira, T. cDNA cloning, expression, and antifungal activity of chitinase from Ficus microcarpa latex: Difference in antifungal action of chitinase with and without chitin-binding domain. *Planta* **2021**, *253*, 120. [CrossRef] [PubMed]

68. Sharma, A.; Arya, S.K.; Singh, J.; Kapoor, B.; Bhatti, J.S.; Suttee, A.; Singh, G. Prospects of chitinase in sustainable farming and modern biotechnology: An update on recent progress and challenges. *Biotechnol. Genet. Eng. Rev.* **2023**, 1–31. [CrossRef] [PubMed]
69. Le, B.; Yang, S.H. Microbial chitinases: Properties, current state and biotechnological applications. *World J. Microbiol. Biotechnol.* **2019**, *35*, 144. [CrossRef]

Disclaimer/Publisher's Note: The statements, opinions and data contained in all publications are solely those of the individual author(s) and contributor(s) and not of MDPI and/or the editor(s). MDPI and/or the editor(s) disclaim responsibility for any injury to people or property resulting from any ideas, methods, instructions or products referred to in the content.

Review

Chitosan-Based Nanoencapsulated Essential Oils: Potential Leads against Breast Cancer Cells in Preclinical Studies

Wen-Nee Tan [1,*], Benedict Anak Samling [1,2], Woei-Yenn Tong [3,*], Nelson Jeng-Yeou Chear [4], Siti R. Yusof [4], Jun-Wei Lim [5,6], Joseph Tchamgoue [7], Chean-Ring Leong [8] and Surash Ramanathan [4]

[1] Chemistry Section, School of Distance Education, Universiti Sains Malaysia, Minden 11800, Penang, Malaysia; benedictsamling@gmail.com
[2] Faculty of Resource Science and Technology, Universiti Malaysia Sarawak, Kota Samarahan 94300, Sarawak, Malaysia
[3] Institute of Medical Science Technology, Universiti Kuala Lumpur, Kajang 43000, Selangor, Malaysia
[4] Centre for Drug Research, Universiti Sains Malaysia, Minden 11800, Penang, Malaysia; nelsonchear@usm.my (N.J.-Y.C.); sryusof@usm.my (S.R.Y.); srama@usm.my (S.R.)
[5] HICoE-Centre for Biofuel and Biochemical Research, Institute of Self-Sustainable Building, Department of Fundamental and Applied Sciences, Universiti Teknologi PETRONAS, Seri Iskandar 32610, Perak Darul Ridzuan, Malaysia; junwei.lim@utp.edu.my
[6] Department of Biotechnology, Saveetha School of Engineering, Saveetha Institute of Medical and Technical Sciences, Saveetha University, Chennai 602105, India
[7] Department of Organic Chemistry, Faculty of Science, University of Yaoundé I, Yaoundé P.O. Box 812, Cameroon; joseph.tchamgoue@facsciences-uy1.cm
[8] Branch Campus Malaysian Institute of Chemical and Bioengineering Technology, Universiti Kuala Lumpur, Alor Gajah 78000, Melaka, Malaysia; crleong@unikl.edu.my
* Correspondence: tanwn@usm.my (W.-N.T.); wytong@unikl.edu.my (W.-Y.T.)

Citation: Tan, W.-N.; Samling, B.A.; Tong, W.-Y.; Chear, N.J.-Y.; Yusof, S.R.; Lim, J.-W.; Tchamgoue, J.; Leong, C.-R.; Ramanathan, S. Chitosan-Based Nanoencapsulated Essential Oils: Potential Leads against Breast Cancer Cells in Preclinical Studies. *Polymers* **2024**, *16*, 478. https://doi.org/10.3390/polym16040478

Academic Editors: Sérgio Paulo Campana-Filho and William Facchinatto

Received: 26 December 2023
Revised: 31 January 2024
Accepted: 5 February 2024
Published: 8 February 2024

Copyright: © 2024 by the authors. Licensee MDPI, Basel, Switzerland. This article is an open access article distributed under the terms and conditions of the Creative Commons Attribution (CC BY) license (https://creativecommons.org/licenses/by/4.0/).

Abstract: Since ancient times, essential oils (EOs) derived from aromatic plants have played a significant role in promoting human health. EOs are widely used in biomedical applications due to their medicinal properties. EOs and their constituents have been extensively studied for treating various health-related disorders, including cancer. Nonetheless, their biomedical applications are limited due to several drawbacks. Recent advances in nanotechnology offer the potential for utilising EO-loaded nanoparticles in the treatment of various diseases. In this aspect, chitosan (CS) appears as an exceptional encapsulating agent owing to its beneficial attributes. This review highlights the use of bioactive EOs and their constituents against breast cancer cells. Challenges associated with the use of EOs in biomedical applications are addressed. Essential information on the benefits of CS as an encapsulant, the advantages of nanoencapsulated EOs, and the cytotoxic actions of CS-based nanoencapsulated EOs against breast cancer cells is emphasised. Overall, the nanodelivery of bioactive EOs employing polymeric CS represents a promising avenue against breast cancer cells in preclinical studies.

Keywords: breast cancer; chitosan; drug delivery; nanoencapsulated essential oils; therapeutics

1. Introduction

Cancer is one of the leading causes of mortality worldwide, accounting for almost ten million deaths in 2020 [1]. Among cancer diseases, breast cancer is the most frequently diagnosed cancer. Breast cancer is a heterogeneous disease generally linked to oestrogen hormones. In addition, it was reported that 5–10% of breast cancer cases are linked to gene alterations. Breast cancer is more common in women between the ages of 65 and 80. However, invasive breast cancer incidence is seen in women under the age of 50. According to studies, early-stage breast cancer can be cured in about 70% of patients, but metastatic breast cancer is typically incurable [2]. Currently, the main treatments for breast cancer include chemotherapy, radiation therapy, and surgery. However, the adverse side effects of conventional therapies have driven researchers to search for alternatives [3].

Regarding the important role of natural products as a source of biologically active constituents, essential oils (EOs) appear as potential candidates for combating human diseases. The use of plant EOs has been evidenced for thousands of years. It was recorded that the Egyptians employed plant EOs for medicinal purposes as early as 4500 BC. Since then, EOs have been produced commercially owing to their widespread use and medicinal value [4,5]. Generally, EOs are a complex mixture of volatile aromatic constituents that are extracted from various plant parts. Due to their wide range of biological effects, EOs are attracting immense scientific attention and may be essential for the treatment of cancer [6,7]. However, EOs present some disadvantages in biomedical applications owing to their volatility, instability, and water insolubility and the heterogeneity in their chemical composition and biological effects [8,9].

In this context, encapsulation may overcome the shortcomings in preserving the bioactive constituents of EOs. Encapsulation can improve the stability, bioavailability, functional properties, and controlled release of bioactive constituents of EOs [8,10]. Additionally, adjusting the pH during encapsulation may enhance the stability of EOs. Encapsulation using biopolymers as carrier agents has attracted immense attention owing to their beneficial attributes. Among the wide range of biopolymers, chitosan (CS) has been widely employed in pharmaceutical settings, including for drug delivery. Generally, CS is the deacetylated form of chitin and is a generally recognized as safe (GRAS) material. It exhibits good properties such as biodegradability, bioavailability, biocompatibility, and non-toxicity [11,12]. Therefore, CS-based nanoencapsulated EOs offer an alternative approach to enhance the physical stability and bioavailability of bioactive EOs. The nanoscale particle size of EOs embedded in CS increased the surface-to-volume ratio, water solubility, and colloidal stability and resulted in a better controlled release of bioactive constituents [13,14]. In this review, we highlight the bioactive attributes of EOs and their constituents against breast cancer cells. The challenges associated with the biomedical applications of EOs are discussed. In addition, we emphasise the benefits of nanoencapsulated EOs, with a specific focus on CS-based nanoencapsulated EOs against breast cancer cells.

2. Essential Oils (EOs)

EOs are volatile organic constituents with low molecular weights extracted from leaves, buds, flowers, seeds, stems, fruits, roots, rhizomes, barks, and tubers [15–18]. Generally, EOs are insoluble in water but soluble in organic solvents [19,20]. Conventionally, EOs are extracted using hydrodistillation, steam distillation, or solvent extraction. Hydrodistillation is regarded as the simplest extraction technique. Plant materials are boiled with distilled water and connected to a Clevenger-type apparatus. Steam distillation, on the other hand, is conducted by passing steam through the plant materials. Hot vapours are then condensed to collect EOs. Solvent extraction involves the usage of organic solvent to macerate the plant materials, followed by filtration and solvent concentration [21,22]. Meanwhile, modern extraction techniques such as supercritical fluid extraction, microwave-assisted extraction, and ultrasonic-assisted extraction have improved the quality and yield of EOs. Shorter extraction time, lower energy consumption, and lesser solvent usage are among the notorious advantages of non-conventional extraction techniques for EOs [23,24]. Supercritical fluid extraction uses carbon dioxide as the solvent to pass through the plant materials. The collected EOs and the extracts are then separated by decompression [25]. Microwave-assisted extraction uses microwaves as the source of energy. It is environmentally friendly and requires a short extraction time [26]. Ultrasonic-assisted hydrodistillation involves the penetration across the plant cells via cavitation. This method improves the extraction efficiency and prevents the degradation of plant materials [27]. Extraction of plant EOs has been the focal point of research interest as it plays a key role in determining the type, amount, and chemical structures of the EO constituents [22].

For decades, EOs have been recognised as source of pharmaceutical agents. They possess a broad spectrum of biological activities, notably antioxidant, anti-inflammatory, anticancer, antibacterial, antiviral, antifungal, antimutagenic, antiparasitic, antimycotic, and antidiabetic activities [15,28,29]. Typically, these biological activities are mainly attributed to the predominance of major constituents. Nonetheless, previous studies have reported that interactions between different EO constituents may lead to additive or synergistic effects. An additive effect is defined as the sum of the individual effects of two or more constituents together. Meanwhile, a synergistic effect is defined as the combined effect of two or more constituents being greater than the sum of all of their individual effects [21]. Thus far, more than 3000 plant EOs have been extracted owing to their attractive biological activities. Nevertheless, the quality and yield of EOs are affected by factors such as plant parts, extraction methods, harvesting seasons, geographical locations, and postharvest storage conditions [30]. In industrial applications, EOs are widely used as flavouring agents in food and beverage, cosmetics, fragrances, and oral products [31]. For example, EOs extracted from citrus and lavender are commonly used in fragrances [32]. In soft drink manufacturing, EOs of cola, cinnamon, and vanilla are often employed [33]. Meanwhile, peppermint EOs are commonly used as the main ingredient in the manufacturing of oral products, such as mouthwash and toothpaste, confectionery, analgesic balms, chewing gums, and tobacco [34].

2.1. Constituents of EOs

EOs comprise more than 300 volatile organic constituents with molecular weights less than m/z 300 [5]. Generally, EOs consist of terpenes and phenylpropanoids [24]. Terpenes and terpenoids have been extensively investigated for their important roles in human health owing to their excellent therapeutic properties [35]. Terpenes are grouped according to the number of isoprene units in their structure. The isoprene units undergo head-to-tail condensation or rearrangement to give an array of terpenes. They are further classified into hemiterpenes (C5), monoterpenes (C10), sesquiterpenes (C15), diterpenes (C20), sesterterpenes (C25), triterpenes (C30), sesquarterpenes (C35), and tetraterpenes (C40). Terpenes are regarded as one of the most prominent groups of plant secondary metabolites. On the contrary, terpenoids are oxygen-containing terpenes that are synthesised via biochemical modifications and reactions [36]. They are categorised into alcohols, aldehydes, epoxides, esters, ether, ketones, and phenols [35].

Among terpenes, hemiterpenes are the simplest. They are regarded as a minor group in plant EOs, with a molecular formula of C_5H_{10}. Hemiterpenes are commonly released from conifers, oaks, and poplars. Some common hemiterpenes found in EOs are angelic acid, isoamyl alcohol, isovaleric acid, senecioic acid, and tiglic acid [35]. Monoterpenes comprise two units of isoprene and have the molecular formula of $C_{10}H_{16}$. They make up approximately 90% of total EO constituents [37]. Monoterpenes primarily control the release of specific odours from plants and are divided into acyclic and cyclic forms. For instance, ocimene and myrcene are acyclic monoterpenes, while limonene and p-cymene are cyclic monoterpenes [38]. Conversely, citral and linalool are common acyclic monoterpenoids, while cyclic monoterpenoids are represented by thymol and eucalyptol [36]. Sesquiterpenes, on the other hand, are less volatile than monoterpenes. Sesquiterpenes are derived from three isoprene units and have the molecular formula of $C_{15}H_{24}$. Similar to other terpenes, sesquiterpenes exist in both acyclic and cyclic forms. Sesquiterpenes and their derivatives are commonly detected in plant EOs. They are receiving much interest owing to their distinctive odour and flavour attributes. Examples of common sesquiterpenes and sesquiterpenoids are α-humulene, β-caryophyllene, patchoulol, and farnesol. Meanwhile, diterpenes, sesterterpenes, triterpenes, sesquarterpenes, and tetraterpenes are rarely detected in EOs due to their low volatility [39,40]. Figure 1 shows the chemical structures of common EO constituents.

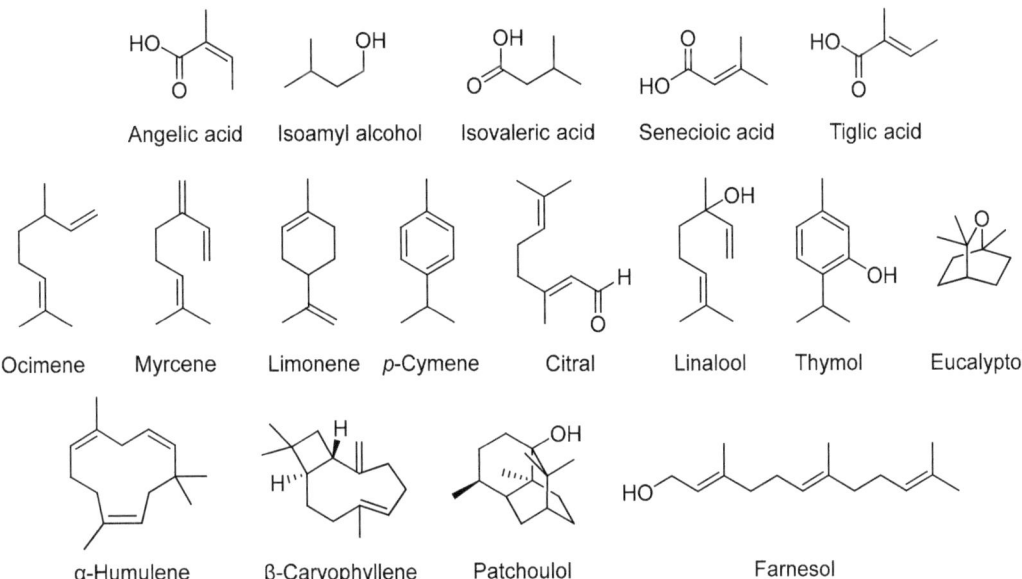

Figure 1. Chemical structures of common EO constituents.

2.2. Cytotoxic EOs against Breast Cancer

Previous studies have reported the cytotoxic nature of EOs against breast cancer cells through different mechanisms of action. EOs trigger the death of cancer cells via apoptosis, necrosis, or cell cycle arrest. This involves the loss of mitochondrial potential, changes in pH gradient, an increase in the cell membrane fluidity, and a decrease in adenosine triphosphate (ATP) synthesis [41]. In addition, the toxicity potential of EOs is suggested to be governed by the type of organism. In eukaryotes, toxicity decreases as the lipophilic components of EOs increase. In prokaryotes, toxicity increases as the lipophilic components of EOs increase [4]. Owing to their promising cytotoxic property, some common EOs were extensively investigated for their in vitro cytotoxic effects on breast cancer cell lines (Table 1). In addition, studies have shown that certain EOs in combination with a chemotherapeutic drug enhanced the cytotoxic activity on cancer cell lines. Thus, only a small dose of a drug is needed when combined with EOs while maintaining the same cytotoxic effect [42]. Common EOs, originating from cinnamon, rose, thyme, chamomile, lavender, jasmine, lemon, agarwood, lemongrass, and citronella, were investigated for their cytotoxic effects against human breast cancer MCF-7 cells. MCF-7 cells are often used in breast cancer studies owing to their ideal characteristics of the mammary epithelium. Physiologically, MCF-7 cells are hormone-responsive breast cancer cells, which express oestrogen receptors (ERs). ERs are nuclear proteins regulating the expression of specific genes in breast cancer development and progression, and approximately 80% of breast cancers are ER-positive. Thus, the MCF-7 cell line is suitable to be used as an experimental model for drug discovery in breast cancer [8].

In a study conducted by Zu and co-workers, thyme (*Thymus vulgaris*) EOs showed the most potent cytotoxic effect against MCF-7 cells with an inhibitory concentration of 0.030% (v/v). Other EOs from cinnamon, rose, chamomile, lavender, jasmine, and lemon showed cytotoxic effects ranging from 0.072 to 0.143% (v/v). The exhibited cytotoxic activity could be attributed to the different constituents present in the EOs, notably terpenes [43]. Lemongrass (*Cymbopogon citratus*) and citronella grass (*Cymbopogon nardus*) are two closely related medicinal herbs native to tropical countries. These aromatic plants have been investigated for their potential against cancer cells due to their excellent pharmacological properties. Lemongrass EOs exhibited a significant inhibitory concentration (IC_{50}) at 0.28% (v/v), while citronella grass EOs showed an IC_{50} at 0.46% (v/v). Previous studies have demonstrated that the amount of major constituents in EOs plays a vital role in biological activities. Citronellal, a major constituent in citronella grass EOs, has been evaluated for its selectivity index (SI) against cancerous MCF-7 cells and non-cancerous cells. The SI in cytotoxicity is a measure used to assess the relative toxicity of a substance to different cell types, typically cancer cells versus non-cancerous cells. A higher SI suggests that a substance is more toxic to cancer cells, indicating potential therapeutic value with lower toxicity to non-cancerous cells [8]. Citronellal showed a high SI of 25.8, indicating its high selectivity towards cancerous MCF-7 cells [44,45]. In addition, agarwood (*Aquilaria* spp.) EOs have been assessed for their cytotoxic potential against MCF-7 cells. It was reported that 44 µg/mL of agarwood EOs could kill 50% of MCF-7 cells. Traditionally, agarwood is mainly used for religious, aromatherapy, and medicinal purposes. The observed cytotoxic effects might be due to the predominance of sesquiterpenoids, namely isoaromadendrene epoxide, agarospirol, and β-guaiene, in the agarwood EOs [46,47]. The ginger plant is popularly used in culinary and folk medicine. Ginger EOs extracted using hydrodistillation have been assessed for cytotoxicity against MCF-7 cells using an MTT assay. The results revealed the cytotoxic potential of ginger EOs against MCF-7 cells with an IC_{50} of 82.6 ± 3.2 µg/mL [48]. Tea tree, scientifically known as *Melaleuca alternifolia*, is highly regarded for its folk uses. A steam-distilled tea tree EO was evaluated for its cytotoxic potential against MCF-7 cells. A terpinen-4-ol-rich tea tree EO showed an IC_{50} of 537 µg/mL after 24 h treatment against MCF-7 cells. In addition, the findings revealed that tea tree EOs induced early-stage apoptosis against MCF-7 cells at concentrations of 100 and 300 µg/mL [49]. *Citrus limon* is a widely known fruit tree from the Rutaceae family. The leaf and branch of *C. limon* were hydrodistilled to give 0.01% (v/w) and 0.005% (v/w) oil yield, respectively. Based on the results, leaf EOs of *C. limon* demonstrated an IC_{50} of 10% (v/v) against MCF-7 cells. In addition, the leaf EOs induced apoptosis through increased expression levels of caspase-8 [50]. In another study, a monoterpenoid-rich lavender EO was studied for its cytotoxic potential against human breast MDA-MB-231 cancer cells. The lavender EO showed a dose-dependent cytotoxic effect on MDA-MB-231 cells with an IC_{50} of 0.259 ± 0.089 µg/mL. It was reported that a high content of eucalyptol is positively correlated with the observed cytotoxicity [51]. Basil is a widely used culinary herb worldwide. In the study conducted by Aburjai in 2020, basil EOs were dominated by linalool (36.26%) and eucalyptol (11.36%). In the MTT assay, basil EOs exhibited an IC_{50} of 432.3 ± 32.2 and 320.4 ± 23.2 µg/mL on MDA-MB-231 and MCF7 cells, respectively. It was hypothesised that the major constituents present in the basil EOs play a main role in the observed cytotoxic effects [52]. In a study by Niksic et al. (2021), thyme EOs were reported to inhibit cancer cell proliferation in a dose-dependent manner. A thymol and *p*-cymene-rich thyme EO was reported to exhibit an LC_{50} value of 60.38 µg/mL in a brine shrimp lethality assay. Meanwhile, the IC_{50} against MCF-7 cancer cells was recorded at 52.65 µg/mL [42]. *Matricaria recutita* (chamomile) is a medicinal herb commonly used in aromatherapy and folk medicine. Chamomile EOs were predominated by terpenoids which accounted for 63.51% of the total oil. In a cytotoxicity assessment, chamomile EOs showed potent cytotoxicity on MDA-MB-231 cells with an IC_{50} of 4 µg/mL. In addition, chamomile EOs reduced the migration and invasion of MDA-MB-231 cells. Based on these findings, it was suggested that the cytotoxic effects of chamomile EOs were due to the inhibition of PI3K/Akt/mTOR signalling pathway in MDA-MB-231 cells [53]. Overall, the demonstrated biological activity might be primarily attributed to the volatile constituents present in the EOs. Nonetheless, the chemical complexity of EOs plays a key role as each constituent contributes to the exhibited bioactivity and may regulate the biological effects of other constituents [54,55].

Table 1. Cytotoxic effects of some common EOs against breast cancer cells.

Breast Cancer Cells	EO	Botanical Source	IC$_{50}$	Reference
MCF-7	thyme	*Thymus vulgaris*	0.030% (v/v)	[43]
	chamomile	*Anthemis nobilis*	0.072% (v/v)	
	rose	*Rosa centifolia*	0.074% (v/v)	
	cinnamon	*Cinnamomum zeylanicum*	0.076% (v/v)	
	jasmine	*Jasminum grandiflora*	0.077% (v/v)	
	lavender	*Lavandula stoechas*	0.142% (v/v)	
	lemon	*Citrus limonum*	0.143% (v/v)	
MCF-7	lemongrass	*Cymbopogon citratus*	0.28% (v/v)	[44]
	citronella grass	*Cymbopogon nardus*	0.46% (v/v)	
MCF-7	agarwood	*Aquilaria* spp.	44 µg/mL	[46]
MCF-7	ginger	*Zingiber officinale*	82.6 ± 3.2 µg/mL	[48]
MCF-7	tea tree	*Melaleuca alternifolia*	537 µg/mL	[49]
MCF-7	lemon	*Citrus limon*	10% (v/v)	[50]
MDA-MB-231	lavender	*Lavandula stoechas*	0.259 ± 0.089 µg/mL	[51]
MCF-7 MDA-MB-231	basil	*Ocimum basilicum*	320.4 ± 23.2 µg/mL 432.3 ± 32.2 µg/mL	[52]
MCF-7	thyme	*Thymus vulgaris*	52.65 µg/mL	[42]
MDA-MB-231	chamomile	*Matricaria recutita*	1.5 µg/mL	[53]

2.3. Limitations of EOs

Even though EOs possess a wide range of bioactivity, their use is limited by their volatility, hydrophobicity, and instability. EOs comprise about 95% volatile and 5% non-volatile constituents [56]. Occasionally, the volatile constituents are affected by many external variables. The quality of EOs may be impacted by environmental variables such as air, heat, and irradiation. In addition, EOs' hydrophobicity has led to their insolubility in a water-based medium. Plant EOs are unstable due to their thermolability. They can readily oxidise or hydrolyse during processing and storage. When exposed to environmental stimuli, they are vulnerable to oxidation, chemical changes, or polymerisation [57]. For example, EOs obtained from citrus trees contain a high concentration of monoterpenes, particularly dextrolimonene. However, they oxidise on contact with air. When dextrolimonene reacts with oxygen in the air, a new constituent is produced, which poses adverse effects [58]. Park and co-workers conducted research on EOs from *Kunzea ambigua*. The EOs were stored under various settings to examine the changes in colour and chemical compositions. A significant colour variation was observed when the EOs of *K. ambigua* were stored at room temperature under a light. On the contrary, the colour of EOs stored at freezing temperature, under refrigeration, or under room temperature without light was more stable. The chemical composition of EOs was generally constant throughout storage at room temperature, under refrigeration, and at freezing temperature in the absence of light. However, the amount of germacrene D, β-caryophyllene, and α-humulene was found to decrease significantly in the EOs of *K. ambigua* when exposed to light, possibly due to isomerisation [59].

3. Nanoencapsulation

Nanoencapsulation is an emerging technique used to encapsulate bioactive constituents (the core material) inside secondary/wall materials (the matrix or shell) to produce nanocapsules [60]. It protects the bioactive constituents from light, heat, air, and moisture. Nanoencapsulation aids in minimising the evaporation of EOs and improving the delivery of bioactive constituents through controlled release. It converts liquid EOs into solid nanoencapsulated EOs and facilitates their applications [10]. The choice of an appropriate wall material is primordial in nanoencapsulation. Lecithin, legumin, gelatin, and albumin are among the proteins that are often employed as wall materials. They are naturally amphipathic, which makes them effective emulsifiers. Moreover, proteins are organic, environmentally friendly, and biodegradable. Bioactive components are also encapsulated in polysaccharides such as dextrin, starch, gums, CS, and alginates. Various wall materials have been used in research to encapsulate bioactive components to increase their bioactivity [61]. Figure 2 shows the nanoencapsulation of EOs with improved properties. For instance, it has been reported that the nanoencapsulation of eugenol with CS increased the antifungal and aflatoxin B1 inhibitory efficacy in stored rice [62]. Poly(lactide-co-glycolide) was used to nanoencapsulate bioactive *Trachyspermum ammi* seed EOs for possible use in the therapy of colon cancer. Based on the in vitro findings, the synthesised nanoencapsulated EO system triggered apoptosis by expressing apoptotic genes in human colon cancer cells [63].

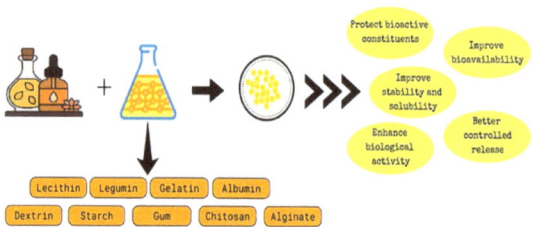

Figure 2. Nanoencapsulation of EOs with improved properties.

3.1. Chitosan (CS)

CS is a naturally occurring polymer made up of N-acetyl-D-glucosamine and D-glucosamine units. It is the second most abundant polysaccharide in nature. In industrial settings, chitin undergoes deacetylation to form CS, wherein the acetamide group is converted into an amino group [64,65]. The resulting CS polymer may vary in length based on the conditions and parameters employed during the deacetylation process. CS typically exhibits molecular weights ranging from 300 to 1000 kDa, influenced by its degree of acetylation. Generally, CS is insoluble in water at neutral pH. However, the protonation of its free amino groups has made CS soluble in dilute acids. This pH-responsive solubility can be advantageous in drug delivery to specific regions of the body with varying pH levels. In addition, the number of amino groups present in CS may impact its mucoadhesion and transfection properties. This can enhance its residence time at the target site compared to other polymers, improving the overall efficiency of drug delivery [66]. Chitin, the precursor to CS, can be obtained from the exoskeletons of crustaceans and insects. CS is highly regarded as a safe and environmentally friendly material. It is non-toxic, biodegradable, and biocompatible. CS breaks down into non-toxic byproducts over time. This characteristic is advantageous in applications where the carrier needs to be gradually eliminated from the body. Owing to these distinctive attributes, CS is used in various applications such as pharmaceuticals, biomedicine, agriculture, and food [64,65]. The general properties of CS are listed in Table 2.

Table 2. General properties of CS.

Property	Description	Reference
Appearance	Semicrystalline of white or slightly yellow	[64]
Solubility	Soluble in diluted acid below pH 6.0. Insoluble in water and organic solvents	[67]
Molecular weight (Mw)	Low Mw: <100 kDa Medium Mw: 100 to 1000 kDa High Mw: >1000 kDa	[68]

In biomedical and pharmaceutical applications, the selection of the right molecular weight, degree of acetylation, and purity of CS are vital. According to the European Pharmacopoeia, the acceptance criteria for impurities present in CS are as follows: total impurities/insoluble $\leq 0.5\%$; heavy metals ≤ 40 ppm; sulphated ash $\leq 1.0\%$; no tolerance on iron and protein. In this aspect, the presence of proteins is critical as it may affect the biological effects of immunity. On the contrary, a high quantity of ash may influence the dissolution and the preparation of efficient CS-based drug delivery systems [66]. However, CS is sensitive to humidity due to its hygroscopic nature. The formation of hydrogen bonding promotes the retention of water on CS and affects its mechanical properties. In addition, CS is reported to thermally degrade at ambient temperature. Thus, low-temperature conditions are suggested for its storage [69]. CS is known to possess anticancer, antimicrobial, antidiabetic, antioxidant, and anti-inflammatory properties [64]. The encapsulation of EOs using CS has been reported to enhance the drug nature after successful drug delivery. In addition, CS has attracted a great deal of attention in drug co-delivery, gene delivery, and tissue engineering [65].

3.2. EO-Loaded CS Nanoparticles

Owing to the favourable attributes of CS, it is widely used in the nanoencapsulation of EOs. Various methods have been employed to encapsulate EOs into CS nanoparticles (Figure 3). Among them, ionic gelation is an economical method for encapsulating plant EOs. The method is eco-friendly and does not require high temperatures. Ionic bonding between the negatively charged cross-linking agents and the positively charged CS is involved in the synthesis of nanoparticles using ionic gelation [70]. In addition, a cross-linking agent such as sodium tripolyphosphate (TPP) is often used in the synthesis due to its biocompatibility and biodegradability attributes [71]. For instance, the ionic gelation method was used in the synthesis of *Cynometra cauliflora* EO-loaded CS nanoparticles. It involved the dissolution of CS in a diluted acetic acid solution containing a surfactant. Subsequently, EOs were added into the CS–surfactant solution followed by TPP. The synthesis was carried out under continuous stirring to facilitate precipitation. The synthesised *C. cauliflora* EO-loaded CS nanoparticles have shown potent cytotoxic effects against human breast cancer MCF-7 cells with an IC_{50} ranging from 3.72 to 17.81 µg/mL after a 72 h incubation period. Meanwhile, they showed cytotoxic effects against human breast cancer MDA-MB-231 cells with an IC_{50} ranging from 16.24 to 17.65 µg/mL after 72 h of treatment. The observed cytotoxicity was significantly enhanced as compared to that of the free EOs of *C. cauliflora* [8].

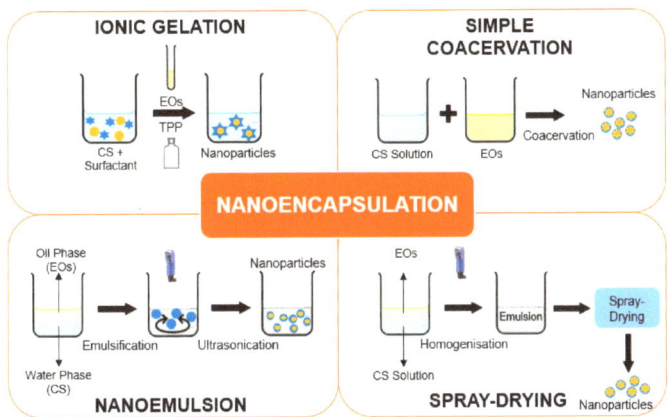

Figure 3. Various methods used to encapsulate EOs into CS nanoparticles.

Another method used for encapsulating reactive and sensitive constituents in a nanoscale setting is coacervation. This method is separated into aqueous (hydrophilic constituents) and organic (lipophilic constituents) phases. In coacervation, however, organic solvents are required. A single polymer is used in simple coacervation. On the contrary, the formation of a matrix enclosing the bioactive constituents from two colloids with opposite charges is required in complex coacervation. Both proteins and polysaccharides are often used as oppositely charged polymers [70]. Compared to other nanoencapsulation methods, coacervation can reach a payload of over 99%. Coacervation regulates the release of bioactive constituents depending on temperature and mechanical stress [72]. In a study by Bastos and co-workers, black pepper EOs were encapsulated utilising the coacervation method. The study showed that the core material retained more than 80% of the terpenes present in the EOs [73].

Nanoemulsion involves a combination of two immiscible liquids. It can be achieved by combining an EO's dispersion phase with water's dispersed medium to create an oil-in-water emulsion [74]. The size of the synthesised EO nanoparticles can range from 10 to 1000 nm. This method serves as a delivery vehicle for targeted mechanisms of action because of the tiny size of the droplets produced in the nanoemulsion. The bioavailability of EOs is increased by nanometric-sized particles and a high surface-to-volume ratio [75]. In nanoemulsion, both high-energy and low-energy processes are appropriate. High-speed homogenisation or high-intensity ultrasonication is used to create high-energy nanoemulsions. This is essential in controlling the zeta potential, retention, and particle size of the synthesised nanoencapsulated system. According to the literature, oil-in-water nanoemulsion was used to encapsulate EOs from cinnamon, rosemary, and oregano. Among the encapsulated EOs, oregano EO nanoemulsions exhibited significant biological effects owing to their smaller droplet size, high encapsulation efficiency, and low separation phase [76].

Spray drying is a simple, inexpensive, and versatile encapsulation method. It is used to enhance the stability of EO constituents that are heat- and light-sensitive. Three main phases are typically involved in spray drying. In experiments, the emulsification of EO constituents occurs in a polymeric solution. Then, the emulsion is homogenised and atomised into a drying medium. A stable matrix is produced as a result of the high temperature used in this method. This method results in better nanoparticle quality and yield. In addition, the synthesised nanoparticles have high solubility and stability [70].

EO-loaded CS nanoparticles are superior in controlled release, stability, and solubility. In numerous investigations, CS nanoparticles were used in the delivery of bioactive EOs [13]. For example, CS nanoparticles have been successfully loaded with the EOs of lemongrass and clove. The synthesised nanoparticles had nanometric size and spherical shape. In comparison to the free EOs, the EO-loaded CS nanoparticles displayed enhanced biological activity [77–79]. In a study conducted by Hesami and co-workers, greater celandine EOs were encapsulated with CS. The encapsulation improved the cytotoxicity of the EOs against human breast cancer cells. The synthesised nanoparticles strongly induced apoptosis in the cancer cells. Hence, CS-based nanoparticles loaded with EOs are a vital source for chemoprevention and cancer suppression [80].

4. Cytotoxic Actions of EO-Loaded CS Nanoparticles against Breast Cancer Cells

CS-based nanoparticles are desirable for delivering bioactive EO constituents in targeted cancer therapy. To deliver the specific bioactive constituents to the target site, the administered particles must be tiny to pass through the microvasculature of cancer cells. Nanoparticles with appropriate particle size can enhance drug penetration and uptake via the cell membrane [81]. In general, EO-loaded CS nanoparticles are able to circumvent immune surveillance and attain target specificity. They penetrate the intracellular space of cancer cells and elude entrapment within endosomes to release bioactive EOs [82]. The effective delivery of bioactive EOs to the intracellular environment is crucial for attaining therapeutic purposes. Occasionally, physical techniques are employed to forcibly transport nanoparticles across the plasma membrane, and the uptake of nanoparticles primarily depends on the inherent endocytic uptake mechanism of the cancer cells [83].

It was reported that EO-loaded CS nanoparticles are able to circulate in the bloodstream for a relatively long time and accumulate at the cancer cell site [66]. This can be achieved through the enhanced permeability and retention (EPR) effect. In regard to this, the physicochemical properties (particle size, surface charge, and shape) of nanoparticles play a vital role in determining the EPR effect. They affect the amount and kinetics of nanoparticle accumulation at the target site. The penetration of EO-loaded CS nanoparticles into the vasculature of cancer cells was believed to occur via a passive targeting mechanism that effectively regulates pharmacokinetics and enhances overall therapeutic efficacy [82]. In addition, EO-loaded CS nanoparticles may deliver the bioactive EO constituents in a site-specific manner while reducing toxicity [66]. The demonstrated cytotoxic effects on cancer cells depend on the type and concentration of EO constituents. They are often regarded as not genotoxic [84]. According to a recent study, bioactive EOs damaged the inner cell membranes and organelles in eukaryotic cells by acting as prooxidants [85]. It was hypothesised that EO constituents reduce the size of cancer cells. They induced the death of cancer cells via apoptosis, necrosis, cell cycle arrest, and disruption of cell organelles [41].

There are several processes that control the release of EOs from CS nanoparticles. These include polymer swelling, diffusion through the polymeric matrix, and erosion or degradation of the polymer. It was postulated that EO-loaded CS nanoparticles release the first burst of bioactive constituents as a result of the polymer swelling, the formation of pores, or the diffusion of the bioactive constituents from the polymer's surface [69]. In a study by Shetta and co-workers, a two-stage release pattern from CS nanoparticles was observed. Peppermint and green tea EOs were encapsulated into CS nanoparticles with an average size of 20 to 60 nm. Under different pH levels, the in vitro results showed a two-stage release pattern from phosphate-buffered saline (PBS) and acetate buffers. The initial burst was noticed for up to 12 h, while the gradual release remained for 72 h. Additionally, the findings showed that the acetate buffer had a higher release rate than the PBS, with full release of 75%. The release kinetics for both buffers was studied and found to follow the Fickian model [86].

Various studies have been conducted to study the cytotoxic effects of EOs loaded with CS nanoparticles against breast cancer cells (Table 3). According to Onyebuchi and Kavaz, an *Ocimum gratissimum* EO encapsulated in CS nanoparticles decreased the viability of MDA-MB-231 cells to 37.44% as compared to the unencapsulated EOs. Blebbing of membrane in breast cancer cells was observed upon treatment with *O. gratissimum* EO-loaded CS nanoparticles, suggesting the occurrence of apoptosis [87]. In a previous study, *Zataria multiflora* EOs were reported to possess antiproliferative properties against breast cancer cells. The EOs of *Z. multiflora* induced the breakdown and oxidation of DNA strands without affecting non-cancerous cells [88]. Encapsulation with CS significantly enhanced the cytotoxic activity against MCF-7 and MDA-MB-231 cells in a dose-dependent manner, with IC_{50} values of 21.2 and 6.2 µg/mL, respectively. In cell morphological study, nanoencapsulated-EO-treated MDA-MB-231 cells showed a pronounced loss in cell membrane structure and a noticeable nuclear fragmentation. Further, the sub-G1 population mass was triggered in a fluorescence-activated cell sorting analysis. In evaluating the MDA-MB-231 cell death mode, the *Z. multiflora* EO-loaded CS nanoparticles exhibited early apoptosis with noticeable chromatin condensation and cell shrinkage. This caused cell accumulation in the G2/M phase, while cells in the G1 phase increased nominally. It was reported that *Z. multiflora* EO-loaded CS nanoparticles triggered the production of intracellular reactive oxygen species (ROS) in the mitochondria, which leads to apoptosis. The observed apoptosis might be due to the oxidation and breakdown of the DNA strands. Based on the findings, an intercalative binding to the DNA helix is observed, suggesting the DNA-binding and -damaging properties of the *Z. multiflora* EO-loaded CS nanoparticles [89]. Traditionally, EOs of *Syzygium aromaticum* are used to treat wounds and burns. It was reported that the EOs possess antibacterial, anti-inflammatory, antioxidant, and anticancer properties. The nanoencapsulation of *S. aromaticum* EOs using CS (IC$_{50}$: 45.89 µg/mL) exhibited significantly enhanced cytotoxic effects against MCF-7 cells as compared to those of its unencapsulated EOs (IC$_{50}$: 172.47 µg/mL) [90]. In a study conducted by Valizadeh et al. (2021), CS nanoparticles containing *S. aromaticum* EOs were studied against human breast cancer MDA-MB-468 cells. The nanoencapsulated EOs showed improved cytotoxicity (IC$_{50}$: 177 µg/mL) when compared to unencapsulated EOs (IC$_{50}$: 243 µg/mL) [91]. EOs from *Citrus aurantium*, *Citrus limon*, and *Citrus sinensis* are predominated by limonene. CS nanoparticles containing *Citrus* EOs were prepared using an ionic gelation method. The synthesised CS-nanoencapsulated *Citrus* EOs exhibited improved cytotoxic properties against cancerous MDA-MB-468 cells. Among the *Citrus* EOs, a nanoencapsulated *C. sinensis* EO demonstrated the most significant cytotoxicity against MDA-MB-468 cells with an IC$_{50}$ value of 23.65 µg/mL [92]. In addition, CS nanoparticles loaded with *Chelidonium majus* have been evaluated for chemotherapeutic potential against breast cancer. The recorded IC$_{50}$ values of the nanoencapsulated leaf and root EOs of *C. majus* against MCF-7 cells were 41.5 and 77.6 µg/mL, respectively. Based on the findings, a nanoencapsulated leaf EO of *C. majus* triggered early apoptosis of MCF-7 cells [80]. The literature revealed that *Cinnamomum verum* EOs possess antiproliferative properties against MCF-7, HeLa, and Raji cells. In a study conducted by Khoshnevisan and colleagues, CS nanoparticles containing *C. verum* EOs were synthesised. They exhibited cytotoxic properties against MDA-MB-468 cells, with an IC$_{50}$ value of 112.35 µg/mL [93]. Previous research has demonstrated the promising anticancer properties of *Curcuma longa* EOs. It was noticed that the unencapsulated EOs did not exert significant cytotoxicity towards MDA-MB-231 and MCF-7 cells. Interestingly, an enhancement in cytotoxicity was observed for nanoencapsulated CS-loaded *C. longa* EOs against MDA-MB-231 (IC$_{50}$: 99.11 µg/mL) and MCF-7 (IC$_{50}$: 82.88 µg/mL) cells. It was postulated that the presence of the CD44 receptor in the MDA-MB-231 and MCF-7 cells plays a key role in the enhanced cytotoxicity. The structural similarity of CS with the natural ligand of the CD44 receptor may facilitate the binding of the nanoencapsulated EOs and promote endocytosis [94]. In a study conducted by Samling and co-workers, the leaf, twig, and fruit EOs of *Cynometra cauliflora* were nanoencapsulated using CS via ionic gelation method. It is worth noting that all nanoencapsulated EOs showed a pronounced enhancement in cytotoxicity against MCF-7 and MDA-MB-231 cells. Among them, the twig EO-loaded CS nanoparticles exhibited the lowest IC$_{50}$ (3.72 µg/mL) against MCF-7 cells after 72 h of treatment. Furthermore, all the nanoencapsulated EOs showed no cytotoxicity against human breast non-cancerous MCF-10A cells [8]. In a recent study, *Cinnamon cassia* EOs were nanoencapsulated to enhance their cytotoxic effects against breast cancer cells. The nanoencapsulated EOs showed better cytotoxicity (IC$_{50}$: 25.24 µg/mL) than unencapsulated EOs (IC$_{50}$: 32.25 µg/mL) against MDA-MB-231 cells. It was postulated that the observed cytotoxic effects might be due to the cell membrane breakdown. The nanoencapsulated EOs of *C. cassia* inhibited the migration and invasion of MDA-MB-231 cells in a dose-dependent manner. In the analysis of ROS, superoxide dismutase (SOD), and malondialdehyde (MDA), the nanoencapsulated EOs showed an increase in ROS and MDA and a decrease in SOD. The intracellular reaction of ROS and peroxidation of lipids might be altered, leading to cell apoptosis. The nanoencapsulated EOs of *C. cassia* lowered the mitochondrial membrane potential and showed an increase in the JC-1 monomer level. They upregulated the expression of caspase-3 and AIF proteins, suggesting the occurrence of cellular apoptosis. In an in vivo study, an inhibition of tumour growth in mice transplanted with breast cancer cells was observed when the nanoencapsulated EOs were administered at a dose of 25 mg/kg. The occurrence of cell apoptosis induced by nanoencapsulated EOs was further confirmed with TUNEL staining with the detection of a significant number of brown apoptotic cells. The nanoencapsulated EOs decreased the expression of the Ki-67 protein, indicating their potential in inhibiting the proliferation of transplanted tumour cells [95].

Table 3. Cytotoxic effects of CS-based nanoencapsulated EOs against breast cancer cells.

Breast Cancer Cells	EOs	Plant Parts	Cell Viability/IC$_{50}$	Reference
MDA-MB-231	*Ocimum gratissimum*	Leaf	Unencapsulated: 44.25% Nanoencapsulated: 37.44%	[87]
MDA-MB-231	*Zataria multiflora*	Aerial	Unencapsulated: 30.5 µg/mL Nanoencapsulated: 6.2 µg/mL	[89]
MCF-7			Unencapsulated: 33.1 µg/mL Nanoencapsulated: 21.2 µg/mL	
MCF-7	*Syzygium aromaticum*	Bud	Unencapsulated: 172.47 µg/mL Nanoencapsulated: 45.89 µg/mL	[90]
MDA-MB-468	*Syzygium aromaticum*	-	Unencapsulated: 243 µg/mL Nanoencapsulated: 177 µg/mL	[91]
MDA-MB-468	*Citrus aurantium*	-	Unencapsulated: 2037.53 µg/mL Nanoencapsulated: 240.44 µg/mL	[92]
	Citrus limon	-	Unencapsulated: 137.03 µg/mL Nanoencapsulated: 40.32 µg/mL	
	Citrus sinensis	-	Unencapsulated: 168.00 µg/mL Nanoencapsulated: 23.65 µg/mL	
MCF-7	*Chelidonium majus* L.	Leaf	Unencapsulated: 90.2 µg/mL Nanoencapsulated: 41.5 µg/mL	[80]
		Root	Unencapsulated: 126.4 µg/mL Nanoencapsulated: 77.6 µg/mL	
MDA-MB-468	*Cinnamomum verum*	-	Unencapsulated: - Nanoencapsulated: 112.35 µg/mL	[93]
MDA-MB-231	*Curcuma longa*	-	Unencapsulated: 329.53 µg/mL Nanoencapsulated: 99.11 µg/mL	[94]
MCF-7			Unencapsulated: 344.60 µg/mL Nanoencapsulated: 82.88 µg/mL	
MCF-7	*Cynometra cauliflora*	Twig	Unencapsulated: NS Nanoencapsulated: 7.69 µg/mL	[8]
		Fruit	Unencapsulated: NS Nanoencapsulated: 17.81 µg/mL	
		Leaf	Unencapsulated: NS Nanoencapsulated: 3.72 µg/mL	
MDA-MB-231		Twig	Unencapsulated: NS Nanoencapsulated: 16.24 µg/mL	
		Fruit	Unencapsulated: NS Nanoencapsulated: 17.65 µg/mL	
		Leaf	Unencapsulated: NS Nanoencapsulated: 16.48 µg/mL	
MDA-MB-231	*Cinnamon cassia*	-	Unencapsulated: 32.25 µg/mL Nanoencapsulated: 25.24 µg/mL	[95]

Note: NS = not significant.

5. Conclusions

EOs, characterised by a variety of volatile constituents with untapped medicinal potential, offer promising prospects for health. Due to the limitations of EOs in specific biomedical applications, this study emphasises the importance of employing nanodelivery with the biopolymer CS. Advances in nanotechnology provide benefits such as enhanced stability, solubility, bioavailability, and controlled release compared to free EOs. Notably, CS-based nanoencapsulated EOs have gained significant attention, becoming a focal point in the development of an alternative drug release system. This approach improves the release profiles of bioactive EOs and enhances selectivity for targeted cells, thereby increasing the effectiveness of EOs against breast cancer cells when embedded in a polymeric matrix. The synergy between nanotechnology and bioactive EOs has the potential to expand their applications in the pharmaceutical and biomedical fields beyond traditional uses. The exploration of CS-based nanoencapsulated EOs signifies a promising pathway with implications for further translation into clinical applications in the treatment of breast cancer.

Author Contributions: Conceptualisation, W.-N.T. and W.-Y.T.; resources, W.-N.T.; data curation, W.-N.T., B.A.S., W.-Y.T. and C.-R.L.; writing—original draft preparation, W.-N.T. and B.A.S.; writing—review and editing, W.-N.T., J.-W.L., N.J.-Y.C., S.R.Y., J.T. and S.R.; project administration, W.-N.T.; funding acquisition, W.-N.T. All authors have read and agreed to the published version of the manuscript.

Funding: The authors would like to acknowledge the Ministry of Higher Education Malaysia for the Fundamental Research Grant Scheme with Project Code FRGS/1/2023/STG01/USM/02/14 and The Royal Society of Chemistry for the RSC Research Fund Grant (R21-3154732300).

Institutional Review Board Statement: Not applicable.

Data Availability Statement: Not applicable.

Conflicts of Interest: The authors declare no conflicts of interest.

References

1. WHO. Cancer. Available online: https://www.who.int/news-room/fact-sheets/detail/cancer/ (accessed on 17 December 2022).
2. Coughlin, S.S. Epidemiology of breast cancer in women. *Adv. Exp. Med. Biol.* **2019**, *1152*, 9–29. [CrossRef] [PubMed]
3. Nanayakkara, A.K.; Boucher, H.W.; Fowler, V.G., Jr.; Jezek, A.; Outterson, K.; Greenberg, D.E. Antibiotic resistance in the patient with cancer: Escalating challenges and paths forward. *CA Cancer J. Clin.* **2021**, *71*, 488–504. [CrossRef]
4. Elshafie, H.S.; Camele, I. An overview of the biological effects of some Mediterranean essential oils on human health. *Biomed. Res. Int.* **2017**, *2017*, 9268468. [CrossRef] [PubMed]
5. El-Tarabily, K.A.; El-Saadony, M.T.; Alagawany, M.; Arif, M.; Batiha, G.E.; Khafaga, A.F.; Elwan, H.A.M.; Elnesr, S.S.; E Abd El-Hack, M. Using essential oils to overcome bacterial biofilm formation and their antimicrobial resistance. *Saudi J. Biol. Sci.* **2021**, *28*, 5145–5156. [CrossRef] [PubMed]
6. Samling, B.A.; Assim, Z.; Tong, W.Y.; Leong, C.R.; Rashid, S.A.; Nik Mohamed Kamal, N.N.S.; Muhamad, M.; Tan, W.N. *Cynometra cauliflora* L.: An indigenous tropical fruit tree in Malaysia bearing essential oils and their biological activities. *Arab J. Chem.* **2021**, *14*, 103302. [CrossRef]
7. Rozman, N.A.S.; Tong, W.Y.; Tan, W.N.; Leong, C.R.; Md Yusof, F.A.; Sulaiman, B. *Homalomena pineodora*, a novel essential oil bearing plant and its antimicrobial activity against diabetic wound pathogens. *J. Essent. Oil Bear. Plants* **2018**, *21*, 963–971. [CrossRef]
8. Samling, B.A.; Assim, Z.; Tong, W.Y.; Leong, C.R.; Rashid, S.A.; Nik Mohamed Kamal, N.N.S.; Muhamad, M.; Tan, W.N. *Cynometra cauliflora* essential oils loaded-chitosan nanoparticles: Evaluations of their antioxidant, antimicrobial and cytotoxic activities. *Int. J. Biol. Macromol.* **2022**, *210*, 742–751. [CrossRef] [PubMed]
9. Guzmán, E.; Lucia, A. Essential oils and their individual components in cosmetic products. *Cosmetics* **2021**, *8*, 114. [CrossRef]
10. Reis, D.R.; Ambrosi, A.; Luccio, M.D. Encapsulated essential oils: A perspective in food preservation. *Future Foods* **2022**, *5*, 100126. [CrossRef]
11. Arabpoor, B.; Yousefi, S.; Weisany, W.; Ghasemlou, M. Multifunctional coating composed of *Eryngium campestre* L. essential oil encapsulated in nano-chitosan to prolong the shelf-life of fresh cherry fruits. *Food Hydrocoll.* **2021**, *111*, 106394. [CrossRef]
12. Nile, S.H.; Baskar, V.; Selvaraj, D.; Nile, A.; Xiao, J.; Kai, G. Nanotechnologies in food science: Applications, recent trends, and future perspectives. *Nanomicro Lett.* **2020**, *12*, 45. [CrossRef]
13. Dupuis, V.; Cerbu, C.; Witkowski, L.; Potarniche, A.V.; Timar, M.C.; Żychska, M.; Sabliov, C.M. Nanodelivery of essential oils as efficient tools against antimicrobial resistance: A review of the type and physical-chemical properties of the delivery systems and applications. *Drug Deliv.* **2022**, *29*, 1007–1024. [CrossRef]
14. Pateiro, M.; Gómez, B.; Munekata, P.E.S.; Barba, F.J.; Putnik, P.; Kovačević, D.B.; Lorenzo, J.M. Nanoencapsulation of promising bioactive compounds to improve their absorption, stability, functionality and the appearance of the final food products. *Molecules* **2021**, *26*, 1547. [CrossRef]
15. Falleh, H.; Ben Jemaa, M.; Saada, M.; Ksouri, R. Essential oils: A promising eco-friendly food preservative. *Food Chem.* **2020**, *330*, 127268. [CrossRef]
16. Lunz, K.; Stappen, I. Back to the roots-an overview of the chemical composition and bioactivity of selected root-essential oils. *Molecules* **2021**, *26*, 3155. [CrossRef]

17. Lee, S.H.; Kim, D.S.; Park, S.H.; Park, H. Phytochemistry and applications of *Cinnamomum camphora* essential oils. *Molecules* **2022**, *27*, 2695. [CrossRef] [PubMed]
18. Ouakouak, H.; Benarfa, A.; Messaoudi, M.; Begaa, S.; Sawicka, B.; Benchikha, N.; Simal-Gandara, J. Biological properties of essential oils from *Thymus algeriensis* Boiss. *Plants* **2021**, *10*, 786. [CrossRef] [PubMed]
19. Vianna, T.C.; Marinho, C.O.; Marangoni Júnior, L.; Ibrahim, S.A.; Vieira, R.P. Essential oils as additives in active starch-based food packaging films: A review. *Int. J. Biol. Macromol.* **2021**, *182*, 1803–1819. [CrossRef] [PubMed]
20. Prince Chidike Ezeorba, T.; Ikechukwu Chukwudozie, K.; Anthony Ezema, C.; Godwin Anaduaka, E.; John Nweze, E.; Sunday Okeke, E. Potentials for health and therapeutic benefits of garlic essential oils: Recent findings and future prospects. *Pharmacol. Res.* **2022**, *3*, 100075. [CrossRef]
21. Basavegowda, N.; Baek, K.-H. Synergistic antioxidant and antibacterial advantages of essential oils for food packaging applications. *Biomolecules* **2021**, *11*, 1267. [CrossRef] [PubMed]
22. Cimino, C.; Maurel, O.M.; Musumeci, T.; Bonaccorso, A.; Drago, F.; Souto, E.M.B.; Pignatello, R.; Carbone, C. Essential Oils: Pharmaceutical applications and encapsulation strategies into lipid-based delivery systems. *Pharmaceutics* **2021**, *13*, 327. [CrossRef]
23. da Silva, L.C.; Viganó, J.; de Souza Mesquita, L.M.; Dias, A.L.B.; de Souza, M.C.; Sanches, V.L.; Chaves, J.O.; Pizani, R.S.; Contieri, L.S.; Rostagno, M.A. Recent advances and trends in extraction techniques to recover polyphenols compounds from apple by-products. *Food Chem. X* **2021**, *12*, 100133. [CrossRef] [PubMed]
24. Ni, Z.-J.; Wang, X.; Shen, Y.; Thakur, K.; Han, J.; Zhang, J.-G.; Hu, F.; Wei, Z.-J. Recent updates on the chemistry, bioactivities, mode of action, and industrial applications of plant essential oils. *Trends Food Sci. Technol.* **2021**, *110*, 78–89. [CrossRef]
25. López-Hortas, L.; Rodríguez, P.; Díaz-Reinoso, B.; Gaspar, M.C.; de Sousa, H.C.; Braga, M.E.M.; Domínguez, H. Supercritical fluid extraction as a suitable technology to recover bioactive compounds from flowers. *J. Supercrit. Fluids* **2022**, *188*, 105652. [CrossRef]
26. Megawati; Fardhyanti, D.S.; Sediawan, W.B.; Hisyam, A. Kinetics of mace (*Myristicae arillus*) essential oil extraction using microwave assisted hydrodistillation: Effect of microwave power. *Ind. Crops Prod.* **2019**, *131*, 315–322. [CrossRef]
27. Tekin, K.; Akalın, M.K.; GulSeker, M. Ultrasound bath-assisted extraction of essential oils from clove using central composite design. *Ind. Crops Prod.* **2015**, *77*, 954–960. [CrossRef]
28. Lammari, N.; Louaer, O.; Meniai, A.H.; Elaissari, A. Encapsulation of essential oils via nanoprecipitation process: Overview, progress, challenges and prospects. *Pharmaceutics* **2020**, *12*, 431. [CrossRef]
29. Stevens, N.; Allred, K. Antidiabetic potential of volatile cinnamon oil: A review and exploration of mechanisms using in silico molecular docking simulations. *Molecules* **2022**, *27*, 853. [CrossRef]
30. Yeshi, K.; Wangchuk, P. Essential oils and their bioactive molecules in healthcare. In *Herbal Biomolecules in Healthcare Applications*; Mandal, S.C., Nayak, A.K., Dhara, A.K., Eds.; Academic Press: Cambridge, MA, USA, 2022; pp. 215–237. [CrossRef]
31. Fuentes, C.; Fuentes, A.; Barat, J.M.; Ruiz, M.J. Relevant essential oil components: A minireview on increasing applications and potential toxicity. *Toxicol. Mech. Methods* **2021**, *31*, 559–565. [CrossRef] [PubMed]
32. Sharmeen, J.B.; Mahomoodally, F.M.; Zengin, G.; Maggi, F. Essential oils as natural sources of fragrance compounds for cosmetics and cosmeceuticals. *Molecules* **2021**, *26*, 666. [CrossRef] [PubMed]
33. Ameh, S.J.; Obodozie-Ofoegbu, O. Essential oils as flavors in carbonated cola and citrus soft drinks. In *Essential Oils in Food Preservation, Flavor and Safety*; Preedy, V.R., Ed.; Academic Press: San Diego, CA, USA, 2016; pp. 111–121. [CrossRef]
34. Gholamipourfard, K.; Salehi, M.; Banchio, E. *Mentha piperita* phytochemicals in agriculture, food industry and medicine: Features and applications. *S. Afr. J. Bot.* **2021**, *141*, 183–195. [CrossRef]
35. Masyita, A.; Sari, R.M.; Astuti, A.D.; Yasir, B.; Rumata, N.R.; Emran, T.B.; Nainu, F.; Simal-Gandara, J. Terpenes and terpenoids as main bioactive compounds of essential oils, their roles in human health and potential application as natural food preservatives. *Food Chem. X* **2022**, *13*, 100217. [CrossRef] [PubMed]
36. Wani, A.R.; Yadav, K.; Khursheed, A.; Rather, M.A. An updated and comprehensive review of the antiviral potential of essential oils and their chemical constituents with special focus on their mechanism of action against various influenza and coronaviruses. *Microb. Pathog.* **2021**, *152*, 104620. [CrossRef] [PubMed]
37. Kashyap, N.; Kumari, A.; Raina, N.; Zakir, F.; Gupta, M. Prospects of essential oil loaded nanosystems for skincare. *Phytomedicine Plus* **2022**, *2*, 100198. [CrossRef]
38. Kang, A.; Lee, T.S. Secondary metabolism for isoprenoid-based biofuels. In *Biotechnology for Biofuel Production and Optimization*; Eckert, C.A., Trinh, C.T., Eds.; Elsevier: Amsterdam, The Netherlands, 2016; pp. 35–71. [CrossRef]
39. Pragadheesh, V.S.; Bisht, D.; Chanotiya, C.S. Terpenoids from essential oils. In *Kirk-Othmer Encyclopedia of Chemical Technology*; John Wiley & Sons: Hoboken, NJ, USA, 2020; pp. 1–22. [CrossRef]
40. Otles, S.; Özyurt, V.H. Biotransformation in the production of secondary metabolites. In *Studies in Natural Products Chemistry*; Atta-ur-Rahman, Ed.; Elsevier: Amsterdam, The Netherlands, 2021; Volume 68, pp. 435–457. [CrossRef]
41. Mansi, S.; Kamaljit, G.; Rupali, J.; Daizy, R.B.; Harminder, P.S.; Ravinder, K.K. Essential oils as anticancer agents: Potential role in malignancies, drug delivery mechanisms, and immune system enhancement. *Biomed. Pharmacother.* **2022**, *146*, 112514. [CrossRef]
42. Haris, N.; Fahir, B.; Emina, K.; Elma, O.; Samija, M.; Bojana, M.; Kemal, D. Cytotoxicity screening of *Thymus vulgaris* L. essential oil in brine shrimp nauplii and cancer cell lines. *Sci. Rep.* **2021**, *11*, 13178. [CrossRef]
43. Zu, Y.; Yu, H.; Liang, L.; Fu, Y.; Efferth, T.; Liu, X.; Wu, N. Activities of ten essential oils towards *Propionibacterium acnes* and PC-3, A-549 and MCF-7 cancer cells. *Molecules* **2010**, *15*, 3200–3210. [CrossRef]

44. Stone, S.C.; Vasconcellos, F.A.; Lenardão, E.J.; do Amaral, R.C.; Jacob, R.G.; Leivas Leite, F.P. Evaluation of potential use of *Cymbopogon* sp. essential oils, (R)-citronellal and N-citronellylamine in cancer chemotherapy. *Int. J. Appl. Res. Nat. Prod.* **2013**, *6*, 11–15.
45. Kumoro, A.C.; Wardhani, D.H.; Retnowati, D.S.; Haryani, K. A brief review on the characteristics, extraction and potential industrial applications of citronella grass (*Cymbopogon nardus*) and lemongrass (*Cymbopogon citratus*) essential oils. *IOP Conf. Ser. Mater. Sci. Eng.* **2021**, *1053*, 012118. [CrossRef]
46. Hashim, Y.Z.; Phirdaous, A.; Azura, A. Screening of anticancer activity from agarwood essential oil. *Pharmacogn. Res.* **2014**, *6*, 191–194. [CrossRef]
47. Wang, M.R.; Li, W.; Luo, S.; Zhao, X.; Ma, C.H.; Liu, S.X. GC-MS study of the chemical components of different *Aquilaria sinensis* (Lour.) gilgorgans and agarwood from different Asian countries. *Molecules* **2018**, *23*, 2168. [CrossRef]
48. Lee, Y. Cytotoxicity evaluation of essential oil and its component from *Zingiber officinale* Roscoe. *Toxicol. Res.* **2016**, *32*, 225–230. [CrossRef]
49. Assmann, C.E.; Cadoná, F.C.; da Silva Rosa Bonadiman, B.; Dornelles, E.B.; Trevisan, G.; da Cruz, I.B.M. Tea tree oil presents in vitro antitumor activity on breast cancer cells without cytotoxic effects on fibroblasts and on peripheral blood mononuclear cells. *Biomed. Pharmacother.* **2018**, *103*, 1253–1261. [CrossRef]
50. Ashmawy, A.; Mostafa, N.; Eldahshan, O. GC/MS analysis and molecular profiling of lemon volatile oil against breast cancer. *J. Essent. Oil Bear. Plants* **2019**, *22*, 903–916. [CrossRef]
51. Boukhatem, M.N.; Sudha, T.; Darwish, N.H.E.; Chader, H.; Belkadi, A.; Rajabi, M.; Houche, A.; Benkebailli, F.; Oudjida, F.; Mousa, S.A. A new eucalyptol-rich lavender (*Lavandula stoechas* L.) essential oil: Emerging potential for therapy against inflammation and cancer. *Molecules* **2020**, *25*, 3671. [CrossRef] [PubMed]
52. Aburjai, T.A.; Mansi, K.; Azzam, H.; Alqudah, D.A.; Alshaer, W.; Abuirjei, M. Chemical compositions and anticancer potential of essential oil from greenhouse-cultivated *Ocimum basilicum* leaves. *Indian J. Pharm. Sci.* **2020**, *82*, 178–183. [CrossRef]
53. An, Z.; Feng, X.; Sun, M.; Wang, Y.; Wang, H.; Gong, Y. Chamomile essential oil: Chemical constituents and antitumor activity in MDA-MB-231 cells through PI3K/Akt/mTOR signaling pathway. *Chem. Biodivers.* **2023**, *20*, e202200523. [CrossRef]
54. Tan, W.; Shahbudin, F.N.; Kamal, N.N.S.N.M.; Tong, W.Y.; Leong, C.R.; Lim, J. Volatile constituents of the leaf essential oil of *Crinum asiaticum* and their antimicrobial and cytotoxic activities. *J. Essent. Oil Bear. Plants* **2019**, *22*, 947–954. [CrossRef]
55. Tan, W.; Tong, W.Y.; Leong, C.R.; Kamal, N.N.S.N.M.; Muhamad, M.; Lim, J.; Khairuddean, M.; Man, M.B.H. Chemical composition of essential oil of *Garcinia gummi-gutta* and its antimicrobial and cytotoxic activities. *J. Essent. Oil Bear. Plants* **2020**, *23*, 832–842. [CrossRef]
56. Wissal, D.; Sana, B.; Sabrine, J.; Nada, B.; Wissem, M. Essential oils' chemical characterization and investigation of some biological activities: A critical review. *Medicines* **2016**, *3*, 25. [CrossRef]
57. Turek, C.; Stintzing, F.C. Impact of different storage conditions on the quality of selected essential oils. *Food Res. Int.* **2012**, *46*, 341–353. [CrossRef]
58. Zhu, Y.; Li, C.; Cui, H.; Lin, L. Encapsulation strategies to enhance the antibacterial properties of essential oils in food system. *Food Control* **2021**, *123*, 107856. [CrossRef]
59. Park, V.; Garland, S.M.; Close, D.C. The influence of temperature, light, and storage period on the colour and chemical profile of Kunzea essential oil (*Kunzea ambigua* (Sm.) Druce). *J. Appl. Res. Med. Aromat. Plants* **2022**, *30*, 100383. [CrossRef]
60. Tiwari, S.; Dubey, N.K. Nanoencapsulated essential oils as novel green preservatives against fungal and mycotoxin contamination of food commodities. *Curr. Opin. Food Sci.* **2022**, *45*, 100831. [CrossRef]
61. Prakash, B.; Kujur, A.; Yadav, A.; Kumar, A.; Singh, P.P.; Dubey, N.K. Nanoencapsulation: An efficient technology to boost the antimicrobial potential of plant essential oils in food system. *Food Control* **2018**, *89*, 1–11. [CrossRef]
62. Das, S.; Singh, V.K.; Dwivedy, A.K.; Chaudhari, A.K.; Deepika, K.; Dubey, N.K. Eugenol loaded chitosan nanoemulsion for food protection and inhibition of Aflatoxin B1 synthesizing genes based on molecular docking. *Carbohydr. Polym.* **2021**, *255*, 117339. [CrossRef] [PubMed]
63. Almnhawy, M.; Jebur, M.; Alhajamee, M.; Marai, K.; Tabrizi, M.H. PLGA-based nano-encapsulation of *Trachyspermum ammi* seed essential oil (TSEO-PNP) as a safe, natural, efficient, anticancer compound in human HT-29 colon cancer cell line. *Nutr. Cancer* **2021**, *73*, 2808–2820. [CrossRef]
64. Sánchez-Machado, D.I.; López-Cervantes, J.; Correa-Murrieta, M.A.; Sánchez-Duarte, R.G.; Cruz-Flores, P.; de la Mora-López, G.S. Chitosan. In *Nonvitamin and Nonmineral Nutritional Supplements*; Nabavi, S.M., Silva, A.S., Eds.; Academic Press: New York, NY, USA, 2019; pp. 485–493. [CrossRef]
65. Zhang, F.; Ramachandran, G.; Mothana, R.A.; Noman, O.M.; Alobaid, W.A.; Rajivgandhi, G.; Manoharan, N. Anti-bacterial activity of chitosan loaded plant essential oil against multi drug resistant *K. pneumonia*. *Saudi J. Biol. Sci.* **2020**, *27*, 3449–3455. [CrossRef]
66. Sharifi-Rad, J.; Quispe, C.; Butnariu, M.; Rotariu, L.S.; Sytar, O.; Sestito, S.; Rapposelli, S.; Akram, M.; Iqbal, M.; Krishna, A.; et al. Chitosan nanoparticles as a promising tool in nanomedicine with particular emphasis on oncological treatment. *Cancer Cell Int.* **2021**, *21*, 318. [CrossRef]
67. Raza, Z.A.; Khalil, S.; Ayub, A.; Banat, I.M. Recent developments in chitosan encapsulation of various active ingredients for multifunctional applications. *Carbohydr. Res.* **2020**, *492*, 108004. [CrossRef]

68. Gonçalves, C.; Ferreira, N.; Lourenço, L. Production of low molecular weight chitosan and chitooligosaccharides (COS): A review. *Polymers* **2021**, *13*, 2466. [CrossRef] [PubMed]
69. Mohammed, M.A.; Syeda, J.T.M.; Wasan, K.M.; Wasan, E.K. An overview of chitosan nanoparticles and its application in non-parenteral drug delivery. *Pharmaceutics* **2017**, *9*, 53. [CrossRef] [PubMed]
70. Chaudhari, A.K.; Singh, V.K.; Das, S.; Dubey, N.K. Nanoencapsulation of essential oils and their bioactive constituents: A novel strategy to control mycotoxin contamination in food system. *Food Chem. Toxicol.* **2021**, *149*, 112019. [CrossRef] [PubMed]
71. Rizeq, B.R.; Younes, N.N.; Rasool, K.; Nasrallah, G.K. Synthesis, bio-applications, and toxicity evaluation of chitosan-based nanoparticles. *Int. J. Mol. Sci.* **2019**, *20*, 5776. [CrossRef] [PubMed]
72. Maqsoudlou, A.; Assadpour, E.; Mohebodini, H.; Jafari, S.M. Improving the efficiency of natural antioxidant compounds via different nanocarriers. *Adv. Colloid Interface Sci.* **2020**, *278*, 102112. [CrossRef] [PubMed]
73. Bastos, L.P.H.; Vicente, J.; dos Santos, C.H.C.; de Carvalho, M.G.; Garcia-Rojas, E.E. Encapsulation of black pepper (*Piper nigrum* L.) essential oil with gelatin and sodium alginate by complex coacervation. *Food Hydrocoll.* **2020**, *102*, 105605. [CrossRef]
74. Espitia, P.J.; Fuenmayor, C.A.; Otoni, C.G. Nanoemulsions: Synthesis, characterization, and application in bio-based active food packaging. *Compr. Rev. Food Sci. Food Saf.* **2019**, *18*, 264–285. [CrossRef] [PubMed]
75. Das, S.; Singh, V.K.; Dwivedy, A.K.; Chaudhari, A.K.; Upadhyay, N.; Singh, A.; Dubey, N.K. Fabrication, characterization and practical efficacy of *Myristica fragrans* essential oil nanoemulsion delivery system against postharvest biodeterioration. *Ecotoxicol. Environ. Saf.* **2020**, *189*, 110000. [CrossRef]
76. Dávila-Rodríguez, M.; López-Malo, A.; Palou, E.; Ramírez-Corona, N.; Jiménez-Munguía, M.T. Antimicrobial activity of nanoemulsions of cinnamon, rosemary, and oregano essential oils on fresh celery. *LWT-Food Sci. Technol.* **2019**, *112*, 108247. [CrossRef]
77. López-Meneses, A.K.; Plascencia-Jatomea, M.; Lizardi-Mendoza, J.; Fernández-Quiroz, D.; Rodríguez-Félix, F.; Mouriño-Pérez, R.R.; Cortez-Rocha, M.O. *Schinus molle* L. essential oil-loaded chitosan nanoparticles: Preparation, characterization, antifungal and anti-aflatoxigenic properties. *LWT-Food Sci. Technol.* **2018**, *96*, 597–603. [CrossRef]
78. Hadidi, M.; Pouramin, S.; Adinepour, F.; Haghani, S.; Jafari, S.M. Chitosan nanoparticles loaded with clove essential oil: Characterization, antioxidant and antibacterial activities. *Carbohydr. Polym.* **2020**, *236*, 116075. [CrossRef] [PubMed]
79. Soltanzadeh, M.; Peighambardoust, S.H.; Ghanbarzadeh, B.; Mohammadi, M.; Lorenzo, J.M. Chitosan nanoparticles encapsulating lemongrass (*Cymbopogon commutatus*) essential oil: Physicochemical, structural, antimicrobial and in-vitro release properties. *Int. J. Biol. Macromol.* **2021**, *192*, 1084–1097. [CrossRef]
80. Hesami, S.; Safi, S.; Larijani, K.; Badi, H.N.; Abdossi, V.; Hadidi, M. Synthesis and characterization of chitosan nanoparticles loaded with greater celandine (*Chelidonium majus* L.) essential oil as an anticancer agent on MCF-7 cell line. *Int. J. Biol. Macromol.* **2022**, *194*, 974–981. [CrossRef]
81. Wang, G.; Li, R.; Parseh, B.; Du, G. Prospects and challenges of anticancer agents' delivery via chitosan-based drug carriers to combat breast cancer: A review. *Carbohydr. Polym.* **2021**, *268*, 118192. [CrossRef] [PubMed]
82. Herdiana, Y.; Wathoni, N.; Shamsuddin, S.; Joni, I.M.; Muchtaridi, M. Chitosan-based nanoparticles of targeted drug delivery system in breast cancer treatment. *Polymers* **2021**, *13*, 1717. [CrossRef] [PubMed]
83. Martens, T.F.; Remaut, K.; Demeester, J.; De Smedt, S.C.; Braeckmans, K. Intracellular delivery of nanomaterials: How to catch endosomal escape in the act. *Nano Today* **2014**, *9*, 344–364. [CrossRef]
84. Russo, R.; Corasaniti, M.T.; Bagetta, G.; Morrone, L.A. Exploitation of cytotoxicity of some essential oils for translation in cancer therapy. *Evid. Based Complement. Altern. Med.* **2015**, *2015*, 397821. [CrossRef]
85. Ahn, C.; Lee, J.H.; Park, M.J.; Kim, J.W.; Yang, J.; Yoo, Y.M.; Jeung, E.B. Cytostatic effects of plant essential oils on human skin and lung cells. *Exp. Ther. Med.* **2020**, *19*, 2008–2018. [CrossRef]
86. Shetta, A.; Kegere, J.; Mamdouh, W. Comparative study of encapsulated peppermint and green tea essential oils in chitosan nanoparticles: Encapsulation, thermal stability, in-vitro release, antioxidant and antibacterial activities. *Int. J. Biol. Macromol.* **2019**, *126*, 731–742. [CrossRef]
87. Onyebuchi, C.; Kavaz, D. Chitosan and N, N, N-trimethyl chitosan nanoparticle encapsulation of *Ocimum gratissimum* essential oil: Optimised synthesis, in vitro release and bioactivity. *Int. J. Nanomed.* **2019**, *14*, 7707–7727. [CrossRef]
88. Salehi, F.; Behboudi, H.; Kavoosi, G.; Ardestani, S.K. Monitoring ZEO apoptotic potential in 2D and 3D cell cultures and associated spectroscopic evidence on mode of interaction with DNA. *Sci. Rep.* **2017**, *7*, 2533. [CrossRef] [PubMed]
89. Salehi, F.; Behboudi, H.; Kavoosi, G.; Ardestani, S.K. Incorporation of *Zataria multiflora* essential oil into chitosan biopolymer nanoparticles: A nanoemulsion based delivery system to improve the in-vitro efficacy, stability and anticancer activity of ZEO against breast cancer cells. *Int. J. Biol. Macromol.* **2020**, *143*, 382–392. [CrossRef] [PubMed]
90. Kamal, I.; Khedr, A.I.M.; Alfaifi, M.Y.; Elbehairi, S.E.I.; Elshaarawy, R.F.M.; Saad, A.S. Chemotherapeutic and chemopreventive potentials of p-coumaric acid–squid chitosan nanogel loaded with *Syzygium aromaticum* essential oil. *Int. J. Biol. Macromol.* **2021**, *188*, 523–533. [CrossRef] [PubMed]
91. Valizadeh, A.; Khaleghi, A.A.; Alipanah, H.; Zarenezhad, E.; Osanloo, M. Anticarcinogenic effect of chitosan nanoparticles containing *Syzygium aromaticum* essential oil or eugenol toward breast and skin cancer cell lines. *BioNanoScience* **2021**, *11*, 678–686. [CrossRef]

92. Alipanah, H.; Farjam, M.; Zarenezhad, E.; Roozitalab, G.; Osanloo, M. Chitosan nanoparticles containing limonene and limonene-rich essential oils: Potential phytotherapy agents for the treatment of melanoma and breast cancers. *BMC Complement. Med. Ther.* **2021**, *21*, 186. [CrossRef]
93. Khoshnevisan, K.; Alipanah, H.; Baharifar, H.; Ranjbar, N.; Osanloo, M. Chitosan nanoparticles containing *Cinnamomum verum* J.Presl essential oil and cinnamaldehyde: Preparation, characterization and anticancer effects against melanoma and breast cancer cells. *Tradit. Integr. Med.* **2022**, *7*, 1. [CrossRef]
94. San, H.H.M.; Alcantara, K.P.; Bulatao, B.P.I.; Chaichompoo, W.; Nalinratana, N.; Suksamrarn, A.; Vajragupta, O.; Rojsitthisak, P.; Rojsitthisak, P. Development of turmeric oil—Loaded chitosan/alginate nanocapsules for cytotoxicity enhancement against breast cancer. *Polymers* **2022**, *14*, 1835. [CrossRef] [PubMed]
95. Xu, X.; Li, Q.; Dong, W.; Zhao, G.; Lu, Y.; Huang, X.; Liang, X. *Cinnamon cassia* oil chitosan nanoparticles: Physicochemical properties and anti-breast cancer activity. *Int. J. Biol. Macromol.* **2023**, *224*, 1065–1078. [CrossRef] [PubMed]

Disclaimer/Publisher's Note: The statements, opinions and data contained in all publications are solely those of the individual author(s) and contributor(s) and not of MDPI and/or the editor(s). MDPI and/or the editor(s) disclaim responsibility for any injury to people or property resulting from any ideas, methods, instructions or products referred to in the content.

Article

Hydrogel Based on Chitosan/Gelatin/Poly(Vinyl Alcohol) for In Vitro Human Auricular Chondrocyte Culture

Carmina Ortega-Sánchez [1,†], Yaaziel Melgarejo-Ramírez [1,†], Rogelio Rodríguez-Rodríguez [2,‡], Jorge Armando Jiménez-Ávalos [2,‡], David M. Giraldo-Gomez [3], Claudia Gutiérrez-Gómez [4], Jacobo Rodriguez-Campos [5], Gabriel Luna-Bárcenas [6], Cristina Velasquillo [7,*], Valentín Martínez-López [7,*] and Zaira Y. García-Carvajal [3,*]

Citation: Ortega-Sánchez, C.;
Melgarejo-Ramírez, Y.;
Rodríguez-Rodríguez, R.;
Jiménez-Ávalos, J.A.; Giraldo-Gomez,
D.M.; Gutiérrez-Gómez, C.;
Rodriguez-Campos, J.;
Luna-Bárcenas, G.; Velasquillo, C.;
Martínez-López, V.; et al.
Hydrogel Based on
Chitosan/Gelatin/Poly
(Vinyl Alcohol) for In Vitro Human
Auricular Chondrocyte Culture.
Polymers 2024, *16*, 479.
https://doi.org/10.3390/
polym16040479

Academic Editors: William
Facchinatto and Sérgio
Paulo Campana-Filho

Received: 12 December 2023
Revised: 3 February 2024
Accepted: 5 February 2024
Published: 8 February 2024

Copyright: © 2024 by the authors.
Licensee MDPI, Basel, Switzerland.
This article is an open access article
distributed under the terms and
conditions of the Creative Commons
Attribution (CC BY) license (https://
creativecommons.org/licenses/by/
4.0/).

1. Laboratorio de Biotecnología, Unidad de Gerociencias, Instituto Nacional de Rehabilitación Luis Guillermo Ibarra Ibarra, Ciudad de México 14389, Mexico; carminaortega52@gmail.com (C.O.-S.); yaazielmr@gmail.com (Y.M.-R.)
2. Biotecnología Médica y Farmacéutica, Centro de Investigación y Asistencia en Tecnología y Diseño del Estado de Jalisco A.C. (CIATEJ), Av. Normalistas No. 800, Col. Colinas de la Normal, Guadalajara 44270, Jalisco, Mexico; rogelio.rodriguez4085@academicos.udg.mx (R.R.-R.); avalos.joar@gmail.com (J.A.J.-Á.)
3. Unidad de Microscopia, Departamento de Biología Celular y Tisular, Facultad de Medicina, Universidad Nacional Autónoma de México, Avenida Universidad 3000, Circuito Interior, Edificio "A" Planta Baja, Ciudad Universitaria, Coyoacán, Ciudad de México 04510, Mexico; davidgiraldo@comunidad.unam.mx
4. División de Cirugía Plástica y Reconstructiva, Hospital General Dr. Manuel Gea González, Ciudad de México 14080, Mexico; dra.claugg8@gmail.com
5. Servicios Analíticos y Metrológicos, Centro de Investigación y Asistencia en Tecnología y Diseño del Estado de Jalisco A.C. (CIATEJ), Av. Normalistas No. 800, Col. Colinas de la Normal, Guadalajara 44270, Jalisco, Mexico; jarodriguez@ciatej.mx
6. Institute of Advanced Materials for Sustainable Manufacturing Tecnológico de Monterrey, Epigmenio González 500, San Pablo, Santiago de Querétaro 76130, Querétaro, Mexico; gabriel.luna@cinvestav.mx
7. Unidad de Ingeniería de Tejidos Terapia Celular y Medicina Regenerativa, Instituto Nacional de Rehabilitación Luis Guillermo Ibarra Ibarra, Ciudad de México 14389, Mexico
* Correspondence: mvelasquillo@inr.gob.mx (C.V.); val_mart76@yahoo.com.mx or vmartinez@inr.gob.mx (V.M.-L.); zgarcia@ciatej.mx (Z.Y.G.-C.)
† These authors contribute in an equal manner.
‡ The research presented in this work corresponds to the Ph.D. research work in CIATEJ, regardless of current affiliation.

Abstract: Three-dimensional (3D) hydrogels provide tissue-like complexities and allow for the spatial orientation of cells, leading to more realistic cellular responses in pathophysiological environments. There is a growing interest in developing multifunctional hydrogels using ternary mixtures for biomedical applications. This study examined the biocompatibility and suitability of human auricular chondrocytes from microtia cultured onto steam-sterilized 3D Chitosan/Gelatin/Poly(Vinyl Alcohol) (CS/Gel/PVA) hydrogels as scaffolds for tissue engineering applications. Hydrogels were prepared in a polymer ratio (1:1:1) through freezing/thawing and freeze-drying and were sterilized by autoclaving. The macrostructure of the resulting hydrogels was investigated by scanning electron microscopy (SEM), showing a heterogeneous macroporous structure with a pore size between 50 and 500 µm. Fourier-transform infrared (FTIR) spectra showed that the three polymers interacted through hydrogen bonding between the amino and hydroxyl moieties. The profile of amino acids present in the gelatin and the hydrogel was determined by ultra-performance liquid chromatography (UPLC), suggesting that the majority of amino acids interacted during the formation of the hydrogel. The cytocompatibility, viability, cell growth and formation of extracellular matrix (ECM) proteins were evaluated to demonstrate the suitability and functionality of the 3D hydrogels for the culture of auricular chondrocytes. The cytocompatibility of the 3D hydrogels was confirmed using a 3-(4,5-dimethylthiazol-2-yl)-2,5-diphenyltetrazolium bromide (MTT) assay, reaching 100% viability after 72 h. Chondrocyte viability showed a high affinity of chondrocytes for the hydrogel after 14 days, using the Live/Dead assay. The chondrocyte attachment onto the 3D hydrogels and the formation of an ECM were observed using SEM. Immunofluorescence confirmed the expression of elastin, aggrecan and type II collagen, three of the main components found in an elastic cartilage extracellular matrix. These results demonstrate the suitability and functionality of a CS/Gel/PVA hydrogel as a 3D support for the auricular chondrocytes culture, suggesting that these hydrogels are a potential biomaterial for cartilage tissue engineering applications, aimed at the regeneration of elastic cartilage.

Keywords: composite hydrogel; three-dimensional hydrogel; chitosan; gelatin; tissue engineering; auricular chondrocytes culture

1. Introduction

Hydrogels are polymeric materials with a hydrophilic structure that allows for the storage of large amounts of water and biological fluids [1–8], and are ideal for in vitro cell cultures, as they provide a supportive matrix that mimics a natural microenvironment [1,9,10]. Three-dimensional (3D) hydrogels provide tissue-like complexities and allow for the spatial orientation of cells, leading to more realistic cellular responses in pathophysiological environments [7,10,11]. Cell behavior in 3D hydrogels is markedly variable, not only between populations of different types of cells, but also between different tissues [5–12]. The cellular microenvironment plays a vital role in cell functions, from controlling morphology to activating a wide range of factors regulating cell growth, proliferation, differentiation and migration [10–12].

Chitosan (CS) is a deacetylated linear polysaccharide, derivative of chitin and composed of variable amounts of attached residues ($\beta 1 \rightarrow 4$) of N-acetyl-2 amino-2-deoxy-D-glucose (glucosamine, GlcN) and 2-amino-residues of 2-deoxy-D-glucose (N-acetyl-glucosamine, GlcNAc) [13,14]. Gelatin (Gel) is a partially hydrolyzed collagen that promotes cell adhesion, proliferation and migration and is also used as a gelling agent carrier for encapsulating bioactive molecules [15,16]. It is mechanically weak and requires chemical modification or combination with other materials to form a hydrogel [17]. Poly(vinyl alcohol) (PVA) is a biodegradable hydrophilic synthetic polymer with excellent biocompatibility [18]. There is a growing interest in developing multifunctional hydrogels [4] using ternary mixtures for biomedical applications [19]. We have previously reported that human adipose-derived mesenchymal stromal cells (AD-hMSCs) displayed chondroinductive properties after being cultured onto bidimensional CS/Gel/PVA hydrogels and following their exposure to chondrogenic stimulation media [20]. The synthesis of 3D hydrogels resulted in open and interconnected macroporous structures, allowing gas and nutrient exchange and improving their mechanical properties. Furthermore, 3D CS/Gel/PVA hydrogels were non-toxic to HT29 cells [21,22] and BRIN-BD11 cells [23], demonstrating their potential use as scaffolds for tissue engineering applications.

Advances in material sciences have the potential to benefit tissue engineering techniques aimed at the treatment of congenital deformities within the field of plastic and reconstructive surgery [24–26]. Three-dimensional hydrogels are used to support the growth of chondrocytes, the cartilage cells, offering an optimal microenvironment similar to native elastic cartilage [25,27–33]. They can positively affect cell morphology, proliferation and differentiation of auricular chondrocytes [32,34–39]. This study aimed to examine the biocompatibility and suitability of human auricular chondrocytes from microtia cultured onto steam-sterilized 3D CS/Gel/PVA hydrogels as scaffolds for tissue engineering applications.

2. Materials and Methods

Ethical Considerations

This study was approved by the Institutional Research Committee and Ethics Committee (Register number: INRLGII 85/19). Pediatric patients were recruited from the microtia clinic at the National Institute of Rehabilitation Luis Guillermo Ibarra Ibarra and from the General Hospital Dr. Manuel Gea González in Mexico City. After informed consent was granted, the microtia cartilage remaining from costal cartilage graft ear reconstruction surgeries was donated.

Chemicals

Low molecular weight chitosan (CS, deacetylation degree ~92.2%), type B gelatin from bovine skin (Gel, Bloom ~75), poly(vinyl alcohol) (PVA, molecular weight ~89 kDa and hydrolysis degree ~99.8%), o-xylene and reagents for the cytotoxicity assay were supplied by Sigma Aldrich (St. Louis, MO, USA). Solvents and reagents were analytical grade.

2.1. Chitosan/Gelatin/Poly(Vinyl Alcohol) Hydrogel Preparation and Sterilization

2.1.1. Ternary Polymer Solutions

CS solution (2.5 wt %) was obtained by dissolving chitosan powder in 0.4 M acetic acid solution and gently stirring it for 12 h at 25 °C. Gel solution (2.5 wt %) was prepared by dissolving gelatin powder in distilled water and gently stirring it for 2 h at 37 °C. PVA solution (2.5 wt %) was prepared by dissolving the powder in distilled water and gently stirring it for 2 h at 80 °C.

2.1.2. Sterilized Hydrogel Preparation

CS/Gel/PVS hydrogels were fabricated by blending the polymer solutions in a 1:1:1 ratio, according to the method previously described by Rodríguez-Rodríguez [21]. The preparation process consisted of four steps: (1) a freeze–thaw cycle, (2) a lyophilization process, (3) chemical treatment with o-xylene as a porogen template, and (4) steam sterilization. The pH of the ternary solutions was adjusted to 4.0, to obtain homogeneous solutions. After that, the polymeric solution was poured into a plastic mold (14 mL) and frozen at −80 °C for 24 h with a subsequent freeze-drying process using a freeze dryer (Telstar LyoQuest, Terrassa, Spain) and a vacuum of 0.2 mbar for 72 h. Dried samples were neutralized using sodium hydroxide solution (NaOH, 0.1 N) for 30 min. Afterwards, to induce porosity, hydrogels were put into the o-xylene solution for 15 min under mechanical agitation at 200 rpm, using an orbital shaker incubator at 37 °C (Luzeren, Hangzhou, China). The CS/Gel/PVA hydrogels were washed with ethyl alcohol from 100 to 0% v/v to remove residual xylene. Finally, hydrogels were sterilized in phosphate-buffered saline (PBS) using a steam autoclave SE 510 (Yamato Scientific, Tokyo, Japan) at 121 °C and 103.4 KPa for 15 min [21].

2.2. Characterization of Chitosan/Gelatin/Poly(Vinyl Alcohol) Hydrogels

2.2.1. Morphological Analysis

Scanning electron microscopy (SEM) analysis was performed to confirm the morphological properties of the steam-sterilized CS/Gel/PVA hydrogels. Freeze-dried hydrogels were mounted on carbon stubs with double-sided adhesive tape without coating. In addition, energy-dispersive X-ray spectrometer analysis (EDS, JSM-7100f, JEOL, Tokyo, Japan) was performed, to determine the elemental component of the hydrogel.

2.2.2. Fourier-Transform Infrared Spectroscopy

Fourier-transform infrared–attenuated total reflectance (FTIR–ATR) was used to identify the structural properties of the CS/Gel/PVA hydrogels. Freeze-dried hydrogels were lyophilized, ground with a grinder, and placed on the ATR surface of the Spectrum GX system (Perkin Elmer, Branford, CT, USA). Spectra were obtained with a resolution of 1 cm^{-1} over 24 scans, ranging from 4000 to 650 cm^{-1}.

2.2.3. Amino Acid Profile

The amino acid profiles of the type B gelatin from bovine skin and the CS/Gel/PVA hydrogels were determined by ultra-performance liquid chromatography (UPLC). Ten milligrams of the sample were hydrolyzed for 24 h with 0.25 N HCl at 110 °C. The hydrolysate was filtered (Whatman No. 2) and dried at 60 °C in a rotary evaporator. The dry hydrolysate was resuspended in 1 mL 0.1 N HCl, and the solution was filtered through a 0.22 µm membrane (AOAC, 2005; Method 982.30). The standards and hydrolysate (10 µL) were derivatized with AccQ-Fluor borate buffer (70 µL) and AccQ-Fluor reagent (20 µL) from Waters. The derivatized was separated and quantified by UPLC Acquity class H with an Acquity diode array detector (Waters Corporation, Milford, MA, USA) and AccQ-Tag ultra C18 column (2.1 x 100 mm). A particle size of 1.7 µm was used in this study. The separation was performed using the following two solvents: (A) AccQ-Tag ultra A/water in proportion 5:95 (v/v) and (B) AccQ-Tag ultra B (Waters®). The gradient elution used was as follows: holding at 0.1% B for 0 to 0.54 min, then 0.1% to 9.1% (B) from 0.54 to 5.74 min, 9.1% to 21.2% (B) from 5.74 to 7.74 min and 21.2% to 59.6% from 7.74 to 8.04. The injection volume was 10 µL, the column temperature was 55 °C and the analysis time was 10 min. Detection was carried out at 260 nm. Data acquisition was performed with Empower 3 system software (WatersCorporation, Milford, MA, USA). Amino acid quantification was carried out using external standards for each amino acid (WatersCorporation, Milford, MA, USA) [40].

2.3. Culture of Chondrocytes in Chitosan/Gelatin/Poly(Vinyl Alcohol) Hydrogels

2.3.1. Isolation and Culture of Human Elastic Chondrocytes

Elastic cartilage remnants were donated from 7- to 10-year-old patients with unilateral congenital microtia who underwent autologous costal cartilage reconstruction. Briefly, auricular remnants were transported in 10% antibiotic–antimycotic/DMEM-F12 media (Gibco®, U.S., New York, NY, USA) under sterile conditions. Tissue remnants were minced and enzymatically predigested with 0.25% trypsin–EDTA (Gibco®, U.S., New York, NY, USA) for 30 min. Afterward, digestion with type II collagenase (3 mg/mL, Worthington Biochemical Corporation), was performed for 5 h at 37 °C under constant agitation. Cell viability was determined using trypan blue. Chondrocytes were seeded onto CS/Gel/PVA hydrogels at a cell density of 2×10^4 cells/cm^2 and kept in DMEM-F12/10% FBS/1% antibiotic–antimycotic supplemented media (Gibco®, U.S., New York, NY, USA) until confluence. At this point, the cell-laden hydrogels were called constructs. Chondrocyte isolation and culture were conducted under standard conditions at 37 °C, 5% CO_2, and 95% relative humidity.

2.3.2. Cytotoxicity Assay

In vitro chondrocyte cytocompatibility with the steam-sterilized CS/Gel/PVA hydrogels was evaluated using supernatant dilutions (extracts), according to ISO 10993-5 [41]. Testing was performed using the (3-(4,5-dimethyl thiazolyl-2) 2,5-diphenyltetrazolium bromide (MTT, Thermo Fisher Scientific, Hillsboro, OR, USA) assay. Sterile hydrogels (~500 mg) were statically incubated in 5 mL of supplemented media (DMEM-F12/10% FBS/1% antibiotic–antimycotic) at 37 °C for 72 h. After this, extracts were diluted in supplemented media to obtain each working concentration tested: 100, 50, 25, 12.5, and 0% v/v.

Auricular chondrocytes were seeded in 96-well plates at a density of 5×10^3 cells per well and cultured for 24 h. Afterwards, cell culture media in each well were replaced with an equal volume of the different hydrogel extract dilutions and incubated for 24 and 72 h in a 5% CO_2 humidified atmosphere. Chondrocytes exposed to 1.0% v/v sodium dodecyl sulfate (SDS, Sigma-Aldrich) were used as positive cytotoxic controls for the same time points. At every time point, extract dilutions were removed from the wells, replaced with 100 µL of MTT solution (0.5 mg/mL) and incubated at 37 °C for 4 h. Formazan crystals were dissolved with isopropanol/dimethyl sulfoxide (DMSO) solution and transferred to a new 96-well plate. Absorbance was measured using a Multiskan GO spectrophotometer (Thermo Scientific, Pittsburgh, PA, USA) at 570 nm, and cell viability was calculated with the following formula:

$$\% \ cell \ viability = \frac{Is}{Ic} \times 100$$

where *Is* corresponds to the absorbance of the cells exposed to the extract, and *Ic* is the absorbance of the non-exposed cells.

2.3.3. Cell Viability Assay

Chondrocyte viability within the CS/Gel/PVA hydrogel constructs was evaluated with a modified protocol from the LIVE/DEAD® viability/cytotoxicity kit (Molecular Probes™ Invitrogen, Carlsbad, CA, USA) [42] after 7 and 14 days. Briefly, 5×10^5 human elastic chondrocytes (third passage) were resuspended in supplemented media (80 µL), injected into 8 mm diameter hydrogel discs using an insulin syringe (U-100 Becton Dickinson, Mexico City, Mexico) and incubated for 1 h at 37 °C and 5% CO_2. After that, constructs were entirely covered with supplemented media (Gibco®, U.S., New York, NY, USA) with periodic media changes every 2–3 days for 7 and 14 days. Assays were performed in triplicate at each time point. Constructs were washed three times with PBS and incubated in Hanks balanced salt solution (HBSS) containing 4 mM calcein-AM and 1 mM ethidium homodimer (EthD-1) for 45 min at 37 °C and 5% CO_2. Then, constructs were examined with an LSM 780 confocal fluorescence microscope (Carl Zeiss, Oberkochen, Baden-Württemberg, Germany). Cell viability was determined using ImageJ software (version 1.52a, Wayne Rasband, National Institutes of Health, Bethesda, MD, USA).

2.3.4. Cell Attachment

Chondrocyte attachment onto CS/Gel/PVA hydrogels was visualized using scanning electron microscopy (SEM) (16538, Electron Microscopy Science, Hatfield, PA, USA). Constructs (5×10^5 cells) were fixed with 2.5% v/v glutaraldehyde (Sigma-Aldrich) in 0.1 M sodium cacodylate (Sigma-Aldrich) at pH 7.2 for 24 h. Samples were washed for 5 min in 0.1 M PBS at pH 7.4 and further dehydrated for 10 min in a serially diluted ethanol solution, ranging from 30% to 99.99%, for 30 min each wash (E7148, Sigma-Aldrich, Merck, Darmstadt, Germany). All specimens were dried with a critical point dryer CO_2 chamber (K850, Quorum Technologies, Kent, UK). Images were acquired with a FIB-SEM Crossbeam 550 field emission microscope (Carl Zeiss, Oberkochen, Baden-Württemberg, Germany) at 5 kV.

2.3.5. Immunofluorescence

After 7 and 14 days in culture, constructs (5×10^5 cells) were fixed with 2% v/v paraformaldehyde (Sigma-Aldrich) for 20 min and included in Tissue-Tek®. Membrane permeabilization was carried out by incubating constructs in 0.1% Triton X-100 in PBS for 1 h at room temperature. Then, blocking was undertaken with 1% bovine serum albumin/PBS for 2 h. Constructs were incubated overnight with primary antibodies against type II collagen (ab34712, 1:100), Sox9 (ab3697, 1:100), aggrecan (ab36861, 1:100), elastin in monolayer (ab21610 1:100) and elastin in constructs (ab9519, 1:100), type I collagen (ab6308, 1:100) and Runx2 (ab76956, 1:100) at 4 °C. The secondary antibodies, anti-Mouse IgG (H+L) (1:300), FITC conjugated (Jackson Immunoresearch, #115–095–003) and goat anti-rabbit, Alexa Fluor 488 (Ab150073) or Alexa Fluor 594 (Ab150120), were incubated for 2 h at 37 °C. Finally, nuclear counterstaining was undertaken with 4′,6-diamidino-2-phenylindole (DAPI). Negative controls were not labeled with primary antibodies. Slides were mounted with fluoroshield™ (Cat: F6937-2ML), and images were taken with an Axio Observer inverted microscope A1 (Carl Zeiss, Oberkochen, Baden-Württemberg, Germany). Autofluorescence was reduced to allow for positive signal identification via confocal microscopy LSM 780 (Carl Zeiss).

Statistical analysis

Data were presented as three independent experiments' mean ± standard deviations (SD) ($n = 3$). Significant differences were considered to be a p-value of < 0.05. A two-way ANOVA, followed by Tukey's test, a multiple comparisons test and a Bonferroni's comparison test, were performed using GraphPad Prism version 9.4.1 for macOS and GraphPad software (San Diego, CA, USA).

3. Results and Discussion

This study aimed to examine the suitability of sterilized CS/Gel/PVA hydrogels as scaffolds for culturing human auricular chondrocytes from microtia patients, in an environment acting as a scaffold, while maintaining the elastic chondrocyte phenotype for auricular tissue engineering applications.

CS/Gel/PVA-based hydrogels have proven to be a versatile platform for supporting various cell types for applications in tissue engineering and wound healing due to their natural bioactivity, biocompatibility, biodegradability, cellular adaptability and cell behavior [20,21,23,43].

Our results show the importance of generating a support biomaterial under controlled conditions with topological and architectural features to improve a 3D micro ambient that favors cell adhesion and interaction for supporting cell growth, for cartilage tissue engineering strategies.

3.1. Chitosan/Gelatin/Poly(Vinyl Alcohol) Hydrogel Preparation

The hydrogels were characterized by their macrostructure, chemical interactions, amino acid profile and ability to promote the growth of auricular chondrocytes in a 3D microenvironment. The CS/Gel/PVA ternary blend was used as part of the formulation hydrogel due to its biocompatibility [20,21,23,43–45] and biodegradability [23,46]. The process followed for preparing hydrogels is outlined in Figure 1. The samples were physically crosslinked by the freeze–thaw/lyophilization process and immersed in NaOH to promote the physical gelation of the deprotonated amino groups of the CS (pKa = 6.3–7.0) [21,38]. The o-xylene induced porosity in the hydrogels, due to its performance as a porogenic template [21,47].

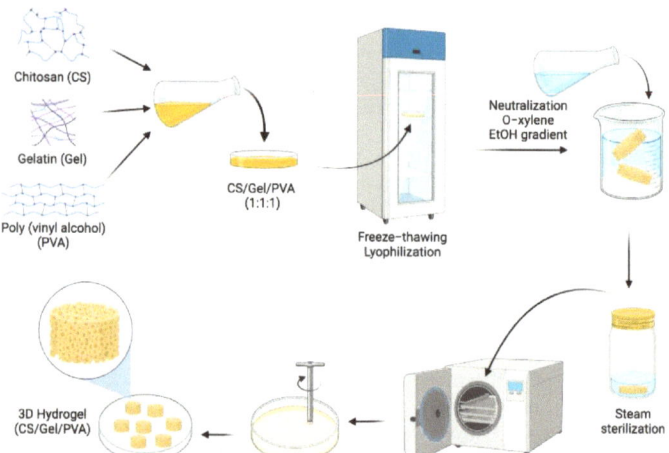

Figure 1. Schematic process of preparing the CS/Gel/PVA hydrogel.

This manufacturing process is feasible, robust, and easily reproducible, providing a reliable method to produce hydrogels with customized shapes and maintaining a pattern in the pore formation [21,46], which is particularly important for cell seeding [47]. The shape of hydrogels for auricular cartilage repair should have sufficient elasticity, flexibility, mechanical strength, and physical stability to support auricular chondrocyte growth [27,48,49].

3.2. Characterization of Chitosan/Gelatin/Poly(Vinyl Alcohol) Hydrogels

3.2.1. Morphological Analysis

The morphological characteristics of the hydrogels were observed. The SEM images revealed that the hydrogels possess a tridimensional (3D) structure with several irregular macroporous, open and closed channels with heterogeneous pore sizes and diameters ranging from 50 to 1500 µm (Figure 2a,b). More elongated and larger pores are present in some cross-sections. These porous structures and the heterogeneity in the size dimensions permit the adequate transport of nutrients, gases and molecules to ensure cell survival and cell functions inside the 3D hydrogel [21]. In our previous research, a different batch of hydrogels was prepared, following the same manufacturing process. Despite the heterogeneity in forming the 3D network and pore dimensions, the manufacturing process allowed control over the formation pattern of interconnected and porous channels, demonstrating the reproducibility of the manufacturing process [21].

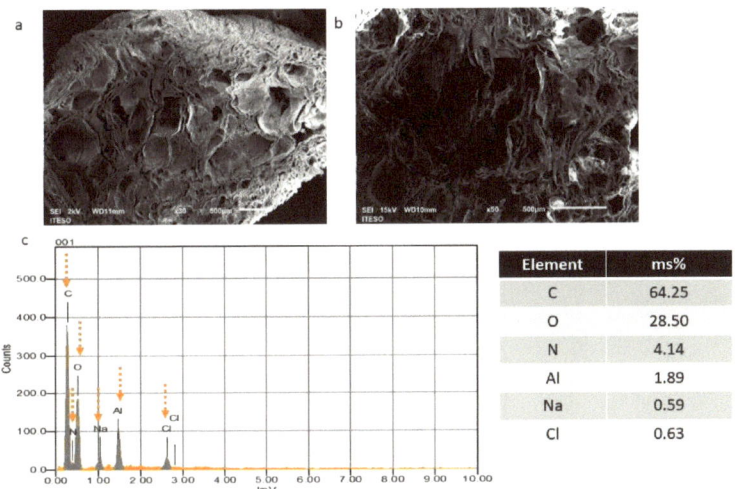

Figure 2. Morphological analysis of CS/Gel/PVA hydrogels. Scanning electron microscope (SEM) images showed heterogeneous channels with pore sizes ranging between 50 and 500 µm. Image magnification: 30× (**a**) and 50× (**b**). Scale bars are 500 µm. Energy-dispersive X-ray spectrometer (EDS) analysis (**c**). Dotted arrows indicate sharp peaks for each element.

Sánchez-Cardona et al. (2021) prepared a hydrogel similar to our research, in terms of their polymer blend proportions of CS/Gel/PVA (1:1:1 w/w) and polymer concentrations (2 wt %). The process preparation consisted of a freeze–thawing (nine cycles at −50 °C for 8 h and 25 °C for 8 h) with a subsequent lyophilization process without neutralization and xylene treatments. Under these conditions, the hydrogels showed homogeneity in the macroporous structure of the hydrogels, with small pore sizes (pore diameter sizes between 0.6 and 265 µm), but the preparation time increased significantly [23].

In addition, energy dispersive X-ray spectroscopy (EDX) analysis was conducted to analyze the elemental composition of the CS/Gel/PVA hydrogels. The EDX spectrum (Figure 2c) reveals sharp peaks of C (64.25%), O (28.50%) and N (4.14%), which can be attributed to the natural polymers (CS and Gel) and PVA (C and O) [50]. The peaks of Cl (0.63%) and Na (0.59%) are due to the PBS (mainly composed of sodium chloride) remaining in the hydrogels after the sterilization process [21]. The Al peak was detected, presumably due to the aluminum foil used to cover and store samples until SEM analysis.

3.2.2. Fourier-Transform Infrared Spectroscopy

The results of the FT-IR spectroscopy were analyzed to understand the chemical interactions between the polymers and to confirm the absence of any unfavorable interactions with the o-xylene treatment of the hydrogels. The spectra of the pristine polymers (CS, Gel, and PVA) were used as controls and compared with the hydrogels (Figure 3). The FTIR spectra were described between 2000 and 600 cm^{-1} peak regions. The characteristic bands (peaks) of pristine polymers are shown in Table 1. The IR spectrum of pristine CS (Figure 3) showed peaks around 890–900 cm^{-1} (corresponding to the CH deformation of the β-glycosidic bond) and 1157 cm^{-1}, corresponding to the saccharide structure [13,14]. The peaks in the range from 1515 to 1570 cm^{-1} were related to the amide II. Peaks corresponding to amide I and II were observed at 1650 and 1322 cm^{-1}, respectively [13,14]. The peaks at 1595 and 1591 cm^{-1} indicated the deacetylation of the chitosan. The peaks in the region between the 1300 and 1200 cm^{-1} were related to the vibrations of NHCO. The peaks between 1000 and 1158 cm^{-1} were related to the vibrational modes of the C–O–C, C–OH and C–C bonds in the ring of the saccharide structure [13]. The skeletal vibrations of the pyranose ring related to NH and OH out of the plane were observed at 600 cm^{-1} [14,45].

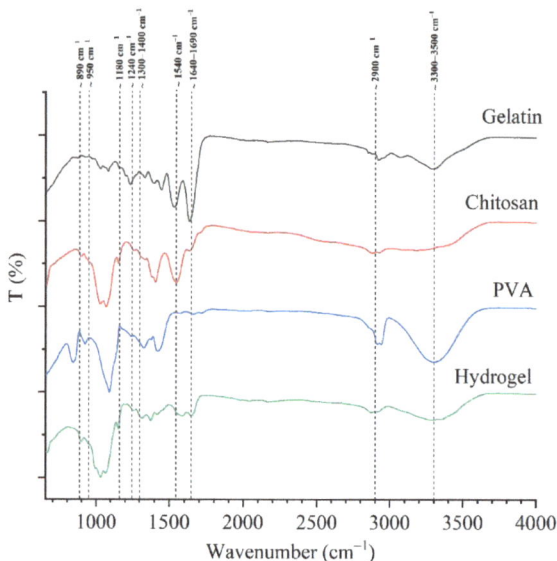

Figure 3. ATR–FTIR spectra of pristine Gel, CS and PVA and of dried CS/Gel/PVA hydrogel. The transmittance is shown on the left axis.

The pristine Gel IR spectrum (Figure 3) showed the following most representative peaks: 1625 cm^{-1} (C=O stretching/hydrogen bonding couple with COO), 1500 cm^{-1} (amide II: N–H bend coupled with C–N stretch), 1435 cm^{-1} (amide II: CH$_2$ bend), 1250 cm^{-1} (amide III: N–H bend), 1100 cm^{-1} (amide III: C–O stretch) and 1000–950 cm^{-1} (amide III: skeletal stretch) [51,52]. The IR spectrum of pristine PVA (Figure 3) showed a strong peak at 1085 cm^{-1} (C–O stretching and O–H bending, and the amorphous sequence of PVA), medium intensity at 1400 cm^{-1} (CH$_2$ bend), 1350 cm^{-1} (scissoring O–H, rocking with C–H, wagging), 915 cm^{-1} (CH$_2$ rocking) and 820 cm^{-1} (C–C stretching) [53].

The FTIR spectrum of the hydrogel exhibited the characteristic bands of the CS, Gel, and PVA. The dominant component of the hydrogel was identified mainly by the amide I (at 1630–1600 cm^{-1}), amide II (at 1540–1350 cm^{-1}) and amide III (at 1250–1300 cm^{-1}) bands present in CS and Gel. The peaks correlated to the saccharide structure (1200–890 cm^{-1}) of CS, and prominent peaks of the PVA correlated to the backbone of the PVA structure (1400–1300 cm^{-1}). The amorphous region (1100–1080 cm^{-1}), the CH$_2$ asymmetric stretching (920 cm^{-1}) and the skeletal vibration of PVA (830 cm^{-1}) were also present.

However, some changes were noticeable in the CS/Gel/PVA hydrogel spectrum, as detailed below:

(1) The characteristic peaks of amide I and II exhibited displacements due to hydrogen bonding interactions between the C=O groups from the gelatin and the OH groups in the PVA and CS. Sánchez Cardona et al. (2021) found similar results, indicating that the three polymers interact through hydrogen bonding between the amino and hydroxyl moieties [23]. According to Rodríguez-Rodríguez et al. (2019), steam sterilization induces the molecular interaction between the amino groups of the CS and Gel and between either primary hydroxyl groups or the carbonyl under high temperatures [21].

(2) The displacement band corresponding to the C=O (1650–1550 cm^{-1}) in CS and Gel suggested interactions between the functional groups of CS and the carboxyl groups in Gel, in agreement with Chang et al. (2009) [21]. Moreover, the peak at ~1550 cm^{-1} diminished in the hydrogel spectra, probably due to the formation of –NH$_3^+$ in the acidic environment and the transition to –NH$_2$ under alkaline conditions [14]. The samples were immersed in NaOH to favor physical gelation throughout the deprotonation of the amino groups of CS (pKa = 6.3–7.0) [21]. The peak at ~1650–1640 cm^{-1} was related to the C=N bond, characteristic of the Schiff's base structure, assuming that a covalent bond formation was maintaining the 3D structure of the hydrogel. Wang et al. (2004) demonstrated the formation of Schiff's base and NH$_{3+}$ in a CS–PVA hydrogel [14]. Corona-Escalera et al. (2022) suggested the formation of covalent bonds by an esterification reaction between free carboxylic and hydroxyl groups in Gel/PVA hydrogels crosslinked with transglutaminase enzyme [54].

(3) Alcohol gradient solutions were used to remove traces of o-xylene. The absence of the characteristic bands of the o-xylene at 743 cm^{-1} in the spectra highlighted the success of the o-xylene extraction process [55]. The effectiveness of xylene as a porogenic template has been demonstrated in CS [55] and CS/Gel/PVA hydrogels [21]. It is well-known that xylene dissolves alcohol during tissue processing in histology [56].

(4) The disappearance of the PVA characteristic peaks at 920 cm^{-1} and 830 cm^{-1} was noticeable, suggesting bonding interactions between CS/Gel and PVA molecules [57]. The vibration of aliphatic ether was associated with the stretching vibration, favoring a physical crosslinking of polymer chains with a hydroxyl group (such as CH$_2$–CHO–CH$_{2-}$) that interacted with other hydroxyl groups of PVA polymer chains, forming bindings such as C–O–C by removing small molecules, such as water molecules [57], due to the xylene/alcohol treatment, since the PVA showed an affinity with o-xylene [58].

Table 1. Main IR peak assignments for the chitosan (CS), gelatin (Gel), and poly(vinyl alcohol) (PVA).

Polymer	Peak Region (Wavenumber cm^{-1})	Assignment	Reference
Chitosan	~1655–1649	Amide I: C=O stretch	[14]
	~1650–1580	NH$_2$ deformation	[14]
	~1570–1510 (1554)	Amide II: N–H deformation and C–N stretching	[14]
	~1625–1500	NH$_3^+$ deformation	[14]
	1560, 1548	Symmetric deformation	
	~1420	O–H and C–H deformation (ring)	[14]
	~1383–1377	CH$_3$ deformation (bend)	[14]
	~1325–1322	Amide III: O–H and C–H deformation	[14]
	~1265–1260	C–H wag (ring)	[14]
	~1200–950	C–O–C and C–O stretching	[13]
	~1160–1150	Primary or secondary alcohol	[14]
	~1080	C–O stretching	[14]
	~900	Aliphatic aldehydes/	[14]
	860–650	CH deformation of the β-glycosidic bond	[13]
Type B gelatin from bovine skin	~1650–1630	Amide I: C=O stretching and N–H stretching /hydrogen bonding coupled with C00-	[51,52]
	~1632	Intermolecular associations with imide residues	[59]
	~1560–1540	Amide II: N–H bend coupled with C–N stretch	[51]
	~1460–1430	Amide II: CH$_2$ bend	[51]
	~1250–1180	Amide III: N–H bend	[51]
	~1130–1022	Amide III: C–O stretch	[51]
	~1200–950	Amide III: skeletal stretch	[51]
	~890–650	Amide III: skeletal stretch	[51]
PVA	~1750–1735	C=O vibration	[53]
	~1640	Water absorption	[60]
	~1460–1400	CH$_2$ bend	[53]
	~1350–1300	Scissoring O–H, rocking with C–H, wagging	[53]
	~1140–1080	C–O (crystallinity)	[53]
		C–O stretching of acetyl group present in the PVA backbone and O–H bending/ the amorphous sequence of PVA	[53]
	~920	CH$_2$ rocking	[53]
	~820	C–C stretching/skeletal vibration	[53]

3.2.3. Amino Acid Profile

The amino acid content of pristine type B gelatin from bovine skin (Sigma-Aldrich) and CS/Gel/PVA hydrogel samples are given in Table 2. The amounts of different amino acids were expressed in grams of amino acid per 100 mg of gelatin sample (g/100 g) and hydrogel. Gelatin and hydrogel samples did not exhibit tryptophan or cysteine because they are typically absent in type I collagen. Gelatin is partially denatured collagen, so the gelatin amino acid composition remains very similar to collagen [61].

Table 2. Amino acid composition of commercial type B gelatin from bovine skin (Sigma-Aldrich) and CS/Gel/PVA hydrogel.

Amino Acid	Pristine Type B Gelatin from Bovine Skin% (g/100 g)	CS/Gel/PVA Hydrogel % (g/100 g)
Histidine (His)	10.38	0.45
Serine (Ser)	1.13	0.08
Arginine (Arg)	0.29	0.01
Glycine (Gly)	14.40	0.73
Aspartic acid (Asp)	ND	ND
Glutamic acid (Glx)	0.98	0.03
Threonine (Thr)	4.59	0.33
Alanine (Ala)	9.07	0.39
Proline (Pro)	16.24	0.82
Lysine (Lys)	0.01	0.01
Tyrosine (Tyr)	ND	ND
Valine (Val)	0.63	0.08
Isoleucine (Ile)	3.09	0.20
Leucine (Leu)	1.54	0.01
Phenylalanine (Phe)	0.01	0.02

Note: ND: not detected.

Collagen contains more glycine than most amino acids but does not contain cysteine or tryptophan (except for type III collagen) [62]. However, the dominant amino acid in pristine bovine gelatin was Proline (Pro, 16.24%), followed by Glycine (Gly, 14.40%) and Histidine (His, 10.38%), which are characteristic of all gelatins [63]. Pro is the amino acid with the highest content during hydrogel preparation. Pro is an amino acid responsible for the stability of the triple helix in the collagen structure [63,64]. The contents of Alanine (Ala, 9.07%), Threonine (Thr, 4.59%), and Isoleucine (Ile 3.09%) were higher than the other amino acids present in the pristine bovine gelatin. Aykin-Dinçer et al. (2017) reported the amino acid profile of different commercial bovine gelatins, with differences in the amino acid profile and contents [63], which may be associated with the sources of skin and the gelatin manufacturing processes [64].

It is notorious that the amino acid content was lower in the hydrogels than in pristine gelatin (Pro 0.82%), Gly (0.73%), His (0.45%), Ala (0.39%), Thr (0.33%) and Ile (0.20%), suggesting chemical interactions between gelatin and the components of the hydrogel during the preparation process. The functional groups of gelatin peptides are involved in H-bonding and NH-bonding formation between the water molecules and the components of the hydrogel (CS and PVA), as corroborated by the FTIR analyses [23,45].

3.3. Culture of Chondrocytes in Chitosan/Gelatin/Poly(Vinyl Alcohol) Hydrogel

The performance of chondrocytes embedded in hydrogels was evaluated by measuring cell cytotoxicity, viability and attachment, and the formation of ECM proteins.

3.3.1. Cytotoxicity Assay

Cytotoxicity is one of the most important properties of biomaterials for tissue engineering applications. To evaluate the cytocompatibility of the steam-sterilized hydrogels, isolated auricular chondrocytes were cultured on the different extracts after 24 and 72 h of incubation, and cell viability was measured using the MTT assay. Figure 4 presents the cell viability in the presence of CS/Gel/PVA hydrogel extracts. All samples showed a cell viability greater than 70%, regardless of the extract concentration. According to ISO10993-5

(ISO10993-5, 1999), the extract medium obtained from the sterilized hydrogels does not contain toxic substances to auricular chondrocyte cells, indicating good cytocompatibility for subsequent assays [21,23,41,44].

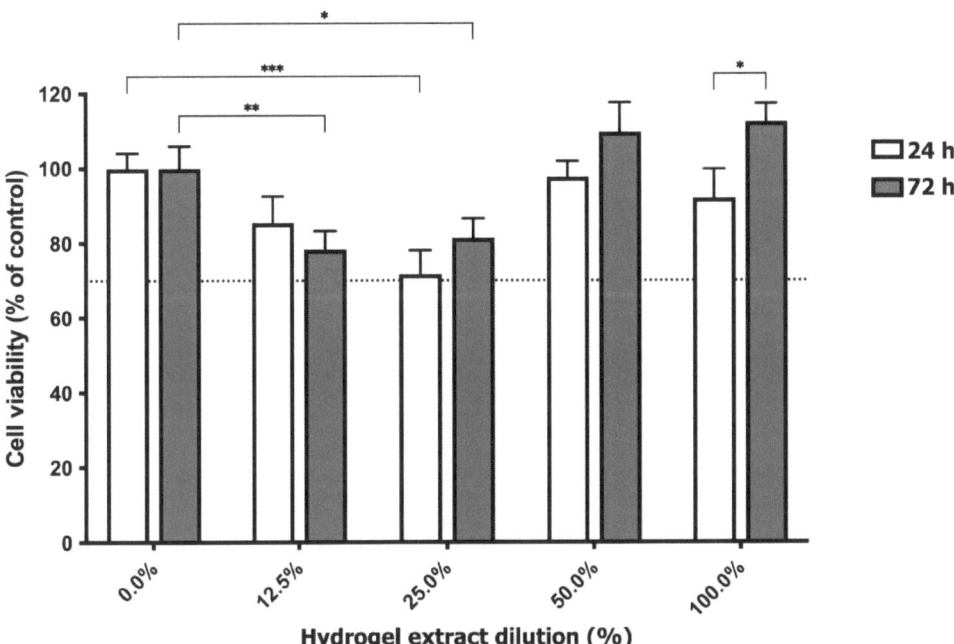

Figure 4. Cell viability of human auricular chondrocytes exposed to CS/Gel/PVA hydrogel extract dilutions. MTT assay was performed to evaluate cellular cytotoxicity 7 and 24 h post-exposure to extract dilutions. Notably, only the treatments marked with asterisks exhibited statistically significant differences from the control group. Therefore, 50 and 100% extract dilutions did not demonstrate statistically significant effects. Data are presented as the mean ± SD of l% of cell viability (n = 3) of three independent experiments. Statistical significances for each condition vs. non-exposed cells, or between conditions, are indicated by * $p < 0.05$, ** $p < 0.01$, and *** $p < 0.001$ (two-way ANOVA and Bonferroni's multiple comparisons test).

However, lower extract concentrations (12.5% and 25%) exhibited a statistically significant decrease in chondrocyte viability compared to the control group (non-exposed cells) at 72 and 24 h incubation, respectively. Conversely, no significant differences were observed at either 24 or 72 h for the highest tested concentrations (50.0% and 100%), compared to non-exposed cells. Interestingly, chondrocytes cultured in 50% and 100% extract concentrations demonstrated the highest viability after 72 h, with a statistically significant increase observed in the 100% extract group between 24 and 72 h. This could be explained by the presence of a higher amount of Gel dissolved in the culture medium than in the other extract dilutions, promoting cell growth and proliferation due to similarities in structure and function with collagen, which is an excellent substrate for cell attachment, proliferation and differentiation [23,65,66].

Previous studies have documented the cytocompatibility of sterilized CS/Gel/PVA hydrogels. Rodríguez-Rodríguez et al. (2020) reported no observable changes in the viability of HT29 cells cultured on extracts prepared via freeze–thawing and autoclaving (the same preparation method but a different batch than those employed in the present study) [21]. Pérez-Díaz et al. (2023) confirmed the biocompatibility of autoclaved 2D CS/Gel/PVA hydrogels by observing the favorable viability of human adipose tissue-derived mesenchymal stromal cells [20]. Massarelli et al. (2021) evaluated the cytocompatibility of NIH/3T3 fibroblasts on autoclaved CS/PVA hydrogels for potential application as a wound dressing [67]. Lastly, Sánchez-Cardona et al. (2021) reported a cell viability greater than 50% of rat pancreatic beta cells (BRIN-BD11) in the presence of CS/Gel/PVA hydrogels that were prepared in a similar way to those used in this study but without NaOH/Xylene/alcohol washes and autoclave sterilization [23].

3.3.2. Cell Viability Assay

Cell viability was investigated by live/dead staining after chondrocytes were encapsulated in the CS/Gel/PVA hydrogels and cultured for 7 and 14 days (Figure 5).

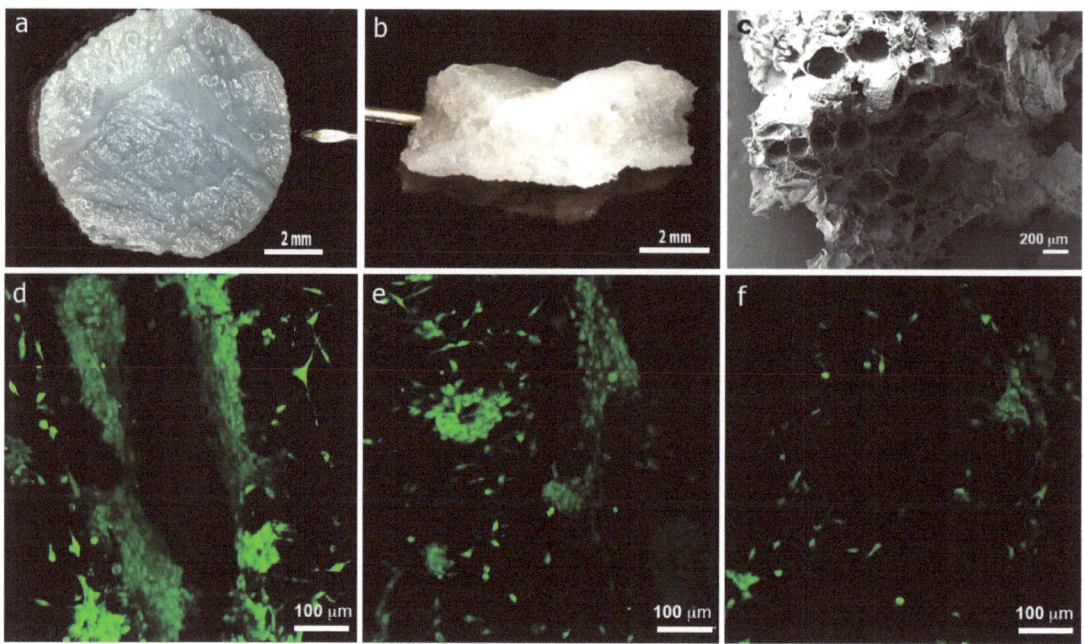

Figure 5. Cell-laden CS/Gel/PVA based hydrogel. Auricular chondrocytes (5×10^5 cells) were injected into eight mm diameter CS/Gel/PVA hydrogel discs (**a**) and a Live/Dead viability test was performed and visualized via confocal microscopy. Hydrogels side view (**b**). SEM micrograph shows porous microstructure (**c**). Viable auricular chondrocytes were visualized near the injection site (**d**) and diffused all through the hydrogel network (**e**,**f**). Scale bars: 2 mm (**a**,**b**), 200 μm (**c**) and 100 μm (**d**–**f**).

The strategy followed for chondrocyte encapsulation was injecting the cell suspension into the hydrogel. It was crucial to know if the structure in the wet state continued to maintain its porous interconnected structure and pore sizes in this condition. The macroscopic appearances of the sterilized hydrogels in a wet state are shown. The hydrogels showed a sponge-like structure with smooth surface topography. The hydrogels showed optimal pore size ranges (greater than 200 μm) and an irregular macrostructure that favors the exchange of nutrients and gases and, therefore, the growth and functionality of chondrocytes (Figure 5a–c).

Auricular chondrocytes embedded in the CS/Gel/PVA construct were identified, confirming that the injected cell suspension favored cell diffusion and distribution into the hydrogel. In all groups, most of the chondrocytes were alive (green signal), and dead cells were observed (red signal) after 7 and 14 days of culture (Figure 5d–f, respectively). The results indicated that the cells in the hydrogels had high cell viability, suggesting that using the CS/Gel/PVA hydrogels for cell encapsulation may be a good strategy. An increased number of viable chondrocytes were found near the injection site, while chondrocyte density diminished in areas far from the injection site (Figure 5d–f).

Confocal images revealed cell proliferation in cell-laden hydrogels with two notable growth patterns: individual cells and aggregated cells (clusters). Similar cell behavior was observed when auricular chondrocytes were encapsulated in fibrin–agarose hydrogels mimicking a 3D environment [68]. For cells that showed individual growth, polygonal morphology was also observed, and the viability was 98.8% by day seven and 98.69% after 14 days of culture (Figure 6a,b). In contrast, clusters showed 99.21% viability by day seven and 98.96% after 14 days of culture, corroborated by the live/dead staining. However, viability between different conditions was not statistically different. Furthermore, the hydrogel stiffness and smooth surface did not influence cell viability. Rodriguez-Rodriguez et al. (2020) demonstrated that sterilized CS/Gel/PVA hydrogels showed higher elastic modulus values than a viscous modulus, indicating that the hydrogels are soft and have predominantly solid viscoelastic behavior [21].

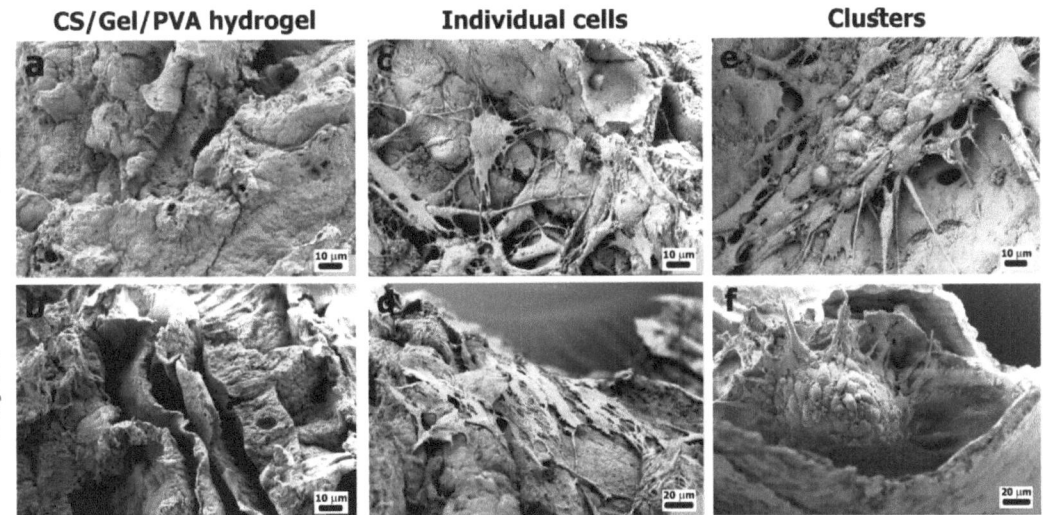

Figure 6. Cell attachment onto CS/Gel/PVA hydrogels. Auricular chondrocytes were cultured onto CS/Gel/PVA hydrogels in standard culture conditions. SEM microscopy was used to visualize cell attachment. Hydrogel without cells after 7 (**a**) and 14 days' (**b**) culture. Individual (**c,d**) and clustered cells growth (**e,f**) after seven and 14 days, respectively. Scale bars 10 μm (**a–c,e**) and 20 μm (**d,f**).

Auricular cartilage comprises chondrocytes and a highly hydrated extracellular matrix, characterized by chondrocytes immersed in an ECM rich in elastic and type II collagen fibers, as well as proteoglycans [69]. The 3D organization of ECM molecules confers elasticity, flexibility, tensile resistance and compression to the elastic cartilage [70]. Several biomaterial scaffold fabrication techniques have been established to process different microstructures with controlled characteristics that mimic the chondrocytes' ECM [71,72]. Characteristics such as porosity, pore size and interconnectivity must be adequate for cells to carry out their primary functions and transport and exchange nutrients to chondrocytes [37]. Lien et al. (2019) demonstrated that a pore size between 250 and 500 μm was optimal for articular chondrocytes to proliferate, to form ECMs and to maintain their phenotype using scaffolds [73].

3.3.3. Cell Attachment

The cell attachment of auricular chondrocytes was evaluated by SEM (Figure 6). The micrographs show CS/Gel/PVA hydrogels without cells in culture conditions after 7 and

14 days (Figure 6a,b). Cell attachment and ECM deposition in cell-laden CS/Gel/PVA-based hydrogels also confirmed the presence of individual cells and clusters observed in Live/Dead staining. Cells remained attached throughout the hydrogel surface. Remarkably, after 14 days of culture, chondrocytes showed more cell growth and ECM deposition than after 7 days. Individual cells showed prolonged filopodia in contact with nearby cells (Figure 6c,d). Clusters possessed a round, nodule-like shape with lamellipodia and filopodia projected towards the periphery of the conglomerates (Figure 6d,e). Results reported by Otto et al. (2018) demonstrated the in vitro cartilage-forming capacity of primary auricular chondrocytes seeded onto gelatin methacryloyl (gelMA)-based hydrogels after 56 days of culture. Chondrocytes displayed a more cluster-like organization of ECM components in a 3D microenvironment, with cartilage-specific matrix deposition and increased ECM production [38].

Initially, the manual cell seeding technique at a high density resulted in the formation of a dense layer condensed in the area where the cell suspension was first seeded. However, later, we determined that cell injection, combined with a 3D interconnected network and the macroporous structure of the hydrogel, provided a better environment for the growth and functionality of chondrocytes [39]. It was found that the surface chemistry provided by the composition of the CS/Gel/PVA polymer, macroporous structure, and stiffness, was favorable for the culture of these particular cells. Also, it is well known that clustered chondrocytes promote differentiation into elastic chondrocytes. Clusters are aggregates where chondrogenesis and osteogenesis take place, in order to form cartilage and bones during embryonic development or repair/regeneration [74,75].

3.3.4. Elastic Cartilage Extracellular Matrix Formation

The expression of elastic cartilage ECM proteins by chondrocytes on monolayer and onto CS/Gel/PVA hydrogels was investigated using immunofluorescence (Figure 7). After in vitro expansion, auricular chondrocytes cultured in monolayer expressed aggrecan (ACAN) a widely distributed proteoglycan in articular hyaline cartilage, elastic cartilage and fibrocartilage (Figure 7a,b), [76]. The expression of elastin (Figure 7e,f), the most abundant fibrous protein in elastic cartilage ECM [48], was also detected. The expression of Sox9, the intrinsic transcription factor related to determining and maintaining chondrogenic lineage, was confirmed. After 14 days in culture, the remaining expression of Sox9 was detected in the nuclei of chondrocytes (Figure 7j); however, COL1 expression was absent at seven days (Figure 7m), and slightly present at 14 days in monolayer culture (Figure 7i,j). However, the expression of type II collagen (COL2), the main collagen in elastic cartilage, was detected by day 7 and significantly increased by day 14 (Figure 7n) Ref [77]. Also, the slight expression of the transcription factor RUNX2 was detected at 7 days (Figure 7m). COL1 and RUNX2 are markers of osteoblast differentiation and cartilage maturation, respectively. Meanwhile, the RUNX2 expression stimulates lineage differentiation toward osteoblasts and contributes to chondrocyte development by regulating maturation into hypertrophic chondrocytes [78]. The expression of these markers demonstrated that auricular chondrocytes from patients with auricular microtia did not lose their phenotype and characteristics as elastic chondrocytes in monolayer culture.

The expression of ECM proteins in 3D CS/Gel/PVA hydrogels showed differences compared to monolayer cultures. Chondrocytes that were densely arranged/distributed around cartilaginous clusters expressed higher levels of ACAN and elastin after 7 and 14 days of culture (Figure 7c,d,g,h). This confirmed the expression of elastic cartilage-specific ECM molecules in the 3D microenvironment. Although the expression of Sox9 was evident by day 7, it decreased by day 14 (Figure 7k,l). Noticeably, COL2 expression increased from day 7 to day 14 (Figure 7o,p). Moreover, Runx2 expression was not detected on CS/Gel/PVA hydrogels, while COL1 had a slight but not relevant expression from day 7 to day 14 (Figure 7k,l).

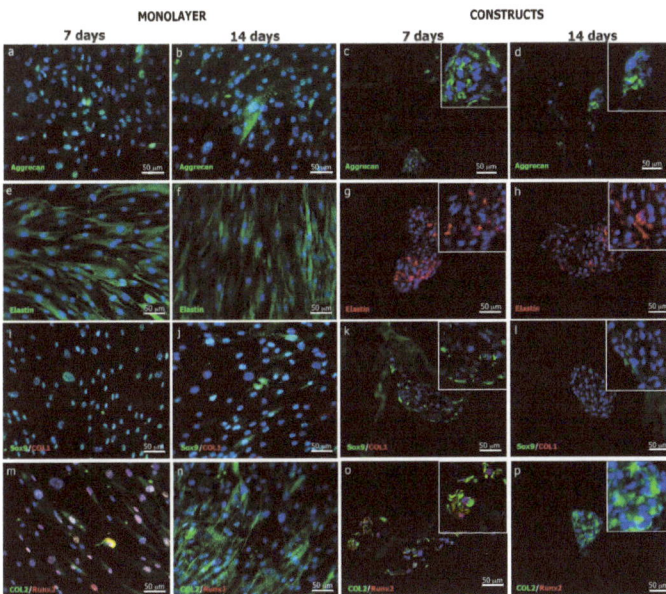

Figure 7. Immunostaining of cartilage markers. Third passage elastic cartilage chondrocytes were cultured in monolayers and onto 3D CS/Gel/PVA hydrogels (constructs) for 7 and 14 days (**d**), respectively. Aggrecan (ACAN) expression was consistent in monolayer (**a**,**b**) and in constructs (**c**,**d**) after 7 and 14 d. Elastin was also expressed in monolayer cells ((**e**,**f**), green signal) and constructs ((**g**,**h**), red signal) at both time frames. Expression of chondrogenic factor SOX9 was detected after 7 and 14 d in monolayer (**i**,**j**) and 7 d constructs (**k**); however, expression diminished in 14 d constructs (**l**). For the osteogenic marker, type I collagen (COL1) expression in monolayer was absent at 7 d (**i**) with a slight increase by 14 d (**j**); however, expression was not detected during 7 or 14 d culture onto 3D CS/Gel/PVA hydrogels (**k**,**l**). Positive expression of type II collagen (COL2), a cartilage ECM marker, was detected in chondrocytes on both monolayer (**m**,**n**) and constructs (**o**,**p**) after 7 and 14 d. Osteogenic factor Runx2 was expressed during the initial stages of monolayer (**m**) or construct (**o**) culture; however, expression was not detected after 14 d culture onto 3D CS/Gel/PVA hydrogels (**n**,**p**). The expression of chondrogenic and osteogenic markers was analyzed using LSM 880 Zeiss confocal microscope. Scale bars are 50 µm.

To our understanding, the preservation of a 3D microenvironment is key to promoting and maintaining chondrogenic phenotypes and the expression of cartilage markers such as elastin, aggrecan and type II collagen [29,30,38,68,79]. The expression of these markers demonstrated that auricular chondrocytes from microtia patients did not lose their phenotype, either in monolayer cultures or after being cultured onto CS/Gel/PVA-based hydrogels. Therefore, these hydrogels can be used as 3D cell supports for potential auricular cartilage regeneration.

4. Conclusions

The CS/Gel/PVA hydrogel showed promising potential for cartilage tissue regeneration, providing the basis for microtia reconstruction based on the culture of auricular chondrocytes from microtia in a 3D environment. The manufacturing process of the CS/Gel/PVA hydrogels is feasible, robust and easily reproducible, providing a reliable method to produce hydrogels with customized shapes and maintaining a pattern in the pore formation, providing a structurally stable, 3D, macroporous and highly interconnected network. The CS/Gel/PVA hydrogels can maintain the phenotype of auricular chondrocytes in culture, avoiding the process of cellular dedifferentiation in fibrocartilage cells usually found in monolayer cultures. Furthermore, long-term cultures of the chondrocyte hydrogel construct revealed a favorable environment that maintains the intrinsic cellular characteristics of elastic cartilage by maintaining its chondral phenotype, cellular cluster formation and elastic cartilage ECM production. These results are an important step toward building bioengineered auricular cartilage tissue.

Author Contributions: Conceptualization: C.O.-S., Y.M.-R., C.V., V.M.-L. and Z.Y.G.-C.; biological material donation and surgical procedures: C.G.-G.; Gel/CS/PVA hydrogel preparation and analysis: R.R.-R. (during his Ph.D.), G.L.-B. (ATR–FTIR) and Z.Y.G.-C.; cell culture and biological analysis: C.O.-S., Y.M.-R., J.A.J.-Á. (during his Ph.D.) and V.M.-L.; cell attachment/SEM analyses: D.M.G.-G.; HPLC assay: J.R.-C.; writing—original draft preparation: C.O.-S., Y.M.-R., C.V., V.M.-L. and Z.Y.G.-C.; writing—review and editing, C.O.-S., Y.M.-R., C.V., V.M.-L. and Z.Y.G.-C.; project administration, Z.Y.G.-C.; funding acquisition: C.V. and Z.Y.G.-C. All authors have read and agreed to the published version of the manuscript.

Funding: This research was supported by the Fondo Sectorial de Investigación en Salud y Seguridad Social, SSA/IMSS/ISSSTE-CONACYT under grant 234073. INRLGII Internal project (85/19). CONACYT-FORDECYT-PRONACES 490754/2020 and SECTEI 183/2019.

Institutional Review Board Statement: The study was conducted under the Declaration of Helsinki and approved by the Institutional Review Board (or Ethics Committee) of Instituto Nacional de Rehabilitación Luis Guillermo Ibarra Ibarra (INR 85/19).

Data Availability Statement: The data presented in this study are available on request from the corresponding author. The data are not publicly available due to intellectual property reasons.

Acknowledgments: The authors are grateful to CONAHCYT for the Ph.D. scholarships of Armando Jiménez Avalos and Rogelio Rodríguez-Rodríguez. The authors are grateful to Reina Araceli Mauricio Sánchez (ATR–FTIR spectroscopy) and Lenin Tamay de Dios (confocal microscopy).

Conflicts of Interest: The authors declare no conflicts of interest. The funders had no role in the design of the study, in the collection, analyses or interpretation of data, in the writing of the manuscript, or in the decision to publish the results.

References

1. Caliari, S.R.; Burdick, J.A. A practical guide to hydrogels for cell culture. *Nat. Methods* **2016**, *13*, 405–414. [CrossRef]
2. Sánchez-Cid, P.; Jiménez-Rosado, M.; Romero, A.; Pérez-Puyana, V. Novel Trends in Hydrogel Development for Biomedical Applications: A Review. *Polymers* **2022**, *14*, 3023. [CrossRef]
3. Yang, Y.; Xu, L.; Wang, J.; Meng, Q.; Zhong, S.; Gao, Y.; Cui, X.; Yang, Y. Recent advances in polysaccharide-based self-healing hydrogels for biomedical applications. *Carbohydr. Polym.* **2022**, *283*, 119161. [CrossRef]
4. Onat, B.; Ulusan, S.; Banerjee, S.; Erel-Goktepe, I. Multifunctional layer-by-layer modified chitosan/poly(ethylene glycol) hydrogels. *Eur. Polym. J.* **2019**, *112*, 73–86. [CrossRef]
5. Lutolf, M.P.; Hubbell, J.A. Synthetic biomaterials as instructive extracellular microenvironments for morphogenesis in tissue engineering. *Nat. Biotechnol.* **2005**, *23*, 47–55. [CrossRef]
6. Maji, S.; Lee, H. Engineering Hydrogels for the Development of Three-Dimensional In Vitro Models. *Int. J. Mol. Sci.* **2022**, *23*, 2662. [CrossRef]
7. El-Husseiny, H.M.; Mady, E.A.; Hamabe, L.; Abugomaa, A.; Shimada, K.; Yoshida, T.; Tanaka, T.; Yokoi, A.; Elbadawy, M.; Tanaka, R. Smart/stimuli-responsive hydrogels: Cutting-edge platforms for tissue engineering and other biomedical applications. *Mater. Today Bio* **2022**, *13*, 100186. [CrossRef]
8. Gong, X.; Mills, K.L. Large-scale patterning of single cells and cell clusters in hydrogels. *Sci. Rep.* **2018**, *8*, 3849. [CrossRef]
9. Hao, D.; Lopez, J.M.; Chen, J.; Iavorovschi, A.M.; Lelivelt, N.M.; Wang, A. Engineering Extracellular Microenvironment for Tissue Regeneration. *Bioengineering* **2022**, *9*, 202. [CrossRef]
10. Gao, J.; Yu, X.; Wang, X.; He, Y.; Ding, J. Biomaterial–Related Cell Microenvironment in Tissue Engineering and Regenerative Medicine. *Engineering* **2022**, *13*, 31–45. [CrossRef]
11. Chan, B.P.; Leong, K.W. Scaffolding in tissue engineering: General approaches and tissue-specific considerations. *Eur. Spine J.* **2008**, *17* (Suppl. 4), 467–479. [CrossRef]
12. Rozario, T.; DeSimone, D.W. The extracellular matrix in development and morphogenesis: A dynamic view. *Dev. Biol.* **2010**, *341*, 126–140. [CrossRef]
13. Kumirska, J.; Czerwicka, M.; Kaczyński, Z.; Bychowska, A.; Brzozowski, K.; Thöming, J.; Stepnowski, P. Application of spectroscopic methods for structural analysis of chitin and chitosan. *Mar. Drugs* **2010**, *8*, 1567–1636. [CrossRef]

14. Wang, T.; Turhan, M.; Gunasekaran, S. Selected properties of pH-sensitive, biodegradable chitosan-poly(vinyl alcohol) hydrogel. *Polym. Int.* **2004**, *53*, 911–918. [CrossRef]
15. Afewerki, S.; Sheikhi, A.; Kannan, S.; Ahadian, S.; Khademhosseini, A. Gelatin-polysaccharide composite scaffolds for 3D cell culture and tissue engineering: Towards natural therapeutics. *Bioeng. Transl. Med.* **2019**, *4*, 96–115. [CrossRef]
16. Klimek, K.; Ginalska, G. Proteins and peptides as important modifiers of the polymer scaffolds for tissue engineering applications—A review. *Polymers* **2020**, *12*, 844. [CrossRef]
17. Echave, M.C.; Burgo, L.S.; Pedraz, J.L.; Orive, G. Gelatin as Biomaterial for Tissue Engineering. *Curr. Pharm. Des.* **2017**, *23*, 3567–3584. [CrossRef]
18. Kumar, A.; Han, S.S. PVA-based hydrogels for tissue engineering: A review. *Int. J. Polym. Mater. Polym. Biomater.* **2017**, *66*, 159–182. [CrossRef]
19. Rodríguez-Rodríguez, R.; Espinosa-Andrews, H.; Velasquillo-Martínez, C.; García-Carvajal, Z.Y. Composite hydrogels based on gelatin, chitosan and polyvinyl alcohol to biomedical applications: A review. *Int. J. Polym. Mater. Polym. Biomater.* **2020**, *69*, 1–20. [CrossRef]
20. Pérez-Díaz, M.A.; Martínez-Colin, E.J.; González-Torres, M.; Ortega-Sánchez, C.; Sánchez-Sánchez, R.; Delgado-Meza, J.; Machado-Bistraín, F.; Martínez-López, V.; Giraldo, D.; Márquez-Gutiérrez, É.A.; et al. Chondrogenic Potential of Human Adipose-Derived Mesenchymal Stromal Cells in Steam Sterilized Gelatin/Chitosan/Polyvinyl Alcohol Hydrogels. *Polymers* **2023**, *15*, 3938. [CrossRef]
21. Rodríguez-Rodríguez, R.; Velasquillo-Martínez, C.; Knauth, P.; López, Z.; Moreno-Valtierra, M.; Bravo-Madrigal, J.; Jiménez-Palomar, I.; Luna-Bárcenas, G.; Espinosa-Andrews, H.; García-Carvajal, Z.Y. Sterilized chitosan-based composite hydrogels: Physicochemical characterization and in vitro cytotoxicity. *J. Biomed. Mater. Res.-Part A* **2020**, *108*, 81–93. [CrossRef]
22. Rodríguez-Rodríguez, R.; García-Carvajal, Z.Y.; Jiménez-Palomar, I.; Jiménez-Avalos, J.A.; Espinosa-Andrews, H. Development of gelatin/chitosan/PVA hydrogels: Thermal stability, water state, viscoelasticity, and cytotoxicity assays. *J. Appl. Polym. Sci.* **2019**, *136*, 47149. [CrossRef]
23. Sánchez-Cardona, Y.; Echeverri-Cuartas, C.E.; López, M.E.L.; Moreno-Castellanos, N. Chitosan/gelatin/pva scaffolds for beta pancreatic cell culture. *Polymers* **2021**, *13*, 2372. [CrossRef]
24. Jessop, Z.M.; Javed, M.; Otto, I.A.; Combellack, E.J.; Morgan, S.; Breugem, C.C.; Archer, C.W.; Khan, I.M.; Lineaweaver, W.C.; Kon, M.; et al. Combining regenerative medicine strategies to provide durable reconstructive options: Auricular cartilage tissue engineering. *Stem Cell Res. Ther.* **2016**, *7*, 19. [CrossRef]
25. Visscher, D.O.; Lee, H.; Van Zuijlen, P.P.; Helder, M.N.; Atala, A.; Yoo, J.J.; Lee, S.J. A photo-crosslinkable cartilage-derived extracellular matrix (ECM) bioink for auricular cartilage tissue engineering. *Acta Biomater.* **2020**, *121*, 193–203. [CrossRef]
26. Ebrahimi, A.; Kazemi, A.; Rasouli, H.R.; Kazemi, M.; Motamedi, M.H.K. Reconstructive surgery of auricular defects: An overview. *Trauma Mon.* **2015**, *20*, e28202. [CrossRef]
27. Melgarejo-Ramírez, Y.; Sánchez-Sánchez, R.; García-Carvajal, Z.; García-López, J.; Gutiérrez-Gómez, C.; Luna-Barcenas, G.; Ibarra, C.; Velasquillo, C. Biocompatibility of Human Auricular Chondrocytes Cultured onto a Chitosan/Polyvynil Alcohol/Epichlorohydrin-Based Hydrogel for Tissue Engineering Application. *Int. J. Morphol.* **2014**, *32*, 1347–1356. [CrossRef]
28. Zeng, J.; Jia, L.; Wang, D.; Chen, Z.; Liu, W.; Yang, Q.; Liu, X.; Jiang, H. Bacterial nanocellulose-reinforced gelatin methacryloyl hydrogel enhances biomechanical property and glycosaminoglycan content of 3D-bioprinted cartilage. *Int. J. Bioprinting* **2023**, *9*, 131–143. [CrossRef]
29. Tang, P.; Song, P.; Peng, Z.; Zhang, B.; Gui, X.; Wang, Y.; Liao, X.; Chen, Z.; Zhang, Z.; Fan, Y.; et al. Chondrocyte-laden GelMA hydrogel combined with 3D printed PLA scaffolds for auricle regeneration. *Mater. Sci. Eng. C* **2021**, *130*, 112423. [CrossRef]
30. Otto, I.; Capendale, P.; Garcia, J.; De Ruijter, M.; Van Doremalen, R.; Castilho, M.; Lawson, T.; Grinstaff, M.; Breugem, C.; Kon, M.; et al. Biofabrication of a shape-stable auricular structure for the reconstruction of ear deformities. *Mater. Today Bio* **2021**, *9*, 100094. [CrossRef]
31. Visscher, D.O.; Gleadall, A.; Buskermolen, J.K.; Burla, F.; Segal, J.; Koenderink, G.H.; Helder, M.N. Design and fabrication of a hybrid alginate hydrogel/poly(ε-caprolactone) mold for auricular cartilage reconstruction. *J. Biomed. Mater. Res. Part B Appl. Biomater.* **2019**, *107*, 1711–1721. [CrossRef]
32. Wang, H.; Zhang, J.; Liu, H.; Wang, Z.; Li, G.; Liu, Q.; Wang, C. Chondrocyte-laden gelatin/sodium alginate hydrogel integrating 3D printed PU scaffold for auricular cartilage reconstruction. *Int. J. Biol. Macromol.* **2023**, *253*, 126294. [CrossRef]
33. Xu, Y.; Xu, Y.; Bi, B.; Hou, M.; Yao, L.; Du, Q.; He, A.; Liu, Y.; Miao, C.; Liang, X.; et al. A moldable thermosensitive hydroxypropyl chitin hydrogel for 3D cartilage regeneration in vitro and in vivo. *Acta Biomater.* **2020**, *108*, 87–96. [CrossRef]
34. Griffin, M.F.; Premakumar, Y.; Seifalian, A.M.; Szarko, M.; Butler, P.E.M. Biomechanical Characterisation of the Human Auricular Cartilages; Implications for Tissue Engineering. *Ann. Biomed. Eng.* **2016**, *44*, 3460–3467. [CrossRef]
35. Yang, Y.; Yao, X.; Li, X.; Guo, C.; Li, C.; Liu, L.; Zhou, Z. Non-mulberry silk fiber-based scaffolds reinforced by PLLA porous microspheres for auricular cartilage: An in vitro study. *Int. J. Biol. Macromol.* **2021**, *182*, 1704–1712. [CrossRef]
36. Wong, C.; Chen, H.; Chiu, H.; Tsuang, H.; Bai, Y.; Chung, J.; Lin, H.; Hsieh, J.; Chen, T.; Yang, L. Facilitating In Vivo Articular Cartilage Repair by Tissue-Engineered Cartilage Grafts Produced from Auricular Chondrocytes. *Am. J. Sports Med.* **2018**, *46*, 713–727. [CrossRef]
37. Loh, Q.L.; Choong, C. Three-dimensional scaffolds for tissue engineering applications: Role of porosity and pore size. *Tissue Eng. Part B Rev.* **2013**, *19*, 485–502. [CrossRef] [PubMed]

38. Otto, I.A.; Levato, R.; Webb, W.R.; Khan, I.M.; Breugem, C.C.; Malda, J. Progenitor cells in auricular cartilage demonstrate cartilage-forming capacity in 3D hydrogel culture. *Eur. Cells Mater.* **2018**, *35*, 132–150. [CrossRef] [PubMed]
39. Cheng, A.; Schwartz, Z.; Kahn, A.; Li, X.; Shao, Z.; Sun, M.; Ao, Y.; Boyan, B.D.; Chen, H. Advances in Porous Scaffold Design for Bone and Cartilage Tissue Engineering and Regeneration. *Tissue Eng. Part B Rev.* **2019**, *25*, 14–29. [CrossRef]
40. Ma, X.; Zhao, D.; Li, X.; Meng, L. Chromatographic method for determination of the free amino acid content of chamomile flowers. *Pharmacogn. Mag.* **2015**, *11*, 176–179. [CrossRef] [PubMed]
41. ISO 10993-5:2009; Biological Evaluation of Medical Devices—Part 5: Tests for In Vitro Cytotoxicity. ISO: Geneva, Switzerland, 2009. Available online: https://www.iso.org/standard/36406.html (accessed on 23 November 2023).
42. Melgarejo-Ramírez, Y.; Sánchez-Sánchez, R.; García-López, J.; Brena-Molina, A.M.; Gutiérrez-Gómez, C.; Ibarra, C.; Velasquillo, C. Characterization of pediatric microtia cartilage: A reservoir of chondrocytes for auricular reconstruction using tissue engineering strategies. *Cell Tissue Bank.* **2016**, *17*, 481–489. [CrossRef]
43. Fan, L.; Yang, H.; Yang, J.; Peng, M.; Hu, J. Preparation and characterization of chitosan/gelatin/PVA hydrogel for wound dressings. *Carbohydr. Polym.* **2016**, *146*, 427–434. [CrossRef] [PubMed]
44. Shamloo, A.; Aghababaie, Z.; Afjoul, H.; Jami, M.; Bidgoli, M.R.; Vossoughi, M.; Ramazani, A.; Kamyabhesari, K. Fabrication and evaluation of chitosan/gelatin/PVA hydrogel incorporating honey for wound healing applications: An in vitro, in vivo study. *Int. J. Pharm.* **2021**, *592*, 120068. [CrossRef] [PubMed]
45. Mhatre, A.; Bhagwat, A.; Bangde, P.; Jain, R.; Dandekar, P. Chitosan/gelatin/PVA membranes for mammalian cell culture. *Carbohydr. Polym. Technol. Appl.* **2021**, *2*, 100163. [CrossRef]
46. López-Velázquez, J.C.; Rodríguez-Rodríguez, R.; Espinosa-Andrews, H.; Qui-Zapata, J.A.; García-Morales, S.; Navarro-López, D.E.; Luna-Bárcenas, G.; Vassallo-Brigneti, E.C.; García-Carvajal, Z.Y. Gelatin–chitosan–PVA hydrogels and their application in agriculture. *J. Chem. Technol. Biotechnol.* **2019**, *94*, 3495–3504. [CrossRef]
47. Norotte, C.; Marga, F.S.; Niklason, L.E.; Forgacs, G. Scaffold-free vascular tissue engineering using bioprinting. *Biomaterials* **2009**, *30*, 5910–5917. [CrossRef] [PubMed]
48. Posniak, S.; Chung, J.H.Y.; Liu, X.; Mukherjee, P.; Wallace, G.G. The importance of elastin and its role in auricular cartilage tissue engineering. *Bioprinting* **2023**, *32*, e00276. [CrossRef]
49. Lee, J.M.; Sultan, M.T.; Kim, S.H.; Kumar, V.; Yeon, Y.K.; Lee, O.J.; Park, C.H. Artificial auricular cartilage using silk fibroin and polyvinyl alcohol hydrogel. *Int. J. Mol. Sci.* **2017**, *18*, 1707. [CrossRef]
50. Almajidi, Y.Q.; Abdullaev, S.S.; Alani, B.G.; Saleh, E.A.M.; Ahmad, I.; Ramadan, M.F.; Al-Hasnawi, S.S.; Romero-Parra, R.M. Chitosan-gelatin hydrogel incorporating polyvinyl alcohol and MnFe double-layered hydroxide nanocomposites with biological activity. *Int. J. Biol. Macromol.* **2023**, *246*, 125566. [CrossRef]
51. Muyonga, J.H.; Cole, C.G.B.; Duodu, K.G. Fourier transform infrared (FTIR) spectroscopic study of acid soluble collagen and gelatin from skins and bones of young and adult Nile perch (*Lates niloticus*). *Food Chem.* **2004**, *86*, 325–332. [CrossRef]
52. Ibrahim, M.; Mahmoud, A.A.; Osman, O.; El-Aal, M.A.; Eid, M. Molecular spectroscopic analyses of gelatin. *Spectrochim. Acta-Part A Mol. Biomol. Spectrosc.* **2011**, *81*, 724–729. [CrossRef] [PubMed]
53. Mansur, H.S.; Sadahira, C.M.; Souza, A.N.; Mansur, A.A.P. FTIR spectroscopy characterization of poly (vinyl alcohol) hydrogel with different hydrolysis degree and chemically crosslinked with glutaraldehyde. *Mater. Sci. Eng. C* **2008**, *28*, 539–548. [CrossRef]
54. Corona-Escalera, A.F.; Tinajero-Díaz, E.; García-Reyes, R.A.; Luna-Bárcenas, G.; Seyfoddin, A.; Padilla-de la Rosa, J.D.; González-Ávila, M.; García-Carvajal, Z.Y. Enzymatic Crosslinked Hydrogels of Gelatin and Poly (Vinyl Alcohol) Loaded with Probiotic Bacteria as Oral Delivery System. *Pharmaceutics* **2022**, *14*, 2759. [CrossRef] [PubMed]
55. Espinosa-García, B.M.; Argüelles-Monal, W.M.; Hernández, J.; Félix-Valenzuela, L.; Acosta, N.; Goycoolea, F.M. Molecularly imprinted Chitosan—Genipin hydrogels with recognition capacity toward o-Xylene. *Biomacromolecules* **2007**, *8*, 3355–3364. [CrossRef]
56. Ananthalakshmi, R.; Ravi, S.; Jeddy, N.; Thangavelu, R.; Janardhanan, S. Natural alternatives for chemicals used in histopathology lab—A literature review. *J. Clin. Diagn. Res.* **2016**, *10*, EE01–EE04. [CrossRef]
57. Sa'adon, S.; Ansari, M.N.M.; Razak, S.I.A.; Anand, J.S.; Nayan, N.H.M.; Ismail, A.E.; Khan, M.U.A.; Haider, A. Preparation and physicochemical characterization of a diclofenac sodium-dual layer polyvinyl alcohol patch. *Polymers* **2021**, *13*, 2459. [CrossRef]
58. Chen, H.L.; Wu, L.G.; Tan, J.; Zhu, C.L. PVA membrane filled β-cyclodextrin for separation of isomeric xylenes by pervaporation. *Chem. Eng. J.* **2000**, *78*, 159–164. [CrossRef]
59. Nkhwa, S.; Lauriaga, K.F.; Kemal, E.; Deb, S. Poly(vinyl alcohol): Physical Approaches to Designing Biomaterials for Biomedical Applications. *Conf. Pap. Sci.* **2014**, *2014*, 403472. [CrossRef]
60. Jipa, I.; Stoica, A.; Stroescu, M.; Dobre, L.; Dobre, T.; Jinga, S.; Tardei, C. Potassium sorbate release from poly(vinyl alcohol)-bacterial cellulose films. *Chem. Pap.* **2012**, *66*, 138–143. [CrossRef]
61. Liu, D.; Nikoo, M.; Boran, G.; Zhou, P.; Regenstein, J.M. Collagen and gelatin. *Annu. Rev. Food Sci. Technol.* **2015**, *6*, 527–557. [CrossRef]
62. Gauza-Włodarczyk, M.; Kubisz, L.; Włodarczyk, D. Amino acid composition in determination of collagen origin and assessment of physical factors effects. *Int. J. Biol. Macromol.* **2017**, *104*, 987–991. [CrossRef] [PubMed]
63. Aykin-Dinçer, E.; Koç, A.; Erbas, M. Extraction and physicochemical characterization of broiler (*Gallus gallus domesticus*) skin gelatin compared to commercial bovine gelatin. *Poult. Sci.* **2017**, *96*, 4124–4131. [CrossRef] [PubMed]

64. Ahmad, T.; Ismail, A.; Ahmad, S.A.; Khalil, K.A.; Kee, L.T.; Awad, E.A.; Sazili, A.Q. Extraction, characterization and molecular structure of bovine skin gelatin extracted with plant enzymes bromelain and zingibain. *J. Food Sci. Technol.* **2020**, *57*, 3772–3781. [CrossRef]
65. Shu, X.Z.; Liu, Y.; Palumbo, F.; Prestwich, G.D. Disulfide-crosslinked hyaluronan-gelatin hydrogel films: A covalent mimic of the extracellular matrix for in vitro cell growth. *Biomaterials* **2003**, *24*, 3825–3834. [CrossRef]
66. Zhang, X.; Qi, L.; Chen, X.G.; Lai, Y.; Liu, K.; Xue, K. Comparative study of alginate and type I collagen as biomaterials for cartilage stem/progenitor cells to construct tissue-engineered cartilage in vivo. *Front. Bioeng. Biotechnol.* **2023**, *10*, 1057199. [CrossRef] [PubMed]
67. Massarelli, E.; Silva, D.; Pimenta, A.; Fernandes, A.; Mata, J.; Armês, H.; Salema-Oom, M.; Saramago, B.; Serro, A. Polyvinyl alcohol/chitosan wound dressings loaded with antiseptics. *Int. J. Pharm.* **2021**, *593*, 120110. [CrossRef]
68. García-Martínez, L.; Campos, F.; Godoy-Guzmán, C.; Del Carmen Sánchez-Quevedo, M.; Garzón, I.; Alaminos, M.; Campos, A.; Carriel, V. Encapsulation of human elastic cartilage-derived chondrocytes in nanostructured fibrin-agarose hydrogels. *Histochem. Cell Biol.* **2017**, *147*, 83–95. [CrossRef]
69. Ferreira, K.D.; Cardoso, L.D.; Oliveira, L.P.; Franzo, V.S.; Pancotti, A.; Miguel, M.P.; Silva, L.A.F.; Vulcani, V.A.S. Histological analysis of elastic cartilages treated with alkaline solution. *Arq. Bras. Med. Vet. Zootec.* **2020**, *72*, 647–654. [CrossRef]
70. Kang, N.; Liu, X.; Guan, Y.; Wang, J.; Gong, F.; Yang, X.; Yan, L.; Wang, Q.; Fu, X.; Cao, Y.; et al. Effects of co-culturing BMSCs and auricular chondrocytes on the elastic modulus and hypertrophy of tissue engineered cartilage. *Biomaterials* **2012**, *33*, 4535–4544. [CrossRef]
71. Huang, Y.; Zhao, H.; Wang, Y.; Bi, S.; Zhou, K.; Li, H.; Zhou, C.; Wang, Y.; Wu, W.; Peng, B.; et al. The application and progress of tissue engineering and biomaterial scaffolds for total auricular reconstruction in microtia. *Front. Bioeng. Biotechnol.* **2023**, *11*, 1089031. [CrossRef]
72. Zopf, D.A.; Flanagan, C.L.; Nasser, H.B.; Mitsak, A.G.; Huq, F.S.; Rajendran, V.; Green, G.E.; Hollister, S.J. Biomechanical evaluation of human and porcine Auricular cartilage. *Laryngoscope* **2015**, *125*, E262–E268. [CrossRef]
73. Lien, S.M.; Ko, L.Y.; Huang, T.J. Effect of pore size on ECM secretion and cell growth in gelatin scaffold for articular cartilage tissue engineering. *Acta Biomater.* **2009**, *5*, 670–679. [CrossRef]
74. Hall, B.K.; Miyake, T. Divide, accumulate, differentiate: Cell condensation skeletal development revisited. *Int. J. Dev. Biol.* **1995**, *39*, 881–893. [PubMed]
75. Fowler, D.A.; Larsson, H.C.E. The tissues and regulatory pattern of limb chondrogenesis. *Dev. Biol.* **2020**, *463*, 124–134. [CrossRef] [PubMed]
76. Hayes, A.J.; Melrose, J. Aggrecan, the primary weight-bearing cartilage proteoglycan, has context-dependent, cell-directive properties in embryonic development and neurogenesis: Aggrecan glycan side chain modifications convey interactive biodiversity. *Biomolecules* **2020**, *10*, 1244. [CrossRef] [PubMed]
77. Wu, Z.; Korntner, S.H.; Mullen, A.M.; Zeugolis, D.I. Collagen type II: From biosynthesis to advanced biomaterials for cartilage engineering. *Biomater. Biosyst.* **2021**, *4*, 100030. [CrossRef] [PubMed]
78. Ba, R.; Kong, L.; Wu, G.; Liu, S.; Dong, Y.; Li, B.; Zhao, Y. Increased Expression of Sox9 during Balance of BMSCs/Chondrocyte Bricks in Platelet-Rich Plasma Promotes Construction of a STable 3-D Chondrogenesis Microenvironment for BMSCs. *Stem Cells Int.* **2020**, *2020*, 5492059. [CrossRef] [PubMed]
79. Chang, C.S.; Yang, C.Y.; Hsiao, H.Y.; Chen, L.; Chu, I.M.; Cheng, M.H.; Tsao, C.H. Cultivation of auricular chondrocytes in poly(ethylene glycol)/poly(ε-caprolactone) hydrogel for tracheal cartilage tissue engineering in a rabbit model. *Eur. Cells Mater.* **2018**, *35*, 350–364. [CrossRef]

Disclaimer/Publisher's Note: The statements, opinions and data contained in all publications are solely those of the individual author(s) and contributor(s) and not of MDPI and/or the editor(s). MDPI and/or the editor(s) disclaim responsibility for any injury to people or property resulting from any ideas, methods, instructions or products referred to in the content.

Article

Thermosensitive Chitosan Hydrogels: A Potential Strategy for Prolonged Iron Dextran Parenteral Supplementation

Emerson Durán [1,2], Andrónico Neira-Carrillo [3], Felipe Oyarzun-Ampuero [4,*] and Carolina Valenzuela [1,*]

1. Departamento de Fomento de la Producción Animal, Facultad de Ciencias Veterinarias y Pecuarias, Universidad de Chile, Santa Rosa 11.735, La Pintana 8820808, Santiago, Chile; emerson.duran@ug.uchile.cl
2. Programa de Doctorado en Ciencias Silvoagropecuarias y Veterinarias, Campus Sur Universidad de Chile, Santa Rosa 11.315, La Pintana 8820808, Santiago, Chile
3. Laboratorios de Materiales Bio-Relacionados (CIMAT) y Síntesis y Caracterización de Polímeros Funcionalizados y Biomoléculas (POLYFORMS), Departamento de Ciencias Biológicas Animales, Facultad de Ciencias Veterinarias y Pecuarias, Universidad de Chile, Santa Rosa 11.735, La Pintana 8820808, Santiago, Chile; aneira@uchile.cl
4. Departamento de Ciencias y Tecnología Farmacéuticas, Facultad de Ciencias Químicas y Farmacéuticas, Universidad de Chile, Santos Dumont 964, Independencia 8380494, Santiago, Chile
* Correspondence: foyarzuna@ciq.uchile.cl (F.O.-A.); cvalenzuelav@u.uchile.cl (C.V.); Tel.: +56-2-29781616 (F.O.-A.); +56-2-2978567 (C.V.)

Citation: Durán, E.; Neira-Carrillo, A.; Oyarzun-Ampuero, F.; Valenzuela, C. Thermosensitive Chitosan Hydrogels: A Potential Strategy for Prolonged Iron Dextran Parenteral Supplementation. *Polymers* **2024**, *16*, 139. https://doi.org/10.3390/polym16010139

Academic Editor: Luminita Marin

Received: 3 November 2023
Revised: 18 December 2023
Accepted: 26 December 2023
Published: 31 December 2023

Copyright: © 2023 by the authors. Licensee MDPI, Basel, Switzerland. This article is an open access article distributed under the terms and conditions of the Creative Commons Attribution (CC BY) license (https://creativecommons.org/licenses/by/4.0/).

Abstract: Iron deficiency anemia (IDA) presents a global health challenge, impacting crucial development stages in humans and other mammals. Pigs, having physiological and metabolic similarities with humans, are a valuable model for studying and preventing anemia. Commonly, a commercial iron dextran formulation (CIDF) with iron dextran particles (IDPs) is intramuscularly administered for IDA prevention in pigs, yet its rapid metabolism limits preventive efficacy. This study aimed to develop and evaluate chitosan thermosensitive hydrogels (CTHs) as a novel parenteral iron supplementation strategy, promoting IDPs' prolonged release and mitigating their rapid metabolism. These CTHs, loaded with IDPs (0.1, 0.2, and 0.4 g of theoretical iron/g of chitosan), were characterized for IM iron supplementation. Exhibiting thermosensitivity, these formulations facilitated IM injection at ~4 °C, and its significant increasing viscosity at 25–37 °C physically entrapped the IDPs within the chitosan's hydrophobic gel without chemical bonding. In vitro studies showed CIDF released all the iron in 6 h, while CTH0.4 had a 40% release in 72 h, mainly through Fickian diffusion. The controlled release of CTHs was attributed to the physical entrapment of IDPs within the CTHs' gel, which acts as a diffusion barrier. CTHs would be an effective hydrogel prototype for prolonged-release parenteral iron supplementation.

Keywords: chitosan; thermosensitive hydrogel; iron deficiency; pig

1. Introduction

Iron deficiency anemia (IDA) is the leading nutritional deficiency in the world, affecting one-third of the global population, especially during critical developmental stages such as childhood, pregnancy, and lactation [1,2]. This deficiency is caused by the high demand for iron during these stages and the low intake of iron in forms that provide higher bioavailability, such as those found in animal-sourced foods, which are expensive [3]. Multiple iron supplementation alternatives have been studied in humans and other mammalians to try and prevent anemia, with unsatisfactory results. In fact, iron deficiency has increased in the world. The use of low-bioavailability sources of non-heme iron (such as ferrous sulfate) in oral supplementation strategies and food fortification is the reason why these strategies have failed. Non-heme iron also causes adverse side effects like gastrointestinal disorders and has unpleasant sensory properties [2,4,5].

In humans, there are few options for parenteral iron supplementation formulations, mainly sodium ferric gluconate and intravenous (IV) iron sucrose. However, their use is not common due to the invasive nature of the procedure, risk of anaphylaxis, and high cost, as these procedures require hospitalization of the patient [6]. IV supplementation is highly effective for iron supplementation, achieving a more efficient increase in hemoglobin compared to oral supplementation, mainly due to its high bioavailability. However, there are risks of infections, anaphylactic reactions, endothelial damage, oxidative stress, and hepcidin overexpression [7].

In pigs, the IM supplementation of 150–200 mg of a commercial formulation of iron dextran (CIDF) in a single dose is used for the prevention of iron deficiency anemia [8], as the condition is highly prevalent in pigs raised in an intensive production system [9,10]. Pigs are considered a relevant and valid animal model to investigate iron deficiency and supplementation due to the physiological and metabolic similarities they have with humans and other omnivorous mammals [11,12]. CIDF is an aqueous dispersion of iron dextran particles (IDPs) with a ferric hydroxide core covered by a dextran shell, which imparts high hydrophilicity, low reactivity, and a nanometric particle size of ~11.5 nm [13]. When CIDF is injected into the muscular tissue, IDPs are rapidly dispersed across the surrounding tissues through simple diffusion mechanisms within muscle fibers, transported by the lymphatic system into the bloodstream, and captured by macrophages, which are responsible for extracting iron and binding it to transferrin for transport to the site of utilization or storage [8]. The preventive administration of CIDF to piglets has been employed for over 70 years. However, it exhibits several disadvantages linked to a rapid metabolism and a high amount of injected iron; a high iron load leads to a substantial increase in blood iron concentration within the first 10 h post-injection [14]. This triggers the overexpression of hepcidin, a hormone that promotes iron efflux from the body and reduces iron absorption, resulting in inefficient utilization of the supplemented iron [15,16]. Furthermore, the accumulation of such a substantial quantity of iron in storage sites induces toxic effects on adjacent tissues, due to its high reactivity potential [16,17].

To mitigate these adverse effects and enhance the effectiveness of injected IDPs, the use of chitosan thermosensitive hydrogels (CTHs) is proposed as a vehicle for the sustained release of IDPs. Chitosan is a biocompatible and biodegradable biopolymer that has been widely utilized to develop sustained release systems in various formats, including IM supplementation [18]. CTHs offer the advantage of being injected as a liquid (sol state) at room temperature (~4–25 °C) and significantly increase the viscosity, achieving a gel state, when interacting with muscular tissues (~36–37 °C) [19]. Several studies demonstrated the various benefits of CTHs for the controlled and/or prolonged release of active agents, which enhances drug retention in situ, thereby extending the duration of the release from the injection site [20,21]. In the present study, we postulate that CTHs possess the capacity to provide IDPs with in situ retention after IM injection. CTHs should act as a barrier, physically prolonging the release from the injection site and promoting a constant and low iron concentration in blood, and be able to prevent hepcidin overexpression and its negative consequences for iron supplementation.

Various molecules have been used to achieve thermosensitive behavior with chitosan, and this study focuses on the use of glycerophosphate (GP) due to its low toxicity, high accessibility, favorable thermosensitive behavior, and outcomes [18]. At room temperature (~4–25 °C), GP functions as an intermediary agent in the chitosan–GP–water system, enabling the maintenance of the sol state (with chitosan suspended) when the formulation is at a neutral pH. These interactions prevent chitosan from losing its cationic potential and settling due to its non-interaction with water [18]. However, as the temperature of the formulation raises to approximately 30 °C, the electrostatic chitosan–GP binding is disrupted, leading to chitosan–chitosan union through hydrogen bonding, which results in the formation of a highly viscous hydrophobic three-dimensional network (gel state) that impedes the outflow of the content from the injection site [18].

For these reasons, it is proposed to develop and study CTH-IDP formulations that maintain the sol state at room temperature (~4–25 °C) and undergo a transition to the gel state at a temperature close to the intramuscular (IM) temperature in mammals (~37 °C).

The loading and entrapment of IDPs for prolonged iron supplementation through CTHs are investigated using a nanometric and hydrophilic active agent, and the outcomes of this study may serve as a model for the development of release studies involving IDPs or similar active agents. Therefore, the objectives of this study were to develop CTH-IDP formulations and to study their potential as a mammalian parenteral iron dextran supplementation strategy.

2. Materials and Methods

2.1. Materials

Iron dextran particles (IDP) obtained from a commercial formulation (CIDF, 20% w/v of iron, obtained from the same batch) were used as the iron source (Veterquímica S.A., Maipú, Chile). For the development of chitosan thermosensitive hydrogels (CTHs), chitosan derived from crab shell (300–350 KDa) with a degree of deacetylation >80% (Sigma-Aldrich, St. Louis, MO, USA) and hydrated disodium glycerophosphate (GP, Sigma-Aldrich, USA) were used. All other reagents were of analytical grade and procured from Merck S.A (Lethabong, South Africa).

2.2. Preparation of CTHs

The CTHs were developed following the procedure described by Sun et al. [22], with some modifications. A 0.2% v/v acetic acid solution in Milli-Q water at 4 °C was prepared, and 1% w/v chitosan was added, which was maintained at 4 °C for 12 h until a transparent and viscous solution (pH of 5.0–5.5) was obtained. GP was prepared at 50% w/v in Milli-Q water and added (0.2 mL/min) to the chitosan solution under magnetic stirring (4 °C), while the pH was constantly monitored until reaching neutral value (7.0 ± 0.5); the final GP concentration was ~6–8% v/v. The obtained CTHs were refrigerated until use.

After the CTHs were formed, IDPs in the form of CIDF were added (using a syringe at a rate of 1 mL/min) and homogenized using a paddle stirrer (1000 rpm for 1 h, OS40-PRO, D-LAB, Beijing, China). Three formulations with increasing iron concentrations (0.1, 0.2, and 0.4 g of theoretical iron/g of chitosan) were developed, referred to as CTH0.1, CTH0.2, and CTH0.4, respectively (CTH0, without iron, was the control).

2.3. CIDF Characterization

CIDF was characterized in terms of pH (AD1020, Adwa, Szeged, Hungary), particle size using dynamic light scattering (DLS, Zetasizer Nano-ZS90, Malvern Instruments, Malvern, UK), zeta potential by laser Doppler anemometry (Zetasizer Nano-ZS90, Malvern Instruments, UK), and viscosity using a rotational viscometer with a no. 1 needle (NDJ-8S, Nirun, Shanghai, China).

2.4. CTH Characterization

2.4.1. Macroscopic Appearance

Digital images were obtained for macroscopic evaluation of the CTH0, CTH0.1, CTH0.2, and CTH0.4 formulations. The images were captured with the CTHs at room temperature (~22 °C) in test tubes from a focal point at 10 cm, focusing on the lower section of the tubes to visualize precipitates and aggregates.

2.4.2. Electron Microscopy

Images of the CTH0, CTH0.1, CTH0.2, and CTH0.4 formulations in a gel state were obtained using scanning electron microscopy (SEM) to provide structural appreciation of the gel state. First, 1 mL of each gelled formulation (previously incubated at 37 °C for 30 min) was frozen at −80 °C and lyophilized (L101, Liotop, São Carlos, Brazil) for 24 h in Eppendorf tubes. Subsequently, the samples were coated with a 10 nm gold film using a Sputter Coater (Cressington model 108, Ted Pella Inc., Redding, CA, USA), and microscopic images were captured using a scanning electron microscope (FEI inspect F50,

Thermo Fisher Scientific, Waltham, MA, USA) equipped with an energy dispersive detector (Ultradry Pathfinder Alpine 129 eV, Thermo Fisher Scientific, USA).

2.4.3. pH

The pH of the CTH0, CTH0.1, CTH0.2, CTH0.4, and CIDF formulations was studied to assess compatibility with muscle tissue. The analysis was conducted on 100 mL samples using a standard pH meter (AD1020, Adwa, Szeged, Hungary).

2.4.4. IDP Content

The concentration of IDPs in the CTH0.1, CTH0.2, CTH0.4, and CIDF formulations was measured using a UV spectrophotometer (UV-5100, Metash, Shanghai, China). A calibration curve for IDPs was obtained (λ = 486 nm, R^2 = 0.99), providing a molar extinction coefficient of 2.9507 mL/mg·cm. This molar extinction coefficient was utilized for IDP quantification in CTHs. Based on this, the necessary dilutions were made for each formulation to achieve a theoretical concentration of 1 mg/mL of IDPs. Results were presented as mg of IDP/mL of formulation.

2.4.5. Viscosity and Sol–Gel Transition Time

To confirm and analyze the thermosensitivity of the CTH0, CTH0.1, CTH0.2, and CTH0.4 formulations, the viscosity was determined in 200 mL samples at 4, 25, and 37 °C using a rotational viscometer (NDJ-8S, Nirun, China). All measurements were carried out with a no. 1 needle, and the unit of measurement used was milliPascal-second (mPa·s). The formulations were incubated (BJPX-200B, Biobase, Jinan, China) for 30 min prior to each experiment, which was conducted at room temperature (~20 °C) immediately following incubation.

To determine the time necessary for the sol–gel transition for these formulations at 37 °C, the tube inversion method described by Wang et al. [23] was used. First, 5 mL of sample was transferred to a sealed glass test tube (13 mL capacity, 1.5 mm diameter) and kept at 4 °C. Simultaneously, a beaker containing distilled water was incubated at 37 °C. The experiment consisted of submerging three-quarters of the test tube into the beaker and measuring the time it takes for the formulation to reach the gel state. The sample was considered to be in a gel state when, upon rotating the tube 180°, the formulation did not flow. The sol–gel transition was monitored every 30 s.

2.4.6. Water–Gel Phase Separation

In order to determine if IDPs are entrapped within them, CTHs, CTH0.1, CTH0.2, and CTH0.4 formulations were centrifuged in the gel state to induce separation between the hydrophobic gel and the aqueous phase of the formulation, thereby revealing the position of the IDPs through macroscopic digital images. Additionally, the same experiment was carried out on the CTH0 and CIDF formulations: 30 mL of the formulations was incubated in Falcon tubes at 37 °C for 30 min and then centrifuged at 3000× g for 30 min using a Thermo Scientific Heraeus Megafuge 16R centrifuge (USA). After centrifugation, the macroscopic appearance of the formulations was assessed in the same manner described in the section on macroscopic appearance.

2.4.7. Fourier-Transform Infrared Spectroscopy

An analysis of the infrared spectrogram was performed on CTH0, CTH0.1, CTH0.2, and CTH0.4 formulations, as well as the chitosan polymer and CIDF (dried at 50 °C for 48 h in an oven), to understand the predominant bonds in the formulations/precursor materials and how IDP content influences the bonds. Fourier-transform infrared spectroscopy (FTIR) analysis was conducted using an ATR/FTIR instrument (Interspec 200-X spectrometer, Tartumaa, Estonia). The formulations were incubated at 37 °C for 30 min before measuring to ensure they were in a gel state. Spectra were obtained by averaging 20 scans in the spectral range of 600–4000 cm^{-1}.

2.4.8. Injectability and Retention Evaluation in Ex Vivo Porcine Tissue

The selected formulations (CTH0.4, due to its higher IDP content, and CIDF as a control) were ex vivo injected to confirm they can be injected in porcine tissue in sol state using a syringe. Digital images were obtained to evaluate and compare the retention of each formulation in the injection site. The selected porcine tissue was top inside round, corresponding to the semitendinosus muscle of the hind limb (common CIDF IM injection site in pigs), of an approximate size of 5 (width) × 5 (length) × 3 (height) cm^3. The porcine tissue was purchased from a supermarket, then sized, and kept refrigerated to be used within the next 12 h. Tissue samples were injected to a depth of 1 cm (using a 3 mL plastic syringe attached to a G21 needle, with an internal diameter of 0.60 mm) with 1 mL of each formulation. The injected tissue samples were then incubated at 37 °C for approximately 1 h, to simulate the temperature of a mammalian in vivo muscle tissue. The images were captured from a focal point at 10 cm, after injection and incubation. To expose the injection site and visualizing the retention of the formulation, the tissue was transversely cut.

2.4.9. In Vitro Iron Release

The quantification of the In vitro iron release from the selected CTH formulation (CTH0.4) and CIDF, a release study using a USP 4 apparatus (Sotax CE7 smart, CY 7 piston pumps, Sotax, Westborough, MA, USA) was carried out. Experiments were conducted in triplicate, utilizing phosphate buffered saline (PBS) as the release medium within each cell (100 mL). A consistent flow rate of 8 mL/min was maintained, and the temperature was regulated at 37 °C. Each cell was equipped with a sample holder containing 200 µL of the formulation, directly positioned above the beads. To minimize turbulence, a 6 mm diameter glass bead was placed at the bottom of each cell, surmounted by a 2 cm layer of 1 mm of diameter glass beads. Sampling was performed at designated intervals (0.5, 1, 2, 4, 6, 8, 12, 24, 48, and 72 h), withdrawing 1 mL of release medium, which was subsequently replaced with the same volume of fresh PBS. Iron content in CTH0.4, CIDF, and collected samples was determined using a Perkin Elmer PINAACLE 900 F Atomic Absorption Spectrophotometer, equipped with an acetylene/air flame for iron quantification. Sample preparation involved digestion with nitric acid and hydrogen peroxide, heating at 90 °C for 2 h. The results were represented as mean values on a cumulative content release curve over time. In vitro drug release data were exposed to mathematical kinetics models (program DDSolver). The Akaike information criteria (AIC), coefficient of determination (R^2), and the model selection criteria (MSC) values were considered for the selection of the model, and then the release data were fitted to different models (zero order, first order, Higuchi, and Korsmeyer–Peppas).

2.4.10. Statistical Analysis

Characterizations of pH, sol–gel transition time, viscosity, IDP content, particle size, zeta potential, and In vitro iron release generated results with continuous and normal data (Shapiro–Wilk test, $p > 0.05$). Characterizations were carried out in triplicate. To determine significant differences, ANOVA ($p < 0.05$) and Tukey's test ($p < 0.05$) were used. Calculations were performed using R software version 4.3.1 (R package, Boston, MA, USA). FTIR analysis was conducted through graphical representation for better visualization and comparison. Macroscopic appearance, electron microscopy, water–gel separation, and injectability analyses were completed using the obtained images.

3. Results and Discussion

3.1. CIDF Characterization

CIDF was characterized by pH and viscosity, which are properties of interest for understanding the use of CIDF as an IM supplement. The pH was 6.38 ± 0.03, which is considered suitable for IM use in mammals (including pigs and humans), as they show a physiological pH close to 7. Therefore, CIDF is unlikely to cause pain upon injection due to the activation of pH-sensitive receptors [24,25]. The viscosity of CIDF significantly decreased with temperature: 23.8 ± 0.2 mPa·s at 4 °C, 12.9 ± 0.2 mPa·s at 25 °C, and 10.2 ± 0.5 mPa·s at 37 °C. This change can be explained by the progressive breaking of hydrogen bonds between the IDPs and the water in the formulation as the energy in the system increases due to the effect of temperature. Therefore, when injected, the temperature of the muscle decreases the viscosity of the CIDF, facilitating its dispersion in the adjacent tissues. The obtained viscosity is in line with the values obtained by other authors (10–25 mPa·s) and is considered suitable for extrusion using a needle with a 21–25 G lumen, commonly used in IM injection in pigs, humans, and other mammals [26].

The average particle size of IDPs in CIDF was 81.9 ± 0.2 nm, which differs from what was described [27], where a particle size of 11.5 nm was obtained using the same method employed in the present study (DLS). This size for these IDP particles prevents diffusion directly into the bloodstream and metabolization through the lymphatic system, acting as a barrier that provides a delayed iron absorption [8]. However, IDP use delays iron absorption but does not add a prolonged or sustained release, resulting in elevated serum iron concentrations within the first 1–10 h post-injection [14]. The zeta potential of IDPs was -0.15 ± 0.56 mV, representing a neutral surface charge value, implying low potential for interaction with membranes and molecules bearing more significant electrical charges [28]. A neutral charge increases the potential for particle aggregation, as they do not electrostatically repel each other [29], which is advantageous for retention within a CTH, promoting accumulation and stability at the injection site.

3.2. CTH Characterization

3.2.1. Macroscopic Appearance

All developed formulations are shown in Figure 1a. Those containing IDPs exhibit a brown/orange coloration, which is attributed to the presence of iron hydroxide in the IDP cores. At room temperature (~22 °C), it is evident that all formulations show a heterogeneous appearance. Moreover, the formation of small clusters can be observed by the naked eye, which may result from the premature gelation of CTHs mediated by nucleation–aggregation processes, where chitosan–chitosan bonds, established via hydrogen bridges at multiple points that simultaneously grow, form small hydrophobic aggregates that increase the formulation's viscosity [18]. The presence of aggregates is initially considered a disadvantage for extrusion through a syringe with a 21–25 G needle due to potential obstructions. Additionally, if this gelation process is triggered at lower temperatures than 20 °C, it also represents a potential disadvantage due to the non-uniform distribution of the active ingredient, which could impede dose uniformity; therefore, the formulations should be injected at lower temperatures.

3.2.2. Electron Microscopy

SEM images of the lyophilized formulations are presented in Figure 1b. The surfaces of the formulations containing IDPs appear to be more heterogeneous than those without it (CTH0), which is most noticeable in the formulation with a higher IDP content (CTH0.4), which also has a rougher surface. This could be attributed to the presence of IDPs interrupting the spatial distribution of CTHs, leading to more discontinuous materials. In an interesting study, Zhao et al. [30] obtained images of CTHs in the gel state (without active principles added) prepared with different types of acid to protonate the amino groups of chitosan, including the same acid used in the present study (acetic acid). They observed highly heterogeneous structures, which align with the simultaneous aggregation of chitosan–chitosan in multiple cores as the temperature increases in CTHs [18]. EDS analysis revealed that the lyophilized CTH0, CTH0.1, CTH0.2, and CTH0.4 formulations exhibit a high presence of oxygen, phosphorus, carbon, and sodium, which are characteristic elements of chitosan and GP. With this strategy, the presence of iron is also observed in the CTH0.1, CTH0.2, and CTH0.4 formulations (~10–20% w/w), indicating the retention of this element in lyophilized CTHs. This could occur through the physical entrapment of IDPs within the CTHs, allowing retention even when water is extracted. Therefore, the three-dimensional networks formed in the gel state of CTHs are heterogeneous due to the gelation of CTHs mediated by the nucleation–aggregation process and the presence of IDPs, being capable of retaining the iron content after water extraction.

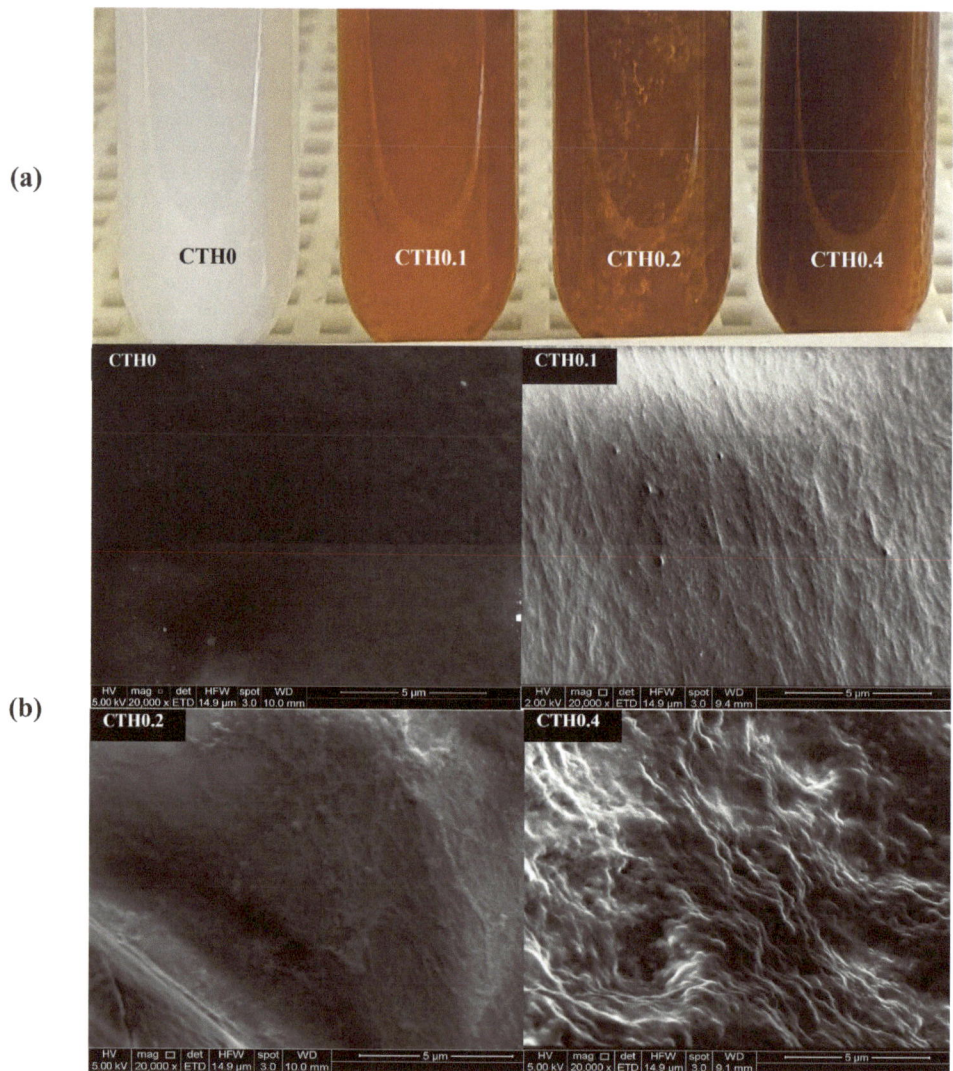

Figure 1. Macroscopic appearance (**a**) and electron microscopy images (**b**) of the thermosensitive chitosan/iron dextran hydrogel formulations (CTHs) with increasing iron concentrations (0.1, 0.2, and 0.4 g of theoretical iron/g of chitosan).

3.2.3. pH

The pH is an important property for the biocompatibility of CTHs, since a change in the proton concentration at the injection site can cause pain in the animal due to the activation of pH-sensitive receptors [24,25]. As shown in Table 1, all CTH formulations have similar pH values and are close to neutrality. These results demonstrate that the addition of IDPs does not have an acidifying effect on the formulations, obtaining pH values close to the physiological range for pigs, humans, and other mammals (~7–7.4) [31]. Therefore, the obtained CTHs show a pH suitable for use as an IM supplement for pigs and other mammals, not requiring pH rectification.

Table 1. pH, sol–gel transition time, iron dextran particle (IDP) content, and viscosity at 4, 25, and 37 °C of the chitosan thermosensitive hydrogel (CTH) formulations with increasing iron concentrations (0.1, 0.2, and 0.4 g of theoretical iron/g of chitosan) and CIDF.

Formulation	pH	Sol–Gel Transition Time (s)	IDP Content (mg/mL)	Viscosity (MPa·s)		
				4 °C	25 °C	37 °C
CTH0	6.88 ± 0.07	90 [a]	-	45 ± 10 [a]	65 ± 13 [a]	2925 ± 108 [b]
CTH0.1	6.83 ± 0.07	120 [b]	2.0 ± 0.3	72 ± 15 [b]	383 ± 33 [b]	2807 ± 284 [b]
CTH0.2	6.62 ± 0.01	120 [b]	4.0 ± 0.8	74 ± 9 [b]	269 ± 40 [c]	3052 ± 421 [b]
CTH0.4	6.68 ± 0.07	300 [c]	13 ± 2	134 ± 14 [c]	447 ± 13 [b]	3060 ± 151 [b]
CIDF	6.38 ± 0.03	-	-	23.8 ± 0.2 [a]	12.9 ± 0.2 [a]	10.2 ± 0.5 [a]

Different letters ([a–c]) indicate significant differences between CTH and CIDF formulations for each of the characterizations ($p < 0.05$).

3.2.4. IDP Content

The IDP content of the CTH formulations is presented in Table 1 and is a direct consequence of the initially added CIDF content (0.1, 0.2, and 0.4 g of theoretical iron/g of chitosan). The IDP content in CTH0.4 is significantly higher than in the other CTH formulations. The iron content in the formulations (in the form of IDPs) could be increased to approach the CIDF content; however, this would result in an excessive increase in the viscosity, due to the high content of IDPs, and a decrease in the concentration of thermosensitive molecules in the formulation.

3.2.5. Viscosity and Sol–Gel Transition Time

The thermosensitivity of CTHs is the most relevant characteristic for IM use, as it enables the smooth injection of the formulation in a sol state and subsequent transition to the gel state within the muscular tissue. As shown in Table 1, the obtained viscosity values demonstrate thermosensitivity in all developed formulations, with similar values at 4 and 25 °C across all groups. However, as the temperature increases to 37 °C, all CTHs exhibit significantly higher and similar viscosity values. Importantly, the transition to the gel state occurs independently of the added IDP content.

The thermosensitivity of these formulations is explained by a series of chemical reactions triggered by the increase in the formulation temperature. At neutral pH, it is not possible to solubilize chitosan in water because it is necessary to protonate its amino groups for the molecule to acquire a polar character and interact with water. However, the addition of GP allows chitosan to not precipitate at neutral pH and remain in a sol state at room temperature (~4–25 °C). This is because GP molecules act as intermediaries in the chitosan–GP–water interaction at neutral pH, allowing chitosan to remain suspended as long as the GP–chitosan interaction between the phosphate and amino groups is maintained. This interaction also prevents the deprotonation of chitosan at neutral pH because its amino groups are shielded by the hydration shell formed by the GP–water interaction [18]. However, increased temperature (higher than 25 °C) promotes the definitive transfer of protons from amino groups to phosphate ions, losing the chitosan–GP interaction, leading to the formation of chitosan–chitosan interactions through hydrogen bonds due to the reduction in chitosan interchain electrostatic repulsion [32]. This proton transference is a direct consequence of a thermosensitive drop in the pKa of chitosan (~−0.025 pK units/°C), promoting neutralization and losing its cationic potential [33]. This chitosan–chitosan union generates non-reversible three-dimensional networks with a nonpolar character, which macroscopically translates into an increase in viscosity, known as the gel state [18].

The significant increase in viscosity between 25 and 37 °C, as observed in Table 1, is advantageous because it allows the formulations to be injected in the sol state close to room temperature (~22 °C), which then substantially increases the viscosity within the muscle tissue. If increased fluidity in IM injection is required, the CTH temperature could be reduced close to 4 °C. This aligns with that previously described in the macroscopic appearance section, where, at room temperature (~22 °C), the formulations appeared viscous and exhibited the formation of small aggregates. This could indicate that the gelation process begins at temperatures <22 °C, with a peak occurring at temperatures higher than 25 °C. The maximum values obtained by the formulations in this study are similar to those reported by [30], ranging from ~2000 to 5000 MPa·s in CTHs synthesized with different materials.

The determination of the time for CTHs to reach the gel state at muscular temperature (~37 °C) is crucial for the development of a potential prolonged-release IDP supplement because the release of the content is likely to be higher prior to the gel state, due to the absence of a force opposing the diffusion of the compound at the injection site. As observed in Table 1, the sol–gel transition time of the CTH formulations increased with the IDP content, being three times longer in CTH0.4 compared to the control. This phenomenon suggests that the reactions responsible for increasing the viscosity require more time to occur in the presence of IDPs and aligns with the fact that the final viscosity values were similar for all formulations (Table 1). This could mean that the formation of a three-dimensional chitosan–chitosan network responsible for the gel state occurs independently of the IDP content, delaying the thermosensitivity but not limiting it. This delay in the transition time could be explained by the presence of non-thermosensitive IDPs and their temporal interference in the chitosan–chitosan binding that leads to the gel state. Finally, these results are consistent with findings from other authors, where transition times of approximately 1–10 min were observed; in those works, the molecular weight and concentration of chitosan were higher [19]. In conclusion, the proposed formulations are thermosensitive, the sol–gel transition time is sensitive to the IDP content, and the tested CTHs show values within appropriate limits for potential IM injection.

3.2.6. Water–Gel Phase Separation

The effective entrapment of IDPs within CTHs in the gel state is necessary to achieve sustained release. To demonstrate this entrapment, the formulations were centrifuged in the gel state to separate the hydrophobic gel and the water in the formulation, forcing the release of IDPs with water since they are highly hydrophilic. This was completed to determine whether the entrapment of IDPs within the chitosan network is sufficient to prevent escape. In Figure 2, the formulations in the gel state before (Figure 2a) and after (Figure 2b) centrifugation are shown. All formulations were separated into two phases, except for CIDF, which keeps the IDPs suspended in water because of its high hydrophilicity. In formulations with phase separation, the upper phase corresponds to the hydrophilic portion of the hydrogel (white arrow, Figure 2b), mainly consisting of GP suspended in water. The lower phase (black arrow, Figure 2b) corresponds to the hydrophobic gel, which was completely separated from the water during centrifugation due to higher density. It is noted that in CTHs containing IDPs, the lower phase (gel) retains the coloration of the initial formulation, indicating that the IDPs (previously suspended) remain in the gel after centrifugation, unlike CIDF. Considering that a chemically attractive interaction between IDPs and chitosan is unlikely due to the respective hydrophilic/hydrophobic characteristics; it is hypothesized that this phenomenon is likely the result of the physical entrapment of IDPs in the chitosan gel. In summary, the action of the CTHs seems to allow IDP retention in chitosan hydrophobic gel at 37 °C.

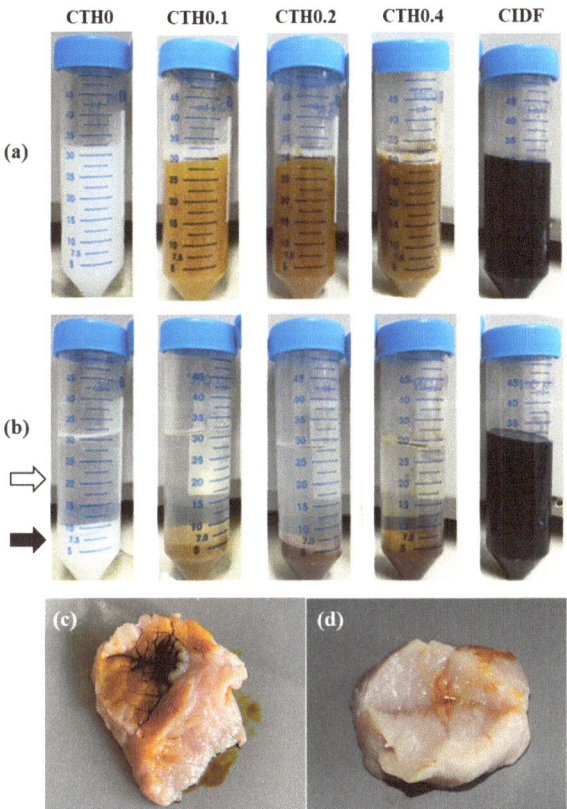

Figure 2. Macroscopic appearance of the chitosan thermosensitive hydrogel (CTH) formulations with increasing iron concentrations (0.1, 0.2, and 0.4 g of theoretical iron/g of chitosan) and the commercial iron dextran formulation (CIDF), showing the separation of the sol–gel phases at 37 °C before (**a**) and after (**b**) centrifugation. The white and black arrows indicate the upper (water) phase and the lower (gel) phase of CTH0, respectively; (**c**,**d**) show the appearance of CIDF and CTH0.4 injected at 37 °C into a piece of pork meat, respectively.

3.2.7. Fourier-Transform Infrared Spectroscopy

The identification of IDP–chitosan interactions and the possible formation of new chemical bonds can help determine the nature of IDP entrapment and anticipate the characteristics of release at the injection site. In Figure 3, the spectra of CTHs in the gel state and the materials used for synthesis (CIDF and chitosan as dry powders) are presented. First, in CTH formulations, an absorption band is observed in the region between 3700 and 3200 cm^{-1}, primarily reflecting O-H bonds from the water molecules highly present in these formulations and N-H interactions from the amino functional group of chitosan [34]. CIDF and CTH formulations show two absorption bands around 3000 cm^{-1}, corresponding to C-H bonds in the aliphatic CH$_2$ and CH$_3$ groups present in the structures of chitosan and dextran [34]. CTH formulations exhibit an absorption band near 1650 cm^{-1}, corresponding to the vibrations of C=O bonds in the NH$_2$ amino group of chitosan monomeric units and O-H bonds from water molecules [34]. In the 1200–1550 cm^{-1} range, CTH formulations, CIDF, and chitosan display absorption bands, mostly corresponding to the characteristic vibrations of O-H, C-H, and C-O bonds inherent to chitosan and dextran structures [34,35]. Likewise, in CTH formulations, within the range of 800–1200 cm^{-1}, there are absorption bands corresponding to -O- and P-O-C bonds characteristic of chitosan and GP structures, respectively [34]. These data appear to reveal that the bonds present in CTH/IDP formulations exhibit absorption bands characteristic of CTHs without IDPs, and there is no evidence of the formation of new bonds or the disappearance of previously existing bonds and specific signals of iron–chitosan interactions, as described by Fahmy and Sarhan et al. [36]. In CT/IDP formulations, there was no observed reduction in the intensity of the absorption band in the 3700–3200 cm^{-1} region or the absorption band near 1650 cm^{-1}, indicating that the bonds of chitosan amino groups show no differences with the addition of iron, and there was no evident appearance of/variation in the characteristic absorption bands of Fe-N or Fe-O interactions [36]. The formation of IDP–chitosan interactions through dextran is unlikely due to dextran's low reactivity, attributed to steric hindrance from its functional groups, and the loss of the cationic potential of chitosan in the gel state. However, it is not possible to rule out the formation of hydrogen bonds between IDPs and chitosan [29]. CTHs containing IDPs exhibit bonds characteristic of CTHs, and there is no evidence of chitosan–IDP interaction.

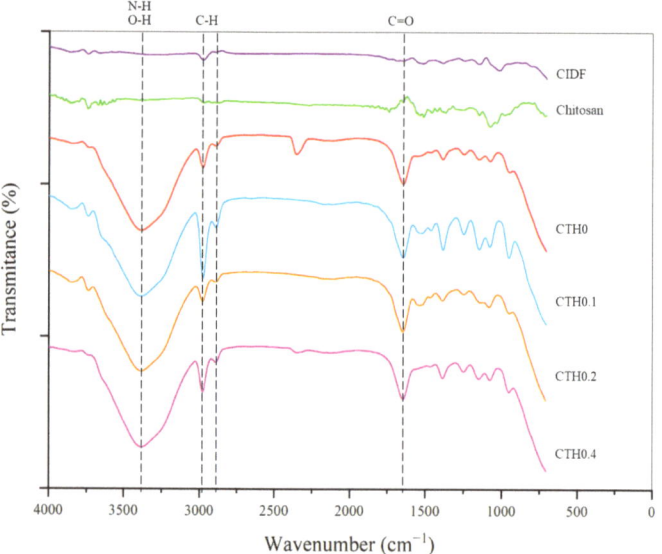

Figure 3. Spectrograms of the chitosan thermosensitive hydrogel (CTH) formulations with increasing iron concentrations (0.1, 0.2, and 0.4 g of theoretical iron/g of chitosan), the commercial iron dextran formulation (CIDF), and precursors. Dotted lines mark the main absorption bands. Data obtained through infrared spectroscopy.

3.2.8. Injectability and Retention Evaluation in Ex Vivo Porcine Tissue

The injectability of the formulations was confirmed through an ex vivo injection test. It was possible to inject the formulation with the higher iron content (CTH0.4) and the gold standard treatment (CIDF) into porcine tissue incubated at 37 °C (Figure 2c,d). CIDF was poorly retained in the injection site (Figure 2c), while CTH0.4 (Figure 2d) was efficiently retained. The low retention of CIDF in injection site can be explained by its high hydrophilicity and low viscosity, which facilitate the rapid diffusion through muscle fibers from the injection site. In contrast, CTH0.4 retention is probably a consequence of the CTH sol–gel transition after its intimate contact with the muscle tissue preheated to 37 °C, triggering the gel formation. The high viscosity in the gel state at 37 °C (Table 1) should promote the retention of IDPs in the injection site. Furthermore, contrasting these results with the water–gel phase separation, it is hypothesized that after the injection of CTHs, the chitosan polymer network forms a viscous gel, concentrating the IDP content at the injection site and prolonging the iron release.

3.2.9. In Vitro Iron Release

With the aim of studying the IDP release from CTHs, we selected a validated methodology commonly used to study formulations with prolonged release (USP apparatus 4). This methodology consists of the supply of a continuous flux of a selected medium in a cell containing the formulation. Although this methodology mimics the dynamic of fluids better than conventional dialysis for parenteral formulations, it is also more demanding due to the constant flux of the medium exposed to the formulation. The release of IDPs from CIDF and CTH0.4 was investigated by quantifying the elemental iron released at 37 °C. Iron release from CIDF reached 100% within 6 h, in contrast to the CTH0.4 formulation, which achieved 40% release over 72 h (Figure 4). This prolonged release from CTH0.4 is attributed to the sol–gel transition of the chitosan thermosensitive hydrogel at the medium's temperature (37 °C), forming a hydrophobic gel that reduces IDP diffusion. In contrast, the rapid release from CIDF is due to the nanometric size and hydrophilic nature of IDPs, facilitating their quick diffusion. CTH gelation creates an effective polymeric barrier that confines IDPs, leading to their prolonged release. The brief duration of iron release observed in this study is consistent with the reported maximum iron concentrations in porcine blood within the first 12 h following CIDF administration [14], which can be attributed to the quick release and subsequent metabolism of IDPs.

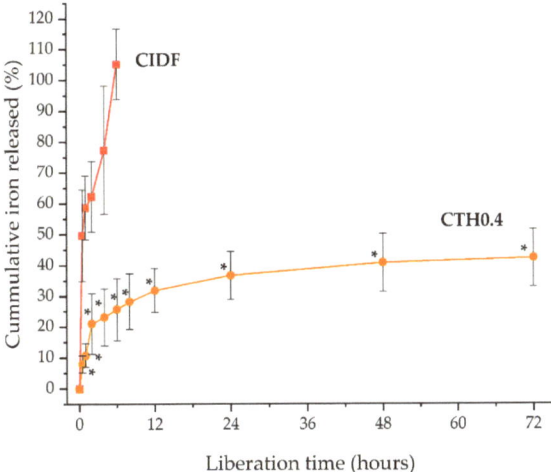

Figure 4. Cumulative iron release from commercial iron dextran formulation (CIDF) and the chitosan thermosensitive hydrogel containing iron dextran particles (CTH0.4) at 37 °C. * Significative differences between formulation values for each measurement time ($p < 0.05$).

The release mechanisms of the CIDF and CTH0.4 formulations were elucidated by fitting the release data to various models (Table 2). The Korsmeyer–Peppas model emerged as the most suitable, indicated by the highest correlation coefficient (R^2), the lowest Akaike information criterion (AIC), and the highest model selection criterion (MSC) [37]. The diffusional exponent (n) values for both CIDF and CTH0.4 (both below 0.5) suggest Fickian diffusion as the predominant release mechanism. This aligns with the rapid liberation observed for CIDF in Figure 4, attributed to constant diffusion driven by the IDP concentration gradient [38]. Conversely, the initial rapid release, followed by a slower release pattern of IDPs from CTH0.4, as shown in Figure 4, implies a similar initial release mechanism to CIDF for non-entrapped IDPs, with a posterior slower release of entrapped IDPs from the gel matrix. The behavior of CTH0.4 could be considered as positive, the initial quick diffusion release may be used for the immediate iron requirements, and the prolonged iron release from the gel could be useful to maintain the biological effect over the time.

Table 2. Kinetic model parameters for CIDF and CTH0.4 release data.

Formulation	Release Model	R^2	AIC	MSC	n
CIDF	Zero order	0.30	52.00	−0.70	
	First order	0.80	44.73	0.50	
	Higuchi	0.81	44.60	0.53	
	Korsmeyer–Peppas	0.88	40.94	1.14	0.27
CTH0.4	Zero order	−0.48	88.92	−0.92	
	First order	−0.17	86.40	−0.69	
	Higuchi	0.60	73.93	0.44	
	Korsmeyer–Peppas	0.91	60.88	1.62	0.33

Figure 5 proposes the interaction dynamics between the different components of the formulation (chitosan, GP, IDPs, and water) in sol and gel states. In the sol state (Figure 5a), the dominant forces are the interactions of water–GP and water–IDP hydrogen bonds, which keep these components suspended, in addition to the electrostatic GP–chitosan bond that prevents the precipitation of the polymer as it is anchored to water through the GP [18]. When transitioning to the gel state (Figure 5b), the increase in temperature generates a decrease in the cationic potential of chitosan, resulting in the loss of the chitosan–GP interaction and the generation of chitosan–chitosan hydrogen bonds, which give origin to the high-viscosity hydrophobic network [18]. Based on the results obtained by and discussed in the present study, it is proposed that IDPs are confined in the chitosan network, which is a physical barrier that stands in the way of diffusion, prolonging IDPs' liberation. Finally, based on all the characterizations included in this article, future studies need to be focused on improving and optimizing the potential shortcoming of our strategy and carrying out an in vivo study in pigs. These improvements should focus on increasing the iron load, improving the CTH homogeneity, and increasing the gelation temperature to avoid any transition at <25 °C, to ensure an adequate injectability in the sol state. In contrast, the irreversible thermosensibility for the obtained materials makes it difficult to use them in high-temperature regions (>25 °C) because the materials can be transformed to the gel state before the injection; in those cases, the CTHs can be refrigerated and injected at a lower temperature.

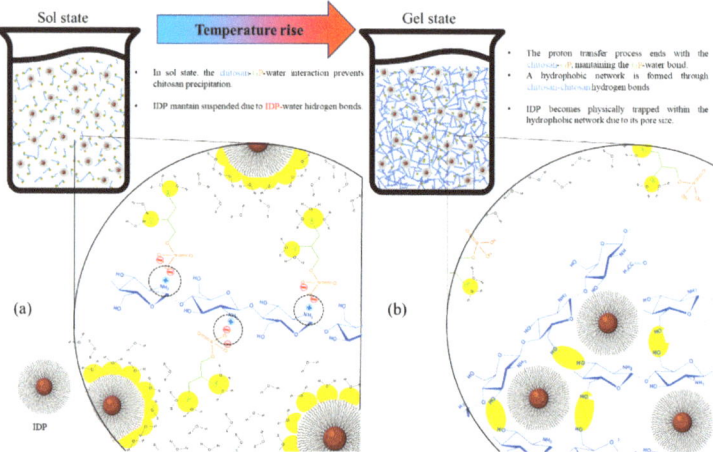

Figure 5. Two-dimensional schematic representation of interactions among chitosan, glycerophosphate (GP), water, and the iron dextran particles (IDPs) in a chitosan thermosensitive hydrogel in sol state (**a**) and gel state (**b**). Dotted circles represent chitosan–GP ionic bonds, and yellow areas represent interactions involving GP–water, chitosan–chitosan, and water–IDP hydrogen bonding.

4. Conclusions

CTHs loaded with IDPs were successfully synthesized, obtaining thermosensitive formulations with neutral pH, suitable for IM injection in a sol state at approximately 4 °C and transitioning into a gel state between 25 and 37 °C. It was determined that IDPs are effectively entrapped within the chitosan polymer network in the gel state, without chemical interactions between chitosan–IDP or GP–IDP. In vitro release studies revealed that CIDF released 100% of its iron content within 6 h, while CTH0.4 released 40% over 72 h, predominantly through Fickian diffusion. The pronounced difference in release profiles was primarily due to the physical confinement of IDPs within the CTH gel, which acted as an additional diffusion barrier. These findings contributed valuable insights on CTHs as a strategy for prolonged iron supplementation and laid the groundwork for developing new prolonged-release micronutrient supplementation approaches for pigs, humans, and other mammals.

Author Contributions: Conceptualization, E.D. and C.V.; methodology, E.D., A.N.-C., F.O.-A., and C.V.; validation, F.O.-A. and C.V.; formal analysis, E.D., F.O.-A., A.N.-C., and C.V.; investigation, E.D.; resources, C.V.; data curation, E.D.; writing—original draft preparation, E.D.; writing—review and editing, E.D., F.O.-A., A.N.-C., and C.V.; visualization, E.D.; supervision, F.O.-A. and C.V.; project administration, C.V.; funding acquisition, C.V. All authors have read and agreed to the published version of the manuscript.

Funding: This study was supported by the grants FONDECYT 1200109 and FONDECYT 1201899.

Institutional Review Board Statement: Not applicable.

Data Availability Statement: Data are contained within the article.

Conflicts of Interest: The authors declare no conflicts of interest.

References

1. Camaschella, C. Iron-deficiency anemia. *N. Eng. J. Med.* **2015**, *372*, 1832–1843. [CrossRef] [PubMed]
2. Gardner, W.; Kassebaum, N. Global, regional, and national prevalence of anemia and its causes in 204 countries and territories, 1990–2019. *Curr. Dev. Nutr.* **2020**, *4*, 830. [CrossRef]
3. Schönfeldt, H.; Hall, N. Determining iron bio-availability with a constant heme iron value. *J. Food Compos. Anal.* **2011**, *24*, 738–740. [CrossRef]
4. Zimmermann, M.; Hurrell, R. Nutritional iron deficiency. *Lancet* **2007**, *370*, 511–520. [CrossRef] [PubMed]
5. Shubham, K.; Anukiruthika, T.; Dutta, S.; Kashyap, A.; Moses, J.; Anandharamakrishnan, C. Iron deficiency anemia: A comprehensive review on iron absorption, bioavailability and emerging food fortification approaches. *Trends Food Sci. Tech.* **2020**, *99*, 58–75. [CrossRef]
6. Nash, C.; Allen, V. The use of parenteral iron therapy for the treatment of postpartum anemia. *J. Obstet. Gyn. Can.* **2015**, *37*, 439–442. [CrossRef] [PubMed]
7. Avni, T.; Bieber, A.; Grossman, A.; Green, H.; Leibovici, L.; Gafter-Gvili, A. The safety of intravenous iron preparations: Systematic review and meta-analysis. *Mayo Clin. Proc.* **2015**, *90*, 12–23. [CrossRef]
8. Svoboda, M.; Drábek, J. Intramuscular versus subcutaneous administration of iron dextran in suckling piglets. *Acta Vet. Brno* **2007**, *76*, 11–15. [CrossRef]
9. Antileo, R.; Figueroa, J.; Valenzuela, C. Characterization of an encapsulated oral iron supplement to prevent iron–deficiency anemia in neonatal piglets. *J. Anim. Sci.* **2016**, *94*, 157–160. [CrossRef]
10. Perri, A.; Friendship, R.; Harding, J.; O'sullivan, T. An investigation of iron deficiency and anemia in piglets and the effect of iron status at weaning on post-weaning performance. *J. Swine Health Prod.* **2016**, *24*, 10–20.
11. Scheinberg, P.; Chen, J. Aplastic anemia: What have we learned from animal models and from the clinic. *Semin. Hematol.* **2013**, *50*, 156–164. [CrossRef] [PubMed]
12. García, Y.; Díaz-Castro, J. Advantages and disadvantages of the animal models v. In vitro studies in iron metabolism: A review. *Animal* **2013**, *7*, 1651–1658. [CrossRef] [PubMed]
13. London, E. The molecular formula and proposed structure of the iron–dextran complex, imferon. *J. Pharm. Sci.* **2004**, *93*, 1838–1846. [CrossRef] [PubMed]
14. Morales, J.; Manso, A.; Martín-Jiménez, T.; Karembe, H.; Sperling, D. Comparison of the pharmacokinetics and efficacy of two different iron supplementation products in suckling piglets. *J. Swine Health Prod.* **2018**, *26*, 200–207. [CrossRef] [PubMed]

15. Starzyński, R.; Laarakkers, C.; Tjalsma, H.; Swinkels, D.; Pieszka, M.; Styś, A.; Mickiewicz, M.; Lipiński, P. Iron supplementation in suckling piglets: How to correct iron deficiency anemia without affecting plasma hepcidin levels. *PLoS ONE* **2013**, *8*, e64022. [CrossRef] [PubMed]
16. Toxqui, L.; Piero, A.; Courtois, V.; Bastida, S.; Sánchez-Muniz, F.; Vaquero, M. Iron deficiency and overload. Implications in oxidative stress and cardiovascular health. *Nutr. Hosp.* **2010**, *25*, 350–365. [PubMed]
17. Lipiński, P.; Starzyński, R.; Canonne-Hergaux, F.; Tudek, B.; Oliński, R.; Kowalczyk, P.; Dziaman, P.; Thibaudeau, O.; Gralak, M.; Smuda, E.; et al. Benefits and risks of iron supplementation in anemic neonatal pigs. *Am. J. Pathol.* **2010**, *177*, 1233–1243. [CrossRef]
18. Supper, S.; Anton, N.; Seidel, N.; Riemenschnitter, M.; Curdy, C.; Vandamme, T. Thermosensitive chitosan/glycerophosphate-based hydrogel and its derivatives in pharmaceutical and biomedical applications. *Expert Opin. Drug Del.* **2014**, *11*, 249–267. [CrossRef]
19. Zhou, H.; Jiang, L.; Cao, P.; Li, J.; Chen, X. Glycerophosphate-based chitosan thermosensitive hydrogels and their biomedical applications. *Carbohydr. Polym.* **2015**, *117*, 524–536. [CrossRef]
20. Zhou, H.; Zhang, Y.; Zhang, W.; Chen, X. Biocompatibility and characteristics of injectable chitosan-based thermosensitive hydrogel for drug delivery. *Carbohydr. Polym.* **2011**, *83*, 1643–1651. [CrossRef]
21. Kolawole, O.; Lau, W.; Khutoryanskiy, V. Chitosan/β-glycerophosphate in situ gelling mucoadhesive systems for intravesical delivery of mitomycin-c. *Int. J. Pharm.* **2019**, *1*, 100007. [CrossRef] [PubMed]
22. Sun, B.; Ma, W.; Su, F.; Wang, Y.; Liu, J.; Wang, D.; Liu, H. The osteogenic differentiation of dog bone marrow mesenchymal stem cells in a thermo-sensitive injectable chitosan/collagen/beta-glycerophosphate hydrogel: In vitro and in vivo. *J. Mater. Sci-mater. Med.* **2011**, *22*, 2111–2118. [CrossRef]
23. Wang, W.; Wat, E.; Hui, P.; Chan, B.; Ng, F.; Kan, C.; Wang, X.; Hu, H.; Wong, E.; Lau, C.; et al. Dual-functional transdermal drug delivery system with controllable drug loading based on thermosensitive poloxamer hydrogel for atopic dermatitis treatment. *Sci. Rep.* **2016**, *6*, 24112. [CrossRef] [PubMed]
24. Żelechowska, E.; Przybylski, W.; Jaworska, D.; Santé-Lhoutellier, V. Technological and sensory pork quality in relation to muscle and drip loss protein profiles. *Eur. Food Res. Technol.* **2012**, *234*, 883–894. [CrossRef]
25. Groves, M. *Parenteral Products: The Preparation and Quality Control of Products for Injection*; Elsevier: Amsterdam, The Netherlands, 2014.
26. Rychen, G.; Aquilina, G.; Azimonti, G.; Bampidis, V.; Bastos, M.; Mantovani, A. Safety and efficacy of iron dextran as a feed additive for piglets. *Efsa J.* **2017**, *15*, e04701. [CrossRef] [PubMed]
27. Juluri, A.; Modepalli, N.; Jo, S.; Repka, M.; Shivakumar, H.; Murthy, S. Minimally invasive transdermal delivery of iron–dextran. *J. Pharm. Sci.* **2013**, *102*, 987–993. [CrossRef] [PubMed]
28. Honary, S.; Zahir, F. Effect of zeta potential on the properties of nano-drug delivery systems-a review (part 1). *Trop. J. Pharm. Res.* **2013**, *12*, 255–264. [CrossRef]
29. Xu, X.Q.; Shen, H.; Xu, J.R.; Xu, J.; Li, X.J.; Xiong, X.M. Core-shell structure and magnetic properties of magnetite magnetic fluids stabilized with dextran. *Appl. Surf. Sci.* **2005**, *252*, 494–500. [CrossRef]
30. Zhao, Q.; Cheng, X.; Ji, Q.; Kang, C.; Chen, X. Effect of organic and inorganic acids on chitosan/glycerophosphate thermosensitive hydrogel. *J. Sol-Gel Sci. Technol.* **2009**, *50*, 111–118. [CrossRef]
31. Rahmati, M.; Samadikuchaksaraei, A.; Mozafari, M. Insight into the interactive effects of β-glycerophosphate molecules on thermosensitive chitosan-based hydrogels. *Bioinspired Biomim. Nanobiomater.* **2016**, *5*, 67–73. [CrossRef]
32. Liu, Y.; Lang, C.; Ding, Y.; Sun, S.; Sun, G. Chitosan with enhanced deprotonation for accelerated thermosensitive gelation with β-glycerophosphate. *Eur. Polym. J.* **2023**, *196*, 112229. [CrossRef]
33. Filion, D.; Lavertu, M.; Buschmann, M. Ionization and solubility of chitosan solutions related to thermosensitive chitosan/glycerol-phosphate systems. *Biomacromolecules* **2007**, *8*, 3224–3234. [CrossRef] [PubMed]
34. Skwarczynska, A.; Kaminska, M.; Owczarz, P.; Bartoszek, N.; Walkowiak, B.; Modrzejewska, Z. The structural (FTIR, XRD, and XPS) and biological studies of thermosensitive chitosan chloride gels with β-glycerophosphate disodium. *J. Appl. Polym. Sci.* **2018**, *135*, 46459. [CrossRef]
35. Deng, A.; Kang, X.; Zhang, J.; Yang, Y.; Yang, S. Enhanced gelation of chitosan/β-sodium glycerophosphate thermosensitive hydrogel with sodium bicarbonate and biocompatibility evaluated. *Mater. Sci. Eng.* **2017**, *78*, 1147–1154. [CrossRef]
36. Fahmy, T.; Sarhan, A. Characterization and molecular dynamic studies of chitosan–iron complexes. *Bull. Mater. Sci.* **2021**, *44*, 142. [CrossRef]
37. Villamizar-Sarmiento, M.G.; Guerrero, J.; Moreno-Villoslada, I.; Oyarzun-Ampuero, F.A. The key role of the drug self-aggregation ability to obtain optimal nanocarriers based on aromatic-aromatic drug-polymer interactions. *Eur. J. Pharm. Biopharm.* **2021**, *166*, 19–29. [CrossRef]
38. Zhang, Y.; Huo, M.; Zhou, J.; Zou, A.; Li, W.; Yao, C.; Xie, S. DDSolver: An add-in program for modeling and comparison of drug dissolution profiles. *AAPS J.* **2010**, *12*, 263–271. [CrossRef]

Disclaimer/Publisher's Note: The statements, opinions and data contained in all publications are solely those of the individual author(s) and contributor(s) and not of MDPI and/or the editor(s). MDPI and/or the editor(s) disclaim responsibility for any injury to people or property resulting from any ideas, methods, instructions or products referred to in the content.

Article

Influence of Chitosan 0.2% in Various Final Cleaning Methods on the Bond Strength of Fiberglass Post to Intrarradicular Dentin

Naira Geovana Camilo [1], Alex da Rocha Gonçalves [1], Larissa Pinzan Flauzino [2], Cristiane Martins Rodrigues Bernardes [1], Andreza Maria Fábio Aranha [2], Priscilla Cardoso Lazari-Carvalho [3], Marco Aurélio de Carvalho [3] and Helder Fernandes de Oliveira [1,*]

1. Department of Endodontics, School of Dentistry, Evangelical University of Goiás, Anápolis 75083-515, GO, Brazil; nairageovana15@gmail.com (N.G.C.); dralexdarocha@gmail.com (A.d.R.G.); cristiane.bernardes@unievangelica.edu.br (C.M.R.B.)
2. Department of Oral Biology, School of Dentistry, University of Cuiabá, Cuiabá 78065-900, MT, Brazil; larissa_pinzan@hotmail.com (L.P.F.); andreza.maria@kroton.com.br (A.M.F.A.)
3. Department of Restorative Sciences, School of Dentistry, Evangelical University of Goiás, Anápolis 75083-515, GO, Brazil; priscilla.lazari@docente.unievangelica.edu.br (P.C.L.-C.); marco.carvalho@docente.unievangelica.edu.br (M.A.d.C.)
* Correspondence: helder.oliveira@docente.unievangelica.edu.br

Citation: Camilo, N.G.; Gonçalves, A.d.R.; Flauzino, L.P.; Bernardes, C.M.R.; Aranha, A.M.F.; Lazari-Carvalho, P.C.; Carvalho, M.A.d.; Oliveira, H.F.d. Influence of Chitosan 0.2% in Various Final Cleaning Methods on the Bond Strength of Fiberglass Post to Intrarradicular Dentin. *Polymers* 2023, *15*, 4409. https://doi.org/10.3390/polym15224409

Academic Editors: Sérgio Paulo Campana-Filho and William Facchinatto

Received: 11 October 2023
Revised: 6 November 2023
Accepted: 7 November 2023
Published: 15 November 2023

Copyright: © 2023 by the authors. Licensee MDPI, Basel, Switzerland. This article is an open access article distributed under the terms and conditions of the Creative Commons Attribution (CC BY) license (https:// creativecommons.org/licenses/by/ 4.0/).

Abstract: The purpose of this study was to analyze the influence of Chitosan 0.2% in various final cleaning methods on the bond strength of fiberglass post (FP) to intrarradicular dentin. Ninety bovine incisors were sectioned to obtain root remnants measuring 18 mm in length. The roots were divided: G1: EDTA 17%; G2: EDTA 17% + PUI; G3: EDTA 17% + EA; G4: EDTA 17% + XPF; G5: Chitosan 2%; G6: Chitosan 2% + PUI; G7: Chitosan 2% + EA; G8: Chitosan 2% +XPF. After carrying out the cleaning methods, the posts were installed, and the root was cleaved to generate two disks from each root third. Bond strength values (MPa) obtained from the micro push-out test data were assessed by using Kruskal–Wallis and Dwass–Steel–Critchlow–Fligner tests for multiple comparisons (α = 5%). Differences were observed in the cervical third between G1 and G8 (p = 0.038), G4 and G8 (p = 0.003), G6 and G8 (p = 0.049), and Control and G8 (p = 0.019). The final cleaning method influenced the adhesion strength of cemented FP to intrarradicular dentin. Chitosan 0.2% + XPF positively influenced adhesion strength, with the highest values in the cervical third.

Keywords: dentin-bonding agent; chelators; fiber post; ultrasonics

1. Introduction

Endodontically managed teeth presenting a considerable coronal structure loss after endodontic treatment imposes the necessity of fiberglass posts (FP) associated with resin cements to restore the biomechanical form and function [1]. The recommendation of these materials is based on characteristics of durability, aesthetics, and operational cost, which make them a very favorable option for the restoration of the teeth in question [2]. However, it must be considered that during the preparation for the intraradical retainer, the formation smear residue resulting from the infected dentin debris, remains of gutta-percha and obturator cement can negatively influence the polymerization processes of resin cement and consequently the final cementation quality of the FP which is essential for successful adhesion [3].

The union of demineralized intrarradicular dentin with the cement and the FP is based on micromechanical retention and therefore, dentine cleaning of the walls of the intraradicular dentin is imperative for ideal retention of the post [4]. In this sense, the efficiency of endodontic irrigation in the sanitization and cleaning stages of the root canal becomes fundamental in reaching areas with difficult access that were untouched during the instrumentation stage, such as isthmuses and lateral canals [5]. Technological advances have allowed irrigating solutions to be agitated within the root canal for more effective smear layer clearance through mechanical, sonic, or ultrasonic agitation methods [6]. Among the solutions tested to remove this smear layer, EDTA and its combinations (mainly sodium hypochlorite) are the most used due to their chelating properties [7]. However, the prolonged use of EDTA can cause erosion in the dentin matrix, thus compromising the bond strength between the post and dentin [8,9]. Thus, there is a search for alternative solutions that are more biocompatible than EDTA in an effort to reduce these possible damages [7,10]. As an alternative, chitosan has been increasingly studied because it is a natural solution, is biocompatible with tissues, and has adequate properties of biodegradability, bioadhesion, and low cytotoxicity [11–16]. The chelating capacity of chitosan has attracted substantial interest due to its strong affinity for various metal ions in acidic pH environments [17,18]. This unique feature has expanded the potential of chitosan as a substitute for EDTA, which not only causes erosion but also poses environmental risks [19]. Previous studies have highlighted the favorable impact of chitosan in minimal concentrations and short-term applications on the demineralization of intraradicular dentin. Furthermore, it presents a cleaning capacity similar to other chelating agents used in clinical practice, such as citric acid and EDTA [7,10,11,17,19].

There is still a notorious absence of consensus about the real influence of post preparation cleaning procedures on the adhesion of FP to radicular dentin [4,20]. So far, there has been a lack of studies evaluating the influence of chitosan in association with various agitation methods, on the adhesion process, and on the biomechanical bond strength of FP. Therefore, it seems opportune to investigate the influence of the final cleaning method on the adhesive force between the post and luting agent and intrarradicular dentin. The null hypotheses assessed were that there would be no difference in the level of bond strength depending on (i) the chelating solution, (ii) the chelator activation method, and (iii) the third of the radicular canal.

2. Materials and Methods

The research was approved by the Animal Ethics Boards of the Universidade Evangélica de Goiás, Brazil (#001/2021). Three hundred extracted bovine lower incisors with fully developed roots, anatomically analogous in size and shape [21–23] were obtained and stored in 0.2% thymol solution (Fitofarma, Goiânia, GO, Brazil). Periapical radiographs were obtained to verify the samples' normality, and only teeth with a unique root canal without obliterations were included in the study. In total, 90 samples were utilized.

2.1. Endodontic Instrumentation and Obturation

The crowns of the teeth were sectioned using a dual-sided diamond disc (KG Sorensen, Sao Paulo, SP, Brazil) perpendicular to its long axis obtaining standardized roots 18 mm in length from the apical end. A #15 K-file (Dentsply Maillefer, Ballaigues, Switzerland) was used to verify the patency of all root canals.

The working length (WL) was established using a #15 K file (Dentsply Maillefer), which was introduced into the root canal until it was visible in the apical foramen. The WL was set 1 mm short of this measurement. To simulate clinical conditions, the root apexes were sealed with flow composite (Top Dam, Dental Products, São Paulo, SP, Brazil). ProTaper® Gold instruments (Dentsply) were utilized for endodontic instrumentation. The channels were instrumented until reaching the instrument F5 (50/0.05). Each instrument was used in the instrumentation of just five root canals through X-Smart Plus endodontic motor (Dentsply), with speed and torque standards established by the manufacturer. During instrumentation, the canals were irrigated with 4 mL of 2.5% sodium hypochlorite (Fitofarma, Goiânia, GO, Brazil). The root canals were irrigated with 17% EDTA (Biodinamica, Ibiporã, PR, Brazil) for 3 min to remove the smear layer.

The roots were subsequently dried using absorbent paper points (Dentsply, Charlotte, NC, USA) and then filled with gutta-percha cones and epoxy resin-based cement (AH Plus;

Dentsply), mixed according to the instructions of the manufacturer using Tagger's hybrid technique. The canal access was sealed with micro-hybrid composite resin (TPH Spectrum, Dentsply Brazil, São Paulo, SP, Brazil). All roots were stored at 37 °C and 100% humidity for 7 days to allow the cement to light cure.

2.2. Post-Space Preparation

After the obturation, heated condensers (Paiva; SS White, Piscataway, NJ, USA) were used to remove the initial portion of the root canal filling mass. The conduits were prepared to a depth of 14 mm using Largo drills #3–5 (Dentsply Maillefer), corresponding to fiber posts of 1.5 mm in diameter (Reforpost #3; Angelus, Londrina, Brazil) [1,21]. Root canals were irrigated with 4 mL of 2.5% NaOCl after each drill change and dried with absorbent paper cones.

2.3. Experimental Groups

The samples were randomly distributed into eight experimental groups and a control group, in accordance with the chelating agent tested and the activation method (Figure 1).

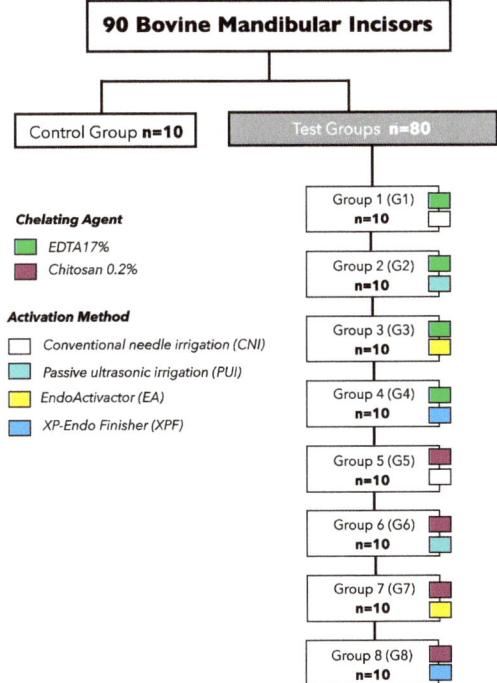

Figure 1. The allocation of experimental groups based on the chelating solution and activation method examined.

2.4. Formulation of Chelating Solutions

The solutions were formulated in a compounding pharmacy (Fitofarma, Troyan, Bulgaria) and were prepared with analytical grade reagents and water purified by a Reverse Osmosis system with Ultraviolet Light (Quimis, Diadema, SP, Brazil) with electrical conductivity lower than 1 µS mm -two. The pH of the solutions was determined with a digital pH meter (Analion, Ribeirão Preto, SP, Brazil). The 0.2% chitosan solution was prepared with 0.2 g of chitosan (ACROS Organics Gell, Belgium; degree of deacetylation

>90%) in 100 mL of 1% acetic acid. The mixture was stirred using a magnetic stirrer at 100 °C in 200 rpm for 2 h [7,11].

CNI

A total volume of 4 mL of 2.5% NaOCl, 4 mL of each chelating agent, and another 4 mL of NaOCl was introduced into the root canals using a 5 mL disposable syringe (Ultradent, Tokyo, Japan) and a 29-gauge needle (NaviTip; Ultradent, Tokyo, Japan). The needle was inserted 1 mm short of the cementoenamel junction (CT) without coming into contact with the canal walls. Each chelating agent was allowed to remain in the canal for a duration of 3 min without undergoing any activation process.

PUI

PUI was conducted in 3 cycles of 20 s each with 2 mL of the solution per cycle. The solutions were passively activated using EMS PM 200 ultrasound (EMS, Nyon, Switzerland) and a E1-Irrisonic tip (Helse, Sao Paulo, Brazil) positioned 1 mm short of the WL, without touching the walls of the root canals, so that it vibrated freely. The ultrasonic unit was adjusted to 10% power following the manufacturer's specifications for the use of the insert [9,24].

EA

Three activation cycles were performed as previously described. The solutions were activated with the EndoActivator system (Dentsply Maillefer) and a medium activator tip (25/0.04), which was inserted 1 mm from the WL for 20 s (each cycle with 2 mL of the solution) at 10,000 cycles per minute.

XPF

Three activation cycles were performed as previously described. The solutions were activated with the XP-Endo Finisher (25/0.00) instrument (FKG Dentaire, La Chaux-de-Fonds, Switzerland), which was inserted 1 mm short of the WL. The instrument operated at a speed of 800 rpm and torque of 1 Ncm. Slow and smooth movements of penetration and withdrawal were performed for 20 seconds (each cycle with 2 mL of the solution). The cleaning methods were completed, and the canals were washed with 4 mL of saline solution and dried with absorbent paper tips.

2.5. Fiber Post Cementation

After applying a thin layer of utilitarian wax on the external surfaces of the roots to prevent lateral polymerization resulting from the photoactivation of the cement, the post underwent a 15-second cleaning with 70% alcohol. Subsequently, the silane (Silane, Angelus) was applied for 1 minute using a micro brush (KG Sorensen, Sao Paulo, SP, Brazil). The self-adhesive resin cement (RelyX U200; 3M-ESPE, St Paul, MN, USA) was manipulated according to the manufacturer's instructions and inserted into each root canal with the assistance of a lentulo spiral instrument (Dentsply Maillefer) and applied to the surface of the fiberglass post. The post was inserted into the canal with appropriate digital pressure, removing excess cement with a clean micro brush (KG Sorensen) after one minute.

Three minutes later, the cement was light-cured using a 1200 mW/cm^2 intensity source (Radii-Cal; SDI, Bayswater, Australia) for 40 seconds on the cervical region, along the long axis of the root, and obliquely on the buccal and lingual surfaces, totaling 120 seconds per root. The dentin-cement-post interface was sealed with composite resin to ensure a hermetic seal of the root canal, ensuring the integrity and stability of the procedure.

2.6. Root Sectioning Procedure

In the meticulous process of root sectioning, each root underwent careful transverse cutting using a double-sided diamond disc (4" diameter × 0.012" thickness × 1/2"; Arbor, Extec, Enfield, CT, USA) mounted on a specialized hard tissue microtome (Isomet 1000, Buehler, Lake Bluff, IL, USA) set at a low speed (400 rpm), ensuring precision and accuracy.

Throughout the procedure, a continuous flow of water provided effective cooling. From different segments of the root—cervical, middle, and apical—two 1 mm thick discs were carefully obtained, yielding a total of 6 discs per root. These dentin discs were precisely cut at distinct measurements: 11 and 12 mm from the root apex for the cervical region, 8 and 9 mm for the middle region, and 5 and 6 mm for the apical region, ensuring comprehensive representation of each third of the root. This standardized method of sectioning, utilizing advanced equipment and precise measurements, guaranteed the consistency and reliability of the obtained samples, forming the foundation for subsequent analyses and evaluation.

2.7. Micro Push-Out Mechanical Test (MPMT)

For conducting the micro push-out test, a specially designed apparatus was employed, crafted with stainless-steel metal bases measuring 3 cm in diameter [25–27]. These bases featured central holes of 2, 3.5, and 4.5 mm (as shown in Figure 2A), accompanied by load applicator tips measuring 1, 1.75, and 2 mm in diameter (as illustrated in Figure 2B). After positioning the set on the base of the mechanical testing machine (Microtensile OM150, Odeme Dental Research, Luzerna, Brazil) (Figure 2C) containing a 10 Kgf load cell, the discs were positioned in the hole of the metal base and the set was aligned to the tip load applicator (Figure 2D). They were then subjected to compression loading in the apex/crown direction at a speed of 0.5 mm/minute until failure occurred. The displacement force values were obtained in KgF which was transformed into Newton. The bond strength, in MPa, was calculated by dividing the force (N) by the area of the adhesive interface. The area of the adhesive interface was calculated by multiplying the height of the disc by the perimeter of the channel lumen, which was considered an ellipse: $A = h \times (\pi (3(a+b) - \sqrt{(3a+b)(a+3b)}))$ where h is the height of the disk, a is the largest radius, and b is the smallest radius.

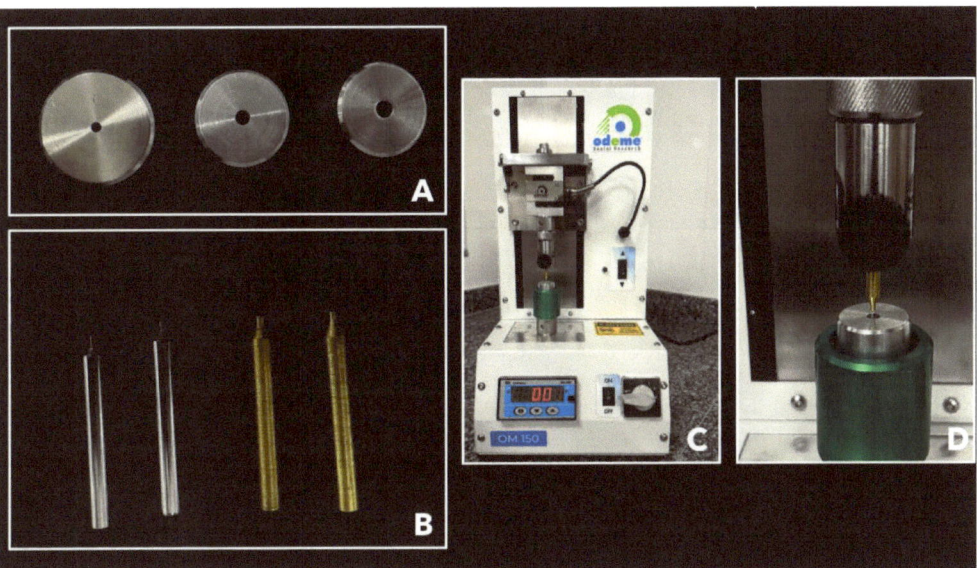

Figure 2. (**A**) Metal bases with 2, 3.5, and 4.5 mm diameter holes in the central region, (**B**) Applicator tips with 1, 1.75, and 2 mm diameter, (**C**) Positioning of the set on the base of the mechanical testing machine (Microtensile OM150, Odeme Dental Research, Brazil), (**D**) Discs aligned in a manner where the load applicator tip precisely matched the orifice in the metallic base.

2.8. Analysis of the Failure Mode by Optical Microscopy

After the mechanical test, each specimen was stored individually in Eppendorf-type microtubes with distilled water for later analysis of the fracture pattern by using 40× optical microscopy without any type of treatment or previous preparation. All samples were analyzed with the aid of an optical microscope (Carl Zeiss, META, Berlin, Germany). The images were processed with the help of the Zeiss LSM Image Browser software version 4.2. Photomicrographs were always obtained with the same increase for all specimens. Failure modes were classified into six categories: Mode 1—adhesive between the post and resin cement, Mode 2—adhesive between resin cement and intrarradicular dentine, Mode 3—mixed, between post, resin cement, and intrarradicular dentine, Mode 4—cohesive in cement, Mode 5—cohesive in the post, and Mode 6—cohesive in dentine (Figure 3).

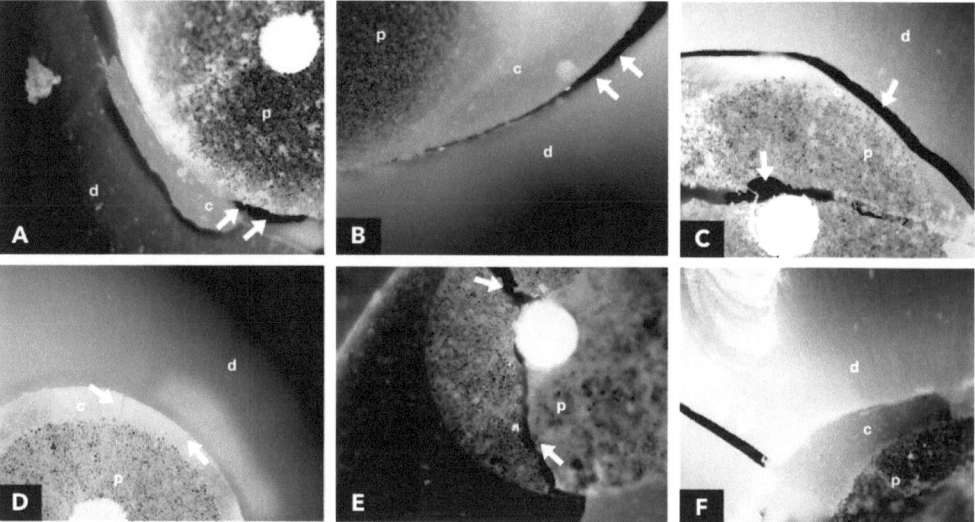

Figure 3. Failure Mode Types: (**A**) Mode 1, (**B**) Mode 2, (**C**) Mode 3, (**D**) Mode 4, (**E**) Mode 5 (**F**) Mode 6. Arrows point to the failure region. d-dentine, c-cement, and p-post.

2.9. Data Analysis

The data analysis was conducted using Jamovi 1.1.9 software (The Jamovi Project, 2019). Various cleaning methods were compared by examining the frequency of failure modes expressed as a percentage within each tested group. The bond strength values (in MPa) obtained from the micro push-out mechanical test underwent rigorous assessment through the Kruskal–Wallis test, followed by the Dwass–Steel–Critchlow–Fligner test for detailed multiple comparisons ($\alpha = 5\%$). Intra-examiner agreement was meticulously assessed using the kappa coefficient, applied to 10% of the sample, ensuring the reliability of the results.

3. Results

The kappa coefficient, standing at 0.86, demonstrated a high level of intra-examiner agreement. Table 1 shows the medians and interquartile ranges for the diverse groups examined.

Table 1. The median and interquartile range (IQR 25–75%), representing the bond strength values in the tested groups categorized by different root thirds.

Experimental Groups	Cervical Third		Middle Third		Apical Third		p-Value
	Median	IQR (25–75%)	Median	IQR (25–75%)	Median	IQR (25–75%)	
Control Group	2.48 [A,a]	1.87–3.81	2.00 [A]	1.58–3.05	2.72 [A]	1.39–3.79	0.387
G1. EDTA 17%	2.54 [A,a]	1.79–3.74	3.42 [A,a]	2.89–4.06	3.88 [A,a]	1.98–5.00	0.116
G2. EDTA 17% + PUI	3.79 [A,a,b]	2.93–5.13	3.76 [A,a]	1.98–5.39	3.43 [A,a]	1.21–4.78	0.387
G3. EDTA 17% + EA	2.71 [A,a,b]	1.80–3.50	2.74 [A,a]	1.63–4.61	2.27 [A,a]	0.94–4.76	1.000
G4. EDTA 17% + XPF	3.16 [A,a]	2.65–3.64	2.46 [A,a]	1.72–3.76	3.09 [A,a]	1.88–3.62	0.796
G5. Chitosan 0.2%	2.90 [A,a,b]	2.33–4.33	2.83 [A,a]	1.91–4.13	2.81 [A,a]	1.54–3.70	0.705
G6. Chitosan 0.2% + PUI	2.45 [A,a]	1.19–3.49	2.79 [A,a]	1.89–3.29	2.52 [A,a]	1.66–3.53	0.212
G7. Chitosan 0.2% + EA	3.91 [A,a,b]	2.34.5.46	2.59 [A,a]	1.22–3.45	3.01 [A,a]	1.26–3.91	0.200
G8. Chitosan 0.2% + XPF	5.35 [A,b]	3.13–6.14	4.01 [A,a]	2.64–5.26	4.11 [A,a]	2.79–4.90	0.350
p-Value	<0.01		0.017		0.159		

Different lowercase letters in the column indicate significant differences ($p < 0.05$). Different uppercase letters in lines indicate significant differences ($p < 0.05$).

During the group analysis, no significant differences were detected among the root thirds ($p > 0.05$). When comparing cleaning methods, variations surfaced in the cervical third between groups G1 and G8 ($p = 0.038$), G4 and G8 ($p = 0.003$), and G6 and G8 ($p = 0.049$), as well as Control and G8 ($p = 0.019$). In the middle third, although the group comparison initially indicated distinctions ($p = 0.017$), these distinctions vanished upon adjustment for multiple comparisons. The comparison between groups G1 and Control ($p = 0.051$) and G8 and Control ($p = 0.053$) was the closest to statistical significance. In the apical third, no notable differences emerged between groups ($p > 0.05$). The failure modes are detailed in Table 2. The predominant failure was Mode 2, constituting 68.3% of the cases, while Mode 6 accounted for 28.7%.

Table 2. Percentage (%) of failure modes in the six different categories.

Experimental Groups	Failures Modes N (%)						TOTAL
	1	2	3	4	5	6	
Control Group	0 (0%)	41 (68.3%)	0 (0%)	0 (0%)	0 (0%)	19 (31.7%)	60 (100%)
G1. EDTA 17%	1 (1.7%)	31 (51.71%)	4 (6.71%)	1 (1.7%)	0 (0%)	23 (38.3%)	60 (100%)
G2. EDTA 17% + PUI	1 (1.7%)	41 (68.3%)	3 (5%)	0 (0%)	0 (0%)	15 (25%)	60 (100%)
G3. EDTA 17% + EA	0 (0%)	40 (66.7%)	0 (0%)	0 (0%)	0 (0%)	20 (33.3%)	60 (100%)
G4. EDTA 17% + XPF	0 (0%)	41 (68.3%)	1 (1.7%)	0 (0%)	0 (0%)	18 (30%)	60 (100%)
G5. Chitosan 0.2%	0 (0%)	35 (58.3%	4 (6.7%)	1 (1.7%)	0 (0%)	20 (33.3%)	60 (100%)
G6. Chitosan 0.2% + PUI	0 (0%)	50 (83.3%	0 (0%)	0 (0%)	0 (0%)	10 (16.7%)	60 (100%)
G7. Chitosan 0.2% + EA	0 (0%)	40 (66.7%)	0 (0%)	0 (0%)	0 (0%)	20 (33.3%)	60 (100%)
G8. Chitosan 0.2% + XPF	0 (0%)	50 (83.3%)	0 (0%)	0 (0%)	0 (0%)	10 (16.7%)	60 (100%)
TOTAL	2 (0.4%)	369 (68.3%)	12 (2.2%)	2 (0.4%)	0 (0%)	155 (28.7%)	540 (100%)

4. Discussion

In the current investigation, exploring the impact of 0.2% chitosan in various final cleaning methods on the adhesion FP to intrarradicular dentin produced convincing results. The study revealed significant disparities in bond strength between groups G8 and G1, as well as between groups G4 and G6, highlighting the substantial influence of the cleaning protocol on adhesion. These findings led to the partial rejection of previously raised null hypotheses, shedding new light on the intricate dynamics of the FP dentin-root bond.

The micro push-out mechanical test, a well-established and widely recognized method for assessing the bond strength between dentin and FP, has been extensively highlighted in prior research [8,20,25]. Its widespread use is attributed to its ability to ensure a more uniform distribution of stresses, minimizing data distortion and reducing the likelihood of premature failures [27,28]. One of its significant advantages lies in enabling the assessment of multiple specimens from the same root, a factor crucial for robust comparative analyses. Moreover, this technique facilitates the exploration of regional variations within the root thirds, providing results that closely mirror real-world clinical scenarios [1,24,29].

The decision to use bovine teeth was based on their easier availability in comparison with human teeth. Additionally, they offer improved standardization of both the teeth's age and the root canal space [21,22,26]. The very similar characteristics between bovine and human teeth, especially in mechanical tests that evaluate the bond strength to dentin and enamel, provide a solid scientific basis, affirming the relevance and suitability of the chosen experimental model [26,30,31].

The choice of the 2.5% sodium hypochlorite solution before using the chelator was based on its suitability, having been previously tested in other studies [24,25,32,33] and proven to be less likely to affect the micromechanical properties of dentin. Higher concentrations were avoided, as these may interfere with experimental results. Previous studies have already demonstrated that increases in NaOCl concentration can cause significant changes in organic and inorganic components, resulting in lower micromechanical resistance due to degradation of the collagen matrix [33,34].

A wide range of materials has been made available on the market for cementing FP [8]. The resin cement used in the study has the capacity to adhere to the tooth structure by two mechanisms: the acidic monomers hybridize the dentin, and the resin chemically interacts with the hydroxyapatite [3,24–26]. Prior research has demonstrated that the presence of the smear layer, which forms during the intrarradicular preparation, can prevent the demineralization process promoted by the cement. This interference adversely affects the adhesive capacity of the cement, compromising its ability to form a strong and enduring bond [35].

EDTA 17% is an important chelator in removing the smear layer [7,10]. However, prolonged use can result in erosive effects on the dentin, leading to a reduction in its microhardness. This can potentially harm the periapical tissues surrounding the tooth [36]. Studies have focused on more biocompatible alternatives to minimize their interference with adhesive and restorative procedures [20,24,25].

Chitosan, derived from the deacetylation process of chitin found in crab and shrimp shells, is a chemical substance of significant interest in dental research. Its cost-effectiveness, biocompatibility, and minimal cytotoxicity have made it a focal point in the exploration of natural polysaccharides for dental applications [7,12–16,37]. In comparisons between cleaning methods, differences in values were found only in the cervical third, as reported in previous investigations [1,21,22].

The findings revealed significant variations in bond strength. The use of 0.2% chitosan + XPF as a final cleaning protocol positively influenced the bond strength of FP to intrarradicular dentin, with the highest values in the cervical third but with no differences for the other thirds (Table 1). Notably, the combination of 0.2% chitosan and XPF positively affected the adhesive strength, especially in the cervical region. It is essential to highlight that the cement's ability to adhere to intrarradicular dentin is influenced by the chosen cleaning technique, as highlighted by numerous studies [11,24–26]. However, it is essential to take into account the fact that a higher concentration of the chelating agent can lead to increased demineralization of the dentin matrix, exposing the collagen and causing a series of inconveniences, including the reduction of microhardness and notably higher incidences of adhesion failures.

These findings align with a previous study [25] that compared the impact of 0.2% chitosan and 17% EDTA utilizing various techniques, including conventional irrigation and PUI. The study revealed higher values for group 6 in the cervical third of the root canal.

Failure modes were analyzed using the serial root sectioning method, from which two slices were obtained per root third, allowing direct inspection of the cementoenamel junction. As for the mode of adhesive failure, it was found that this most frequently occurred, followed by cohesive failure in the dentin (Table 2). This corroborates previous studies that demonstrated greater fragility at the cement-dentin interface [8,21,22,24,25], which can be justified by the presence of remaining obturators that adhered to the intraradicular dentin walls and within the dentinal tubules [4].

Certain limitations of the present study deserve attention in future research efforts. Notably, the samples were not exposed to mechanical and thermal conditions, which could better reproduce oral conditions, offering more authentic results [38]. Clinical trials are essential to confirm the results of this study and evaluate the effectiveness of new materials and methods for cleaning intrarradicular dentin post-retainer preparation, particularly in the context of FP-related restorations.

5. Conclusions

According to the methods used, it can be concluded that:

1. the use of 0.2% chitosan + XPF as a final cleaning method positively influenced the bond strength between FP and intrarradicular dentin, with higher bond strength values in the cervical third.

Author Contributions: Conceptualization: H.F.d.O., N.G.C. and A.d.R.G. Data Curation: C.M.R.B., L.P.F. and A.M.F.A. Formal analysis: H.F.d.O., P.C.L.-C. and M.A.d.C. Investigation and Methodology: H.F.d.O., N.G.C. and M.A.d.C. Project administration: H.F.d.O. Resources: H.F.d.O., N.G.C. and A.d.R.G. Supervision: H.F.d.O. and M.A.d.C. Writing-review and editing: H.F.d.O. All authors have read and agreed to the published version of the manuscript.

Funding: This research received no external funding.

Institutional Review Board Statement: The study was approved by the Animal Use Ethics Committee (CEUA) of the Universidade Evangélica de Goiás, Anápolis, Brazil (#001/2021).

Data Availability Statement: The data presented in this study are available on request from the corresponding author.

Conflicts of Interest: The authors state no conflict of interest.

References

1. Menezes, M.S.; Queiroz, E.C.; Campos, R.E.; Martins, L.R.; Soares, C.J. Influence of endodontic sealer cement on fibreglass post bond strength to root dentine. *Int. Endod. J.* **2008**, *41*, 476–484.
2. Baena, E.; Flores, A.; Ceballos, L. Influence of root dentin treatment on the push-out bond strength of fiber posts. *Odontology* **2017**, *105*, 170–177.
3. Akman, M.; Eldeniz, A.U.; Ince, S.; Guneser, M.B. Push-out bond strength of a new post system after various post space treatments. *Dent. Mater. J.* **2016**, *35*, 876–880. [CrossRef]
4. Oliveira, L.V.; Maia, T.S.; Zancopé, K.; Menezes, M.S.; Soares, C.J.; Moura, C.C.G. Can intra-radicular cleaning protocols increase the retention of fiberglass posts? A systematic review. *Braz. Oral. Res.* **2018**, *32*, e16.
5. Haapasalo, M.; Shen, Y.; Wang, Z.; Gao, Y. Irrigation in endodontics. *Br. Dent. J.* **2014**, *216*, 299–303.
6. Gu, L.S.; Kim, J.R.; Ling, J.; Choi, K.K.; Pashley, D.H.; Tay, F.R. Review of contemporary irrigant agitation techniques and devices. *J. Endod.* **2009**, *35*, 791–804.
7. Silva, P.V.; Guedes, D.F.; Pécora, J.D.; da Cruz-Filho, A.M. Time-dependent effects of chitosan on dentin structures. *Braz. Dent. J.* **2012**, *23*, 357–361.

8. Barreto, M.S.; Rosa, R.A.; Seballos, V.G.; Machado, E.; Valandro, L.F.; Kaizer, O.B.; Só, M.V.R.; Bier, C.A.S. Effect of Intracanal Irrigants on Bond Strength of Fiber Posts Cemented with a Self-adhesive Resin Cement. *Oper. Dent.* **2016**, *41*, e159–e167.
9. Mancini, M.; Cerroni, L.; Iorio, L.; Dall'Asta, L.; Cianconi, L. FESEM evaluation of smear layer removal using different irrigant activation methods (EndoActivator, EndoVac, PUI and LAI). An in vitro study. *Clin. Oral. Investig.* **2018**, *22*, 993–999.
10. Sen, B.H.; Ertürk, O.; Pişkin, B. The effect of different concentrations of EDTA on instrumented root canal walls. *Oral. Surg. Oral. Med. Oral. Pathol. Oral. Radiol. Endod.* **2009**, *108*, 622–627.
11. Silva, P.V.; Guedes, D.F.; Nakadi, F.V.; Pécora, J.D.; Cruz-Filho, A.M. Chitosan: A new solution for removal of smear layer after root canal instrumentation. *Int. Endod. J.* **2013**, *46*, 332–338. [CrossRef] [PubMed]
12. Akncbay, H.; Senel, S.; Ay, Z.Y. Application of chitosan gel in the treatment of chronic periodontitis. *J. Biomed. Mater. Res. B Appl. Biomater.* **2007**, *80*, 290–296. [CrossRef] [PubMed]
13. Bounegru, A.V.; Bounegru, I. Chitosan-Based Electrochemical Sensors for Pharmaceuticals and Clinical Applications. *Polymers* **2023**, *15*, 3539. [CrossRef] [PubMed]
14. Desai, N.; Rana, D.; Salave, S.; Gupta, R.; Patel, P.; Karunakaran, B.; Sharma, A.; Giri, J.; Benival, D.; Kommineni, N. Chitosan: A Potential Biopolymer in Drug Delivery and Biomedical Applications. *Pharmaceutics* **2023**, *15*, 1313. [CrossRef] [PubMed]
15. Herdiana, Y. Chitosan Nanoparticles for Gastroesophageal Reflux Disease Treatment. *Polymers* **2023**, *15*, 3485. [CrossRef]
16. Kurita, K. Chemistry and application of chitin and chitosan. *Polym. Degrad. Stab.* **1998**, *59*, 117–120. [CrossRef]
17. Da Cruz-Filho, A.M.; Bordin, A.R.V.; Souza-Flamini, L.E.; Guedes, D.; Saquy, P.C.; Silva, R.G.; Pécora, J. Analysis of the shelf life of chitosan stored in different types of packaging, using colorimetry and dentin microhardness. *Restor. Dent. Endod.* **2017**, *42*, 87–94. [CrossRef]
18. Kurita, K. Chitin and chitosan: Functional biopolymers from marine crustaceans. *Mar. Biotechnol.* **2006**, *8*, 203–226. [CrossRef]
19. Spanó, J.C.; Silva, R.G.; Guedes, D.F.; Sousa-Neto, M.D.; Estrela, C.; Pécora, J.D. Atomic absorption spectrometry and scanning electron microscopy evaluation of concentration of calcium ions and smear layer removal with root canal chelators. *J. Endod.* **2009**, *35*, 727–730. [CrossRef]
20. Elnaghy, A.M. Effect of QMix irrigant on bond strength of glass fibre posts to root dentine. *Int. Endod. J.* **2014**, *47*, 280–289. [CrossRef]
21. Renovato, S.R.; Santana, F.R.; Ferreira, J.M.; Souza, J.B.; Soares, C.J.; Estrela, C. Effect of calcium hydroxide and endodontic irrigants on fibre post bond strength to root canal dentine. *Int. Endod. J.* **2013**, *46*, 738–746. [CrossRef] [PubMed]
22. Guedes, O.A.; Chaves, G.S.; Alencar, A.H.; Borges, A.H.; Estrela, C.R.; Soares, C.J.; Estrela, C. Effect of gutta-percha solvents on fiberglass post bond strength to root canal dentin. *J. Oral. Sci.* **2014**, *56*, 105–112. [CrossRef] [PubMed]
23. Santana, F.R.; Soares, C.J.; Silva, J.A.; Alencar, A.H.; Renovato, S.R.; Lopes, L.G.; Gonzaga, L.; Carlos, E. Effect of Instrumentation Techniques, Irrigant Solutions and Artificial accelerated Aging on Fiberglass Post Bond Strength to Intraradicular Dentin. *J. Contemp. Dent. Pract.* **2015**, *16*, 523–530. [PubMed]
24. Boggian, L.C.; Silva, A.V.; Santos, G.R.; Oliveira, G.F.; Silva, W.L.; Nery Neto, I.; Guedes, O.A.; Estrela, C. Effect of intra-radicular cleaning protocols after post-space preparation on marginal adaptation of a luting agent to root dentin. *J. Oral. Sci.* **2023**, *65*, 81–86. [CrossRef] [PubMed]
25. Dorileo, M.C.G.O.; Guiraldo, R.D.; Lopes, M.B.; de Almeida Decurcio, D.; Guedes, O.A.; Aranha, A.M.F.; Gonini, Á.H.B.A., Jr. Effect of 0.2% Chitosan Associated with Different Final Irrigant Protocols on the Fiber Post Bond Strength to Root Canal Dentin of Bovine Teeth: An Study. *Open Dent. J.* **2022**, *16*, e2205310. [CrossRef]
26. Soares, C.J.; Pereira, J.C.; Valdivia, A.D.; Novais, V.R.; Meneses, M.S. Influence of resin cement and post configuration on bond strength to root dentine. *Int. Endod. J.* **2012**, *45*, 136–145. [CrossRef]
27. Soares, C.J.; Santana, F.R.; Castro, C.G.; Santos-Filho, P.C.; Soares, P.V.; Qian, F.; Armstrong, S.R. Finite element analysis and bond strength of a glass post to intraradicular dentin: Comparison between microtensile and push-out tests. *Dent. Mater.* **2008**, *24*, 1405–1411. [CrossRef]
28. Goracci, C.; Tavares, A.U.; Fabianelli, A.; Monticelli, F.; Raffaelli, O.; Cardoso, P.C.; Tay, F.; Ferrari, M. The adhesion between fiber posts and root canal walls: Comparison between microtensile and push-out bond strength measurements. *Eur. J. Oral. Sci.* **2004**, *112*, 353–361. [CrossRef]
29. Guedes, O.A.O.H.; Santana, G.; Raineri Capeletti, L.; De Araújo, E.C.R.; De Almeida, E.C.; Decurcio, D. Effect of Endodontic Retreatment Protocols on Bond Strength of Fiberglass Post to Root Canal Dentine: An In-vitro Study. *J. Clin. Diagn. Res.* **2021**, *15*, 11–14. [CrossRef]
30. Dong, C.C.; McComb, D.; Anderson, J.D.; Tam, L.E. Effect of mode of polymerization of bonding agent on shear bond strength of autocured resin composite luting cements. *J. Can. Dent. Assoc.* **2003**, *69*, 229–234.
31. Nakamichi, I.; Iwaku, M.; Fusayama, T. Bovine teeth as possible substitutes in the adhesion test. *J. Dent. Res.* **1983**, *62*, 1076–1081. [CrossRef] [PubMed]
32. Pascon, F.M.; Kantovitz, K.R.; Sacramento, P.A.; Nobre-dos-Santos, M.; Puppin-Rontani, R.M. Effect of sodium hypochlorite on dentine mechanical properties. A review. *J. Dent.* **2009**, *37*, 903–908. [CrossRef] [PubMed]
33. Tartari, T.; Bachmann, L.; Zancan, R.F.; Vivan, R.R.; Duarte, M.A.H.; Bramante, C.M. Analysis of the effects of several decalcifying agents alone and in combination with sodium hypochlorite on the chemical composition of dentine. *Int. Endod. J.* **2018**, *51*, e42–e54. [CrossRef] [PubMed]

34. Doğan, H.; Qalt, S. Effects of chelating agents and sodium hypochlorite on mineral content of root dentin. *J. Endod.* **2001**, *27*, 578–580. [CrossRef]
35. Gerth, H.U.; Dammaschke, T.; Züchner, H.; Schäfer, E. Chemical analysis and bonding reaction of RelyX Unicem and Bifix composites—A comparative study. *Dent. Mater.* **2006**, *22*, 934–941. [CrossRef]
36. De-Deus, G.; Paciornik, S.; Mauricio, M.H. Evaluation of the effect of EDTA, EDTAC and citric acid on the microhardness of root dentine. *Int. Endod. J.* **2006**, *39*, 401–407. [CrossRef]
37. Kurita, K.; Shimada, K.; Nishiyama, Y.; Shimojoh, M.; Nishimura, S.I. Nonnatural Branched Polysaccharides: Synthesis and Properties of Chitin and Chitosan Having alpha-Mannoside Branches. *Macromolecules* **1998**, *31*, 4764–4769. [CrossRef]
38. Bitter, K.; Paris, S.; Pfuertner, C.; Neumann, K.; Kielbassa, A.M. Morphological and bond strength evaluation of different resin cements to root dentin. *Eur. J. Oral. Sci.* **2009**, *117*, 326–333. [CrossRef]

Disclaimer/Publisher's Note: The statements, opinions and data contained in all publications are solely those of the individual author(s) and contributor(s) and not of MDPI and/or the editor(s). MDPI and/or the editor(s) disclaim responsibility for any injury to people or property resulting from any ideas, methods, instructions or products referred to in the content.

Article

Supramolecular Responsive Chitosan Microcarriers for Cell Detachment Triggered by Adamantane

Lixia Huang [1], Yifei Jiang [1], Xinying Chen [2,*], Wenqi Zhang [2], Qiuchen Luo [2], Siyan Chen [1], Shuhan Wang [3], Fangqing Weng [1] and Lin Xiao [2,*]

[1] Hubei Key Laboratory of Purification and Application of Plant Anti-Cancer Active Ingredients, School of Chemistry and Life Sciences, Hubei University of Education, Wuhan 430205, China; huanglixia@hue.edu.cn (L.H.); jiangyifei2423@163.com (Y.J.); chensiyan0911@163.com (S.C.); wengfq.hue@foxmail.com (F.W.)
[2] School of Biomedical Engineering, Shenzhen Campus of Sun Yat-sen University, Shenzhen 518107, China; zhangwq87@mail2.sysu.edu.cn (W.Z.); luoqch3@mail2.sysu.edu.cn (Q.L.)
[3] Shenzhen Institute for Drug Control, Shenzhen Testing Center of Medical Devices, Shenzhen 518057, China; wangshuhan@szidc.org.cn
* Correspondence: xiaolin23@mail.sysu.edu.cn (L.X.); chxiny@mail2.sysu.edu.cn (X.C.)

Citation: Huang, L.; Jiang, Y.; Chen, X.; Zhang, W.; Luo, Q.; Chen, S.; Wang, S.; Weng, F.; Xiao, L. Supramolecular Responsive Chitosan Microcarriers for Cell Detachment Triggered by Adamantane. *Polymers* 2023, *15*, 4024. https://doi.org/10.3390/polym15194024

Academic Editors: Sérgio Paulo Campana-Filho and William Facchinatto

Received: 10 September 2023
Revised: 30 September 2023
Accepted: 3 October 2023
Published: 8 October 2023

Copyright: © 2023 by the authors. Licensee MDPI, Basel, Switzerland. This article is an open access article distributed under the terms and conditions of the Creative Commons Attribution (CC BY) license (https://creativecommons.org/licenses/by/4.0/).

Abstract: Supramolecular responsive microcarriers based on chitosan microspheres were prepared and applied for nonenzymatic cell detachment. Briefly, chitosan microspheres (CSMs) were first prepared by an emulsion crosslinking approach, the surface of which was then modified with β-cyclodextrin (β-CD) by chemical grafting. Subsequently, gelatin was attached onto the surface of the CSMs via the host–guest interaction between β-CD groups and aromatic residues in gelatin. The resultant microspheres were denoted CSM-g-CD-Gel. Due to their superior biocompatibility and gelatin niches, CSM-g-CD-Gel microspheres can be used as effective microcarriers for cell attachment and expansion. L-02, a human fetal hepatocyte line, was used to evaluate cell attachment and expansion with these microcarriers. After incubation for 48 h, the cells attached and expanded to cover the entire surface of microcarriers. Moreover, with the addition of adamantane (AD), cells can be detached from the microcarriers together with gelatin because of the competitive binding between β-CD and AD. Overall, these supramolecular responsive microcarriers could effectively support cell expansion and achieve nonenzymatic cell detachment and may be potentially reusable with a new cycle of gelatin attachment and detachment.

Keywords: chitosan; microcarrier; cell culture; cell detachment; host–guest interaction

1. Introduction

Microcarriers are 100 to 300 micron supporting matrices that permit the growth of adherent cells in bioreactor systems. They have a larger surface area to volume ratio in comparison to single cell monolayers, enabling cost-effective cell production and expansion [1–4]. Microcarriers are composed of a solid matrix that must be separated from expanded cells during downstream processing stages.

Most of the current cellular microcarriers need to harvest cells by tedious downstream operations after cell expansion, and cells adherent to the surface of microcarriers are usually isolated by pancreatic enzyme digestion because microcarriers are mainly used for the culture of anchorage-dependent cells. However, the efficiency of pancreatic enzyme digestion is limited. It is difficult to dissociate cells completely from microcarriers with regular pancreatic enzyme digestion, while overdigestion may cause irreversible damage to cells and reduce total cellular yield and function [5]. Therefore, constructing smart microcarriers with stimuli-responsiveness to achieve nonenzymatic automatic desorption of cells has become one of the main directions of current microcarrier technology. A variety of nonenzymatic responsive microcarriers have been reported during past decades [6–8]. As an example, Akbari et al. reported multifunctional temperature-responsive microcarriers (cytoGel) made of an interpenetrating hydrogel network composed of poly(N-isopropylacrylamide) (PNIPAM), poly(ethylene glycol) diacrylate (PEGDA), and gelatin methacryloyl (GelMA) [7]. Cell detachment was achieved by cooling the system to room temperature with up to 70% detachment efficiency. Quantitative analysis with a flow cytometer indicated that more than 90% of cell viability was obtained for these thermally detached cells. Zhao et al. developed near-infrared (NIR) light-responsive graphene oxide hydrogel microcarriers for controlled cell culture and release [8]. After exposure to NIR light, the cell-laden microcarriers underwent rapid shrinkage and the cells were released. The cell release was enhanced as the irradiation power intensity increased. However, to obtain a comparable cell viability with enzyme-digested cells, the power intensity should be lower than 0.5 W cm^{-2} and the irradiation period should be less than 15 s.

Chitosan (CS) is a deacetylation product of chitin. It has good biocompatibility and biodegradability and a variety of biological activities, such as coagulation, antibacterial activity, antitumor activity, and immunomodulatory function. It has the potential to be modified due to the large amount of free amino groups. Based on these advantages, chitosan has been widely used in biomedical fields such as tissue engineering and drug release [9–11]. We previously developed highly porous chitosan microspheres as microcarriers for 3D cell culture, in which adhesion and growth of L-02 cells not only took place on their external surface but also within the internal pores of the microcarriers, allowing multidirectional cell–cell interactions [1]. In this work, responsive microcarriers with the abilities of in vitro cell expansion and nonenzymatic detachment were explored based on chitosan microspheres and surface modification. First, chitosan microspheres (CSMs) were prepared through a conventional emulsification crosslinking approach, where glutaraldehyde was used as a crosslinking agent. By crosslinking the amino groups, glutaraldehyde forms covalent bonds that make it difficult for water molecules to penetrate the chitosan microspheres, resulting in increased hydrophobicity and improved mechanical properties. This makes the chitosan microspheres more resistant to harsh environments, such as high pH or temperature, making them ideal candidates for use in various applications, such as drug delivery systems, biomaterials, and tissue engineering [12–14]. Second, CSMs were surface-modified with β-cyclodextrin (β-CD) with chemical grafting and subsequently modified with gelatin, which binds β-CD groups via host–guest interactions with its aromatic residues. The modification of gelatin is favorable for cell attachment and expansion when the microspheres are used as microcarriers. Finally, due to the relatively higher affinity of adamantane (AD) to β-CD than gelatin, it is possible to detach the gelatin molecules from the microsphere surface with the addition of AD, which may thus cause detachment of the cells anchored to gelatin. Compared with the reported nonenzymatic responsive microcarriers, this biocompatible microcarrier system is based on gentle supramolecular interactions, enabling cell detachment in a simple and mild manner without changing any external environmental conditions, which is favorable for maintaining cell viability [15,16].

2. Materials and Methods

2.1. Material

Chitosan (degree of deacetylation > 95%) was purchased from Sigma Aldrich, and its molecular weight (MV) was 1.02×10^6 Da. Surfactants siban 80 (S80), Tween 60 (T60), acetic acid, petroleum ether, glutaraldehyde, and ethanol were purchased from Shanghai Sinopharm Chemical Reagent Co., Ltd.(Shanghai, China), and 1-(3-dimethylaminopropyl)-3-ethylcarbodiimide salt (EDC), N-hydroxysuccinimide (NHS), carboxymethyl β-cyclodextrin (CMCD), gelatin, and adamantane (AD) were purchased from Shanghai Aladdin Biochemical Technology Co., Ltd. (Shanghai, China). All reagents were of analytical grade and used without further purification.

Human fetal hepatocytes L-02 were obtained from Tongji Medical College, Huazhong University of Science and Technology, and cultured in DMEM supplemented with 10%

fetal bovine serum and 1% antibiotics (100 mg/mL streptomycin and 100 U/mL penicillin) at 37 °C in a humidified atmosphere containing 5% CO_2. The cell culture grade reagents used were all procured from Gibico products from Thermo Fisher Scientific (Waltham, MA, USA).

2.2. Preparation of Chitosan Microspheres

Chitosan microspheres were prepared with a conventional emulsification crosslinking method [1]. Briefly, 0.6 g of chitosan powder was dispersed in 30 mL of deionized water under magnetic stirring at 400 rpm to form a white turbid liquid. Then, 0.3 g of acetic acid was added dropwise to the suspension to facilitate the dissolution of chitosan, and a transparent 2% (w/v) chitosan solution was obtained. The chitosan solution was then mixed with 90 mL of petroleum ether containing 4.8 g of Span 80 and 0.2 g of Tween 60. The mixture was subjected to emulsification under stirring at 400 rpm at 40 °C for 2 h, after which 1.5 mL of 25 wt% glutaraldehyde aqueous solution was added to the emulsion and crosslinked for 3 h. The crosslinked emulsion was then centrifuged, and the precipitates were collected. After being washed with absolute ethanol and deionized water 3 times each, the precipitates of chitosan microspheres (CSMs) were freeze-dried and stored at room temperature for subsequent studies.

2.3. Preparation of Cyclodextrin-Grafted Chitosan Microspheres (CSM-g-CD)

Chitosan microspheres (CSMs) were surface-modified with β-cyclodextrin groups with a reaction between the amino groups of chitosan and the carboxyl groups of carboxymethyl-β-cyclodextrin (CMCD) mediated with hydrochloride (EDC) and N-hydroxysuccinimide (NHS). Briefly, 0.45 g of CMCD was dissolved in 10 mL of deionized water, and then 0.396 g EDC and 0.242 g NHS were added to the CMCD solution and reacted at 25 °C with stirring for 12 h. The activated CMCD solution was then added gradually to the CSM suspension containing 0.12 g CSM and reacted at 25 °C with stirring at 100 rpm for 24 h. The products of cyclodextrin-grafted chitosan microspheres (CSM-g-CD) were obtained after centrifugation and washing the precipitates with deionized water, followed by freeze-drying.

2.4. Preparation of Gelatin-Modified Responsive Microspheres (CSM-g-CD-Gel)

To obtain responsive microspheres, gelatin was attached to the surface of CSM-g-CD through dynamic host–guest interactions between β-cyclodextrin (β-CD) groups and aromatic residues in gelatin. Briefly, 0.4 g of gelatin was dissolved in 10 mL of deionized water under stirring at 100 rpm at 37 °C to obtain a 4% (w/v) gelatin solution, which was sterilized by filtration through a 0.45 μM filter. Then, after autoclaving, 0.5 g of CSM-g-CD microspheres was added to the gelatin solution and reacted under stirring at room temperature for 3 h. The resultant reaction mixture was centrifuged at 2000 rpm for 5 min, and the precipitates were retained. The unreacted gelatin molecules on the microsphere surface were removed by washing with deionized water. The modified responsive microspheres were labeled CSM-g-CD-Gel.

2.5. Physical Characterization of Chitosan Microspheres

2.5.1. SEM

Chitosan microsphere samples in the wet state were frozen in a −20 °C freezer and then dried in a vacuum freeze dryer. Samples were surface gold sprayed for 100 s, and the morphologies of the appearance of the samples at an accelerating voltage of 20.0 kV were observed using an environmental scanning electron microscope (Quanta 200, FEI, Hillsboro, The Netherlands) and photographed for preservation.

2.5.2. FTIR

The CSM, CSM-g-CD, CSM-g-CD-Gel, and AD-treated CSM-g-CD-Gel as well as gelatin were tested using the single reflection level smart ATR accessory of a Fourier transform micro infrared spectrometer (Vertex 70, Bruker, Germany). β-CD powder and AD powder were then tested using FTIR tableting, and the sample powder was thoroughly ground with an appropriate amount of KBr in a quartz mortar and poured into a tableting machine to squeeze into tablets. The transmittance of the samples at different wavenumbers was scanned in the range of 500~4000 cm^{-1}, and infrared spectra were obtained and aligned based on data mapping.

2.5.3. 2D NMR

A ^{1}H NOESY NMR test was performed on CSM-g-CD-Gel at 37 °C to confirm the interaction between the aromatic residues of gelatin and β-CD groups in CSM-g-CD-Gel.

2.5.4. Cytocompatibility Evaluation

For the nondirect contact cytotoxicity test, the CSM leaching solution was prepared according to ISO10993-5. After autoclaving, CSM and CSM-g-CD-Gel were soaked in cell culture medium at different concentrations (0.5, 1.0, and 2.0 mg/mL) for 72 h at 37 °C. The supernatant was collected as the extract after centrifugation at 10,000 rpm for 5 min. The extract was sterilized by filtration through a 0.22 μm filter before use. L-02 cells were seeded in 96-well plates at a density of 1×10^4 cells/well and incubated overnight at 37 °C in a 5% CO_2 atmosphere. Then, the aspirated and discarded medium was replaced by the extract solution, and the cell culture medium was used as a negative control. After 1, 2, and 3 days of culture, CCK-8 (Dojindo, Kumamoto, Japan) reagent was added to the plates at 10 μL/well and incubated at 37 °C for 30 min after gentle shaking for 5 min. The OD value of the solution in the well plate was measured at 450 nm using a multifunctional microplate reader (Multiskan, Waltham, MA, Thermo Fisher Scientific, USA). For direct contact cytotoxicity evaluation, cells were seeded in a mixture of microsphere suspensions at concentrations of 0.5, 1.0, and 2.0 mg/mL, and the cytotoxicity evaluation was performed according to the procedure described above. Cell viability was calculated by referring to the following formula, where OD_S, OD_B, and OD_N are the OD values of the sample, blank control, and negative control, respectively.

$$\text{cell viability} = \frac{OD_S - OD_B}{OD_N - OD_B} \times 100\% \tag{1}$$

2.5.5. Cell Adhesion and Distribution on Microspheres

To evaluate the adhesion of cells to the above-prepared microspheres, CSM and CSM-g-CD-Gel were added to the cell culture medium to prepare a suspension of 2.0 mg/mL. The microsphere suspension (1 mL) was first mixed with 500 μL of cell culture medium containing 4×10^4 L-02 cells for adequate contact. After that, the mixture was gently transferred to a 24-well plate and incubated at 37 °C in a 5% CO_2 atmosphere. The adhesion of L-02 cells to the microsphere surface was examined under a light microscope after incubation for 4, 24, and 48 h. To further observe the growth and distribution of cells on the microsphere surface, cells were fluorescently stained. Nuclei of L-02 cells were stained using Hoechst 33,258 dye with blue fluorescence excitation using a mercury lamp as the excitation light source and a UV filter. The cell membrane of L-02 cells was stained using DIL dye, with a green filter for red fluorescence excitation. The cells were then incubated for 4, 24, and 48 h before observation under a fluorescence microscope.

2.5.6. Responsive Detachment of Cells from CSM-g-CD-Gel Microspheres

L-02 cells were seeded onto responsive microspheres (CSM-g-CD-Gel) and cultured at 37 °C in a 5% CO_2 atmosphere for 48 h. Adamantane (AD) solution at a concentration of 1.5% (w/v) was added to the culture and incubated for 5 min and 30 min, respectively. Then, the adhesion of cells on the microspheres was observed under a light microscope. To further verify the detachment of cells, cells were stained with Hoechst 33,258 dye and observed under a fluorescence microscope.

2.6. Statistical Analysis

Quantitative data are expressed as the arithmetic mean ± standard deviation (SD). All quantitative results were obtained from at least triplicate samples. The difference between groups was tested with t test. $p < 0.05$ was considered statistically significant, and $p < 0.001$ was considered highly statistically significant.

3. Results and Discussion

3.1. Morphological Analysis of the Microspheres

The surface morphologies of the microspheres were observed with SEM, and the results are shown in Figure 1. Figure 1A,C,E are the SEM images of CSM, and Figure 1B,D,F are the SEM images of CSM-g-CD-Gel. The surface of the CSM microspheres is smooth and flat, while the surface of the CSM-g-CD-Gel microspheres is relatively rough. It can be more clearly seen in the partially enlarged image (Figure 1F) that the surface of CSM-g-CD-Gel was wrapped with a thin layer of substance. It is speculated that this layer of substance might be gelatin, which attached to the microsphere surface by forming inclusion complexes with β-cyclodextrin groups through host–guest interactions.

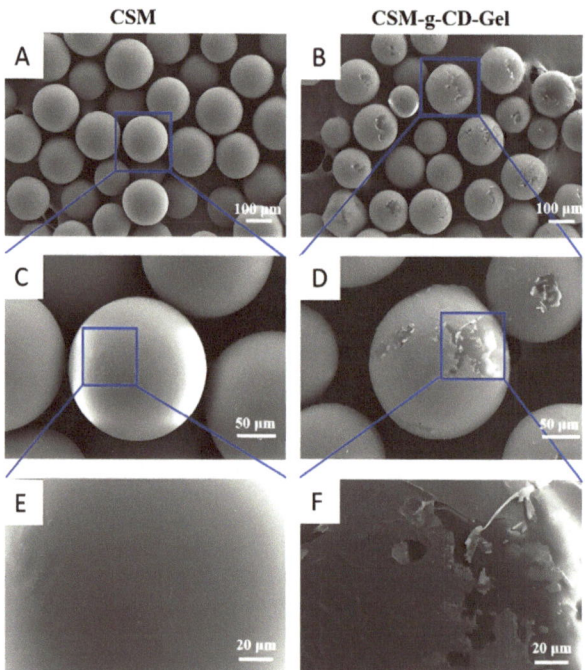

Figure 1. SEM images of CSMs (**A,C,E**) and responsive microspheres (CS-g-CD-Gel) (**B,D,F**); (**C,E**) and (**D,F**) are partially enlarged images of (**A**) and (**B**), respectively.

3.2. FTIR Analysis of the Microspheres

To verify the grafting of β-cyclodextrin groups on the surface of chitosan microspheres as well as the subsequent modification of gelatin, the materials were characterized using FTIR, and the spectra are shown in Figure 2. In the spectrum of CSM, the large absorption peak at 3434 cm^{-1} is attributed to the stretching vibration of -OH and -NH groups on the molecular chain of chitosan. The peak at 2875 cm^{-1} is assigned to the -CH$_2$ stretching vibration. The absorption peak at 1642 cm^{-1} is due to the C=O stretching vibration of the amide I bond, while the absorption peak at 1566 cm^{-1} is due to the N-H stretching vibration of the amide II bond [9]. The peak at 1382 cm^{-1} is assigned to the N-H absorption of the amide III bond. The peak at 1157 cm^{-1} is the stretching vibration of the glucoside C-O-C bond between the chitosan monomers and glucose. The peaks at 1080 cm^{-1} and 1028 cm^{-1} are the characteristic stretching peaks of polysaccharide C-OH. In comparison to the spectrum of CSM, the amide I peak of CSM-g-CD at 1642 cm^{-1} was shifted to 1635 cm^{-1}, and the peak at 1566 cm^{-1} disappeared, which resulted from the formation of a new amide bond between the amino groups on chitosan and the carboxyl groups on CMCD. The formation of new absorption peaks at 1417 cm^{-1} and 1247 cm^{-1}, derived from CMCD [10], was also observed in the spectrum of CSM-g-CD, indicating that CMCD was successfully grafted onto the chitosan microspheres to form CSM-g-CD. It was further found that in the spectrum of CSM-g-CD-Gel, new peaks at 577 cm^{-1} and 494 cm^{-1} were present, which were obtained from gelatin [17]. Combined with the SEM results, it can be concluded that gelatin successfully reacted onto the microspheres to form CSM-g-CD-Gel. In addition, the CSM-g-CD-Gel microspheres were also subjected to IR analysis after adamantane (AD) treatment. As shown in Figure 2, compared with CSM-g-CD-Gel, the peak intensities at 577 cm^{-1} and 494 cm^{-1} of gelatin were obviously weakened, and a new peak appeared at 455 cm^{-1}, which might be the skeleton C-C-C bond deformation vibration of AD according to the literature reports [18]. This result suggests that gelatin molecules on the surface of the CSM-g-CD-Gel microspheres were replaced by AD molecules because of the higher affinity between the AD and CD groups than between gelatin and CD [19].

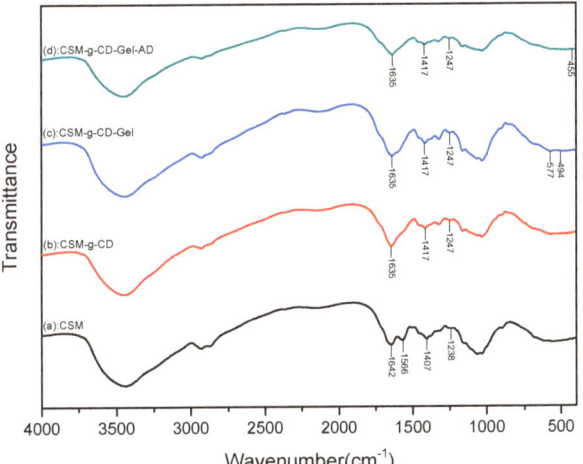

Figure 2. FTIR spectra of CSM (a), CSM-g-CD (b), CSM-g-CD-Gel (c), and CSM-g-CD-Gel-AD (d). CSM: chitosan microspheres; CSM-g-CD: CSM modified with β-CD groups; CSM-g-CD-Gel: CSM-g-CD modified with gelatin; CSM-g-CD-Gel-AD: CSM-g-CD-Gel after AD treatment.

3.3. 2D NMR Analysis

Two-dimensional (2D) ^{1}H NOESY NMR was used to verify the binding of gelatin to the microspheres via the formation of inclusion complexes between aromatic residues in gelatin and β-CD groups on the surface of the microspheres. As shown in Figure 3, multiple symmetrical cross-signal regions can be seen in the spectrum of CSM-g-CD-Gel, indicating that the aromatic hydrogens of gelatin and the alkane hydrogens of β-CD are spatially close to each other, and their atomic nuclei produce a NOE effect, thereby demonstrating that the aromatic residues of gelatin formed inclusion complexes with the β-CD groups through host–guest interactions; thus, gelatin molecules were successfully linked to the surface of the microspheres [20].

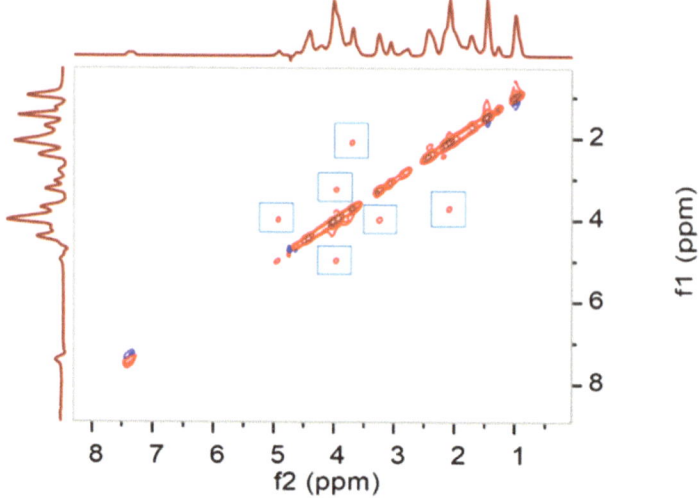

Figure 3. The 2D ^{1}H NOESY spectrum of CSM-g-CD-Gel.

3.4. Cytocompatibility

To analyze the cytocompatibility of chitosan microspheres and responsive microspheres, CCK-8 was employed for cytotoxicity experiments. L-02 cells were seeded on CSMs and CSM-g-CD-Gel at concentrations of 0.5 mg/mL, 1.0 mg/mL, and 2.0 mg/mL. As shown in Figure 4, in general, the cell viability of all the samples with different concentrations at different time intervals was more than 100%, which indicated that both the CSM and CSM-g-CD-Gel microspheres did not cause toxic effects on the cells and had good cytocompatibility. Notably, it was found that the cell viability on the responsive microspheres was slightly higher than that on the CSMs. Meanwhile, it was also found that the cell viability on CSM-g-CD-Gel had a tendency to increase with the material concentration. This may be related to the fact that gelatin on responsive microspheres can promote cell adhesion and growth, which has been well-recognized previously [21–23].

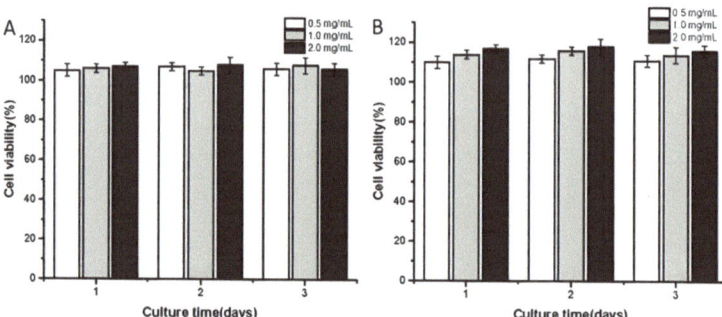

Figure 4. CCK-8 cytotoxicity analysis of chitosan microspheres and smart microspheres (**A**) CSM; (**B**) CSM-g-CD-Gel.

3.5. Cell Adhesion and Distribution on CSM-g-CD-Gel

The adhesion and distribution of L-02 cells on CSM and CSM-g-CD-Gel are shown in Figure 5. It was observed under a light microscope that, with increasing incubation time, the number of cells adhering to the surface of the microspheres gradually increased. In the first 4 h, only a small number of cells adhered to the microsphere surface, and the number of cells began to increase at 24 h. When the incubation time increased to 48 h, most of the microsphere surface was covered with cells. The number of cells on the surface of the CSM-g-CD-Gel was greater than that on the surface of the CSM, especially at 4 h; that is, at the cell attachment stage, a large number of cells adhered to the surface of the responsive microspheres. This is probably because there are gelatin molecules on the surface of the responsive microspheres that promote cell adhesion. This is also consistent with the results of the cytotoxicity tests.

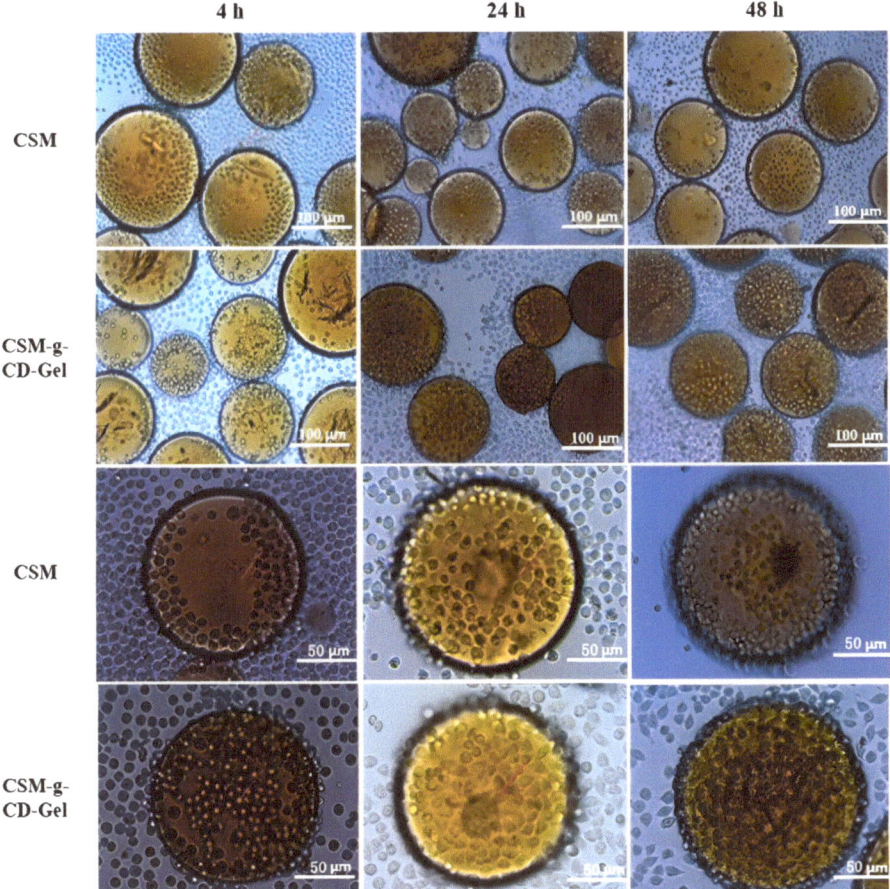

Figure 5. Optical microscopy images of the attachment and growth of L-02 cells cultured on CSM and CSM-g-CD-Gel for 4 h, 24 h, and 48 h. The scale bar for the upper two rows of images is 100 μm. The scale bar for the lower two rows of images is 50 μm.

To further analyze the cell distribution as well as the cell morphology on CSM-g-CD-Gel, the cells on the microspheres were fluorescently stained for nuclei and cell membranes, and the results observed under a fluorescence microscope are shown in Figure 6. In general, the results were consistent with those observed under an ordinary light microscope, and the number of cells adhering to the microspheres increased as the incubation time increased. At the beginning of the culture (4 h), the cells were distributed only at the edges of the microspheres, and subsequently, the cells began spreading, migrating, and growing over the microspheres. At 48 h of culture, the cells had spread over the microspheres. Meanwhile, it could also be seen that the cells had an intact structure and morphology from the blue fluorescence-labeled nucleus and the red fluorescence-labeled cell membrane. These results further indicate that the CSM-g-CD-Gel microspheres are favorable for cell adhesion and proliferation and can maintain the morphology of cells. It is well-known that collagens are crucial structural components of the extracellular matrix that provide cells with abundant adhesive sites and affect cell fate. Peptide sequences such as RGD in collagens can act as specific ligands to integrins, a family of transmembrane proteins of cells that facilitate cell adhesion, growth, proliferation, and communication [24–26]. Gelatin, as a product of partial hydrolysis of collagens, possesses remarkable biological effects similar to collagens and thus has been widely applied in the biomedical field. The specific peptide sequences retained by gelatin can also act as anchoring sites for cells and promote cell adhesion, growth, proliferation, and communication [27,28]. Therefore, through the interaction between gelatin and the cells, CSM-g-CD-Gel microcarriers displayed superior abilities to support cell adhesion and expansion in vitro.

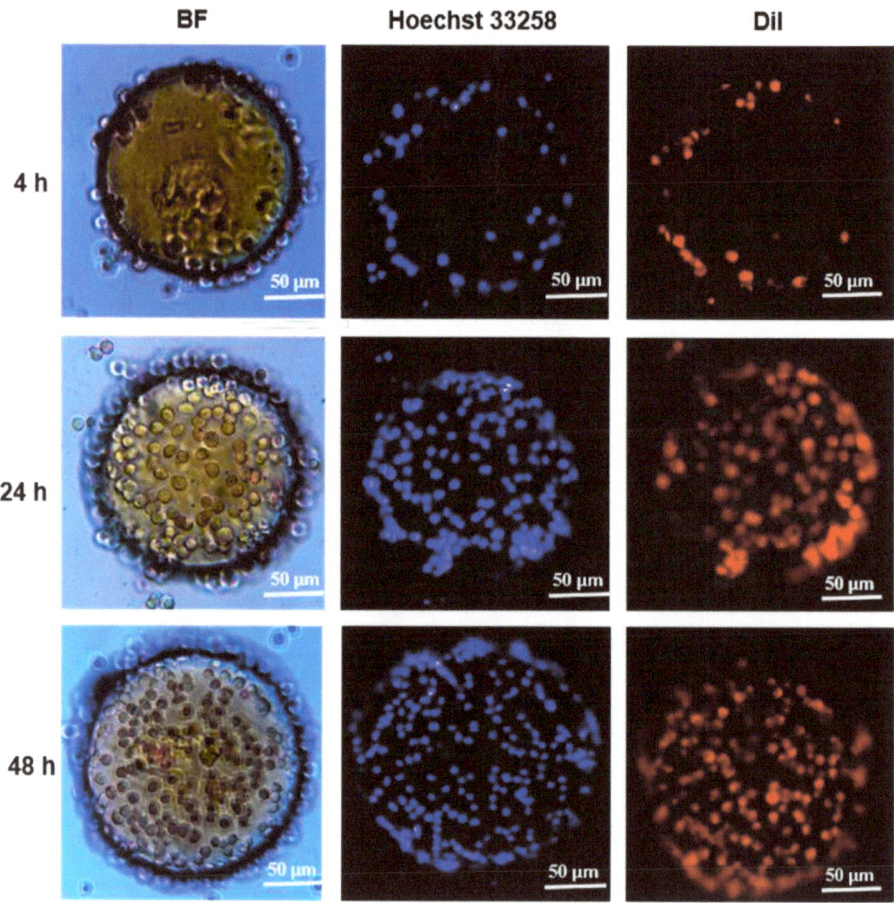

Figure 6. Fluorescence microscopy image of cell morphology and distribution on CSM-g-CD-Gel microspheres for 4 h, 24 h, and 48 h. Hoechst 33,258 stained nucleus; DiI stained cell membrane. The scale bar is 50 μm.

3.6. Responsive Detachment of Cells

Cell detachment from the CSM-g-CD-Gel microspheres was achieved by adding AD molecules to the culture. The results are shown in Figure 7. When the cells were cultured on the microspheres for 48 h, the cells spread well on both CSM and CSM-g-CD-Gel (Figure 7A,B). When AD molecules were added and reacted for 5 min and 30 min, the number of cells on CSM was not significantly different from that before AD treatment, as shown in Figure 7D,G. In comparison, the number of cells on CSM-g-CD-Gel decreased after 5 min of AD treatment, as shown in Figure 6E. Furthermore, after 30 min of AD treatment, the number of cells on the CSM-g-CD-Gel microspheres was significantly decreased (Figure 7H), indicating that a large number of cells had detached from the microspheres. To further verify the cell detachment, the nuclei of the cells on microspheres were stained to show the decrease in cell number on microspheres more clearly after AD treatment, as shown in Figure 7C,F,I. The AD-responsive cell detachment from the CSM-g-CD-Gel microspheres is due to the competitive binding of AD molecules to the CD cavities, leading to the release of gelatin molecules, which tightly bind cells. The cells detached together with the gelatin released from the microspheres. The above results indicate that the CSM-g-CD-Gel microspheres can be used as smart microcarriers for cell culture and enzyme-free cell detachment.

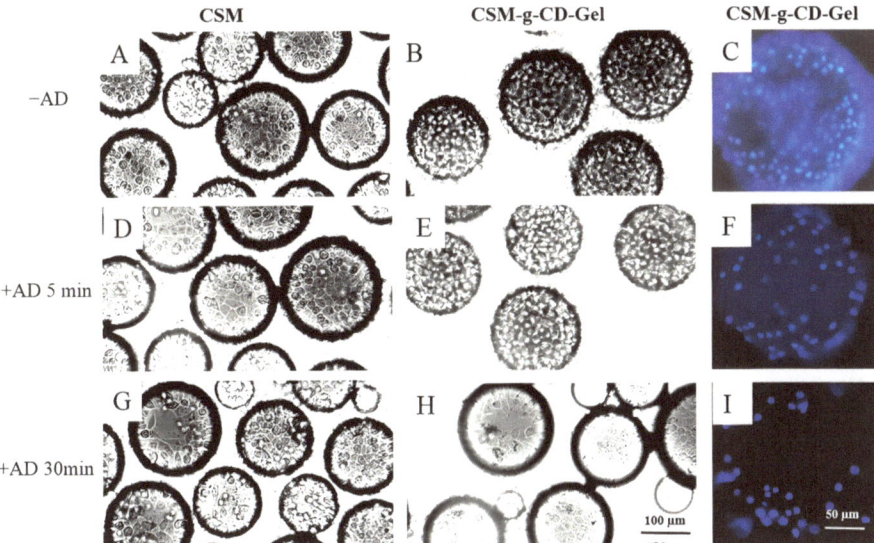

Figure 7. Cell detachment from CSM-g-CD-Gel after treatment with amantadine (**A,D**). The scale bar for (**A,B,D,E,G,H**) is 100 μm. The scale bar for (**C,F,I**) is 50 μm.

4. Conclusions

The construction of responsive microcarriers based on chitosan microspheres and their application in cell culture and enzyme-free detachment were studied in this work. First, chitosan microspheres were prepared with the emulsification crosslinking method, and through optimization of the preparation conditions, solid chitosan microspheres (CSMs) with spherical shapes and particle sizes ranging from 100 to 300 μm were obtained. This was followed by the successful grafting of β-cyclodextrin (β-CD) groups on the surface of CSM via an EDC/NHS-mediated acylation reaction and further gelatin modification to obtain CSM-g-CD-Gel utilizing the host–guest interactions between β-CD cavities and the aromatic residues of gelatin. SEM, FTIR, and 2D 1H NMR were employed to characterize the microspheres. The CSM-g-CD-Gel microspheres exhibited good cytocompatibility, allowing cells to attach and grow over the microsphere surface and maintain cell morphology. Finally, nonenzymatic detachment of cells together with gelatin from the CSM-g-CD-Gel microspheres can be achieved by the addition of AD based on the competitive binding of

AD molecules to the CD cavities. Overall, CSM-g-CD-Gel microspheres may potentially be applied as smart microcarriers for 3D cell culture and enzyme-free detachment.

Author Contributions: Conceptualization, L.H. and L.X.; data curation, W.Z., Q.L. and F.W.; formal analysis, L.H., Y.J., Q.L. and S.C.; funding acquisition, L.H., F.W. and L.X.; investigation, L.H., Y.J., X.C., W.Z., S.W. and F.W.; methodology, Y.J., X.C., W.Z., Q.L., S.C. and L.X.; project administration, L.H. and L.X.; resources, X.C., S.W. and L.X.; software, L.H. and Y.J.; supervision, L.H. and L.X.; validation, Y.J., X.C., S.C. and L.X.; writing—original draft, L.H. and Y.J.; writing—review and editing, X.C., S.W. and L.X. All authors have read and agreed to the published version of the manuscript.

Funding: This research was funded by the Open Foundation of Hubei Key Laboratory of Purification and Application of Plant Anti-Cancer Active Ingredients (Hubei University of Education) (HLPAI2023012), the National Natural Science Foundation of China (52173151), Hubei Provincial Natural Science Foundation (2020CFB491,2023AFB152), and Science Research Program of the Education Department of Hubei Province (Q20203002).

Institutional Review Board Statement: Not applicable.

Informed Consent Statement: Not applicable.

Data Availability Statement: Data will be made available upon request.

Conflicts of Interest: The authors declare no conflict of interest.

References

1. Huang, L.; Xiao, L.; Jung Poudel, A.; Li, J.; Zhou, P.; Gauthier, M.; Liu, H.; Wu, Z.; Yang, G. Porous chitosan microspheres as microcarriers for 3D cell culture. *Carbohydr. Polym.* **2018**, *202*, 611–620. [CrossRef]
2. Huang, L.; Abdalla, A.M.E.; Xiao, L.; Yang, G. Biopolymer-based microcarriers for three-dimensional cell culture and engineered tissue formation. *Int. J. Mol. Sci.* **2020**, *21*, 1895.
3. Chen, X.; Chen, J.; Tong, X.; Mei, J.; Chen, Y.; Mou, X. Recent advances in the use of microcarriers for cell cultures and their ex vivo and in vivo applications. *Biotechnol. Lett.* **2020**, *42*, 1–10.
4. Wei, F.; Zhang, X.; Cui, P.; Gou, X.; Wang, S. Cell-based 3D bionic screening by mimicking the drug-receptor interaction environment in vivo. *J. Mater. Chem. B. Mater. Biol. Med.* **2021**, *3*, 9. [CrossRef]
5. Niederau, C.; Fronhoffs, K.; Klonowski, H.; Schulz, H.U. Active pancreatic digestive enzymes show striking differences in their potential to damage isolated rat pancreatic acinar cells. *J. Lab. Clin. Med.* **1995**, *125*, 265–275.
6. Derakhti, S.; Safiabadi-Tali, S.H.; Amoabediny, G.; Sheikhpour, M. Attachment and detachment strategies in microcarrier-based cell culture technology: A comprehensive review. *Mater. Sci. Eng. C Mater. Biol. Appl.* **2019**, *103*, 109782.
7. Dabiri, S.M.H.; Samiei, E.; Shojaei, S.; Karperien, L.; Khun Jush, B.; Walsh, T.; Jahanshahi, M.; Hassanpour, S.; Hamdi, D.; Seyfoori, A.; et al. Multifunctional Thermoresponsive Microcarriers for High-Throughput Cell Culture and Enzyme-Free Cell Harvesting. *Small* **2021**, *17*, 2103192. [CrossRef]
8. Wang, J.; Chen, G.; Zhao, Z.; Sun, L.; Zou, M.; Ren, J.; Zhao, Y. Responsive graphene oxide hydrogel microcarriers for controllable cell capture and release. *Sci. China Mater.* **2018**, *61*, 1314–1324. [CrossRef]
9. Li, X.; Ji, X.; Chen, K.; Ullah, M.W.; Li, B.; Cao, J.; Xiao, L.; Xiao, J.; Yang, G. Immobilized thrombin on X-ray radiopaque polyvinyl alcohol/chitosan embolic microspheres for precise localization and topical blood coagulation. *Bioact. Mater.* **2021**, *6*, 2105–2119.
10. Agarwal, T.; Chiesa, I.; Costantini, M.; Lopamarda, A.; Tirelli, M.C.; Borra, O.P.; Varshapally, S.V.S.; Kumar, Y.A.V.; Koteswara Reddy, G.; De Maria, C.; et al. Chitosan and its derivatives in 3D/4D (bio) printing for tissue engineering and drug delivery applications. *Int. J. Biol. Macromol.* **2023**, *246*, 125669.
11. Ansari, M.T.; Murteza, S.; Ahsan, M.N.; Hasnain, M.S.; Nayak, A.K. Chitosan as a responsive biopolymer in drug delivery. In *Chitosan in Drug Delivery*; Academic Press: Cambridge, MA, USA, 2022; pp. 389–410.
12. Barbosa, O.; Ortiz, C.; Berenguer-Murcia, Á.; Torre, R.; Rodrigue, R.; Fernandez-Lafuent, R. Glutaraldehyde in bio-catalysts design: A useful crosslinker and a versatile tool in enzyme immobilization. *RSC Adv.* **2014**, *4*, 1583–1600. [CrossRef]
13. Migneault, I.; Dartiguenave, C.; Bertrand, M.J.; Waldron, K.C. Glutaraldehyde: Behavior in aqueous solution, reaction with proteins, and application to enzyme crosslinking. *BioTechniques* **2004**, *37*, 790–802. [CrossRef] [PubMed]
14. Monsan, P. Optimization of glutaraldehyde activation of a support for enzyme immobilization. *J. Mol. Catal Chem.* **1978**, *3*, 371–384. [CrossRef]
15. Zhang, J.; Cui, Z.; Field, R.; Moloney, M.G.; Rimmer, S.; Ye, H. Thermo-responsive microcarriers based on poly(N-isopropylacrylamide). *Eur. Polym. J.* **2015**, *67*, 346–364.
16. Tamura, A.; Kobayashi, J.; Yamato, M.; Okano, T. Thermally responsive microcarriers with optimal poly(N-isopropylacrylamide) grafted density for facilitating cell adhesion/detachment in suspension culture. *Acta Biomater.* **2012**, *8*, 3904–3939. [CrossRef]

17. Hamza, M.F.; Wei, Y.; Althumayri, K.; Fouda, A.; Hamad, N.A. Synthesis and Characterization of Functionalized Chitosan Nanoparticles with Pyrimidine Derivative for Enhancing Ion Sorption and Application for Removal of Contaminants. *Materials.* **2022**, *15*, 4676. [CrossRef] [PubMed]
18. Zhang, W.; Wang, Y.; Guo, J.; Huang, C.; Hu, Y. Polysaccharide supramolecular hydrogel microparticles based on carboxymethyl β-cyclodextrin/chitosan complex and EDTA-chitosan for controlled release of protein drugs. *Polym. Bull.* **2022**, *79*, 6087–6097. [CrossRef]
19. Ndlovu, S.P.; Ngece, K.; Alven, S.; Aderibigbe, B.A. Gelatin-based hybrid scaffolds: Promising wound dressings. *Polymers* **2021**, *13*, 2959. [CrossRef] [PubMed]
20. Niquini, F.M.; Moura, A.L.S.; Machado, P.H.; Oliveira, K.M.; Correa, R.S. Synthesis, infrared and molecular structure of adamantane-1-ammonium picrate monohydrate: A derivative of the antiviral symmetrel. *Crystallogr. Rep.* **2020**, *65*, 879–884. [CrossRef]
21. Sugane, K.; Maruoka, Y.; Shibata, M. Self-healing epoxy networks based on cyclodextrin–adamantane host–guest interactions. *J. Polym. Res.* **2021**, *28*, 423. [CrossRef]
22. Xiao, L.; Xu, W.; Huang, L.; Liu, J.; Yang, G. Nanocomposite pastes of gelatin and cyclodextrin-grafted chitosan nanoparticles as potential postoperative tumor therapy. *Adv. Compos. Hybrid. Mater.* **2023**, *6*, 15. [CrossRef]
23. Liang, J.; Guo, Z.; Timmerman, A.; Grijpma, D.; Poot, A. Enhanced mechanical and cell adhesive properties of photo-crosslinked PEG hydrogels by incorporation of gelatin in the networks. *Biomed. Mater.* **2019**, *14*, 024102. [CrossRef] [PubMed]
24. Pang, X.; He, X.; Qiu, Z.; Zhang, H.; Xie, R.; Liu, Z.; Gu, Y.; Zhao, N.; Xiang, Q. Targeting integrin pathways: Mechanisms and advances in therapy. *Sig. Transduct. Target. Ther.* **2023**, *8*, 1.
25. Boraschi-Diaz, I.; Wang, J.; Mort, J.; Komarova, S. Collagen Type I as a Ligand for Receptor-Mediated Signaling. *Front. Phys.* **2017**, *5*, 12.
26. Azreq, M.A.E.; Naci, D.; Aoujit, F. Collagen/β1 integrin signaling up-regulates the ABCC1/MRP-1 transporter in an ERK/MAPK-dependent manner. *Mol. Biol. Cell.* **2012**, *23*, 3437–3684. [CrossRef]
27. Zaman, M.H. Understanding the Molecular Basis for Differential Binding of Integrins to Collagen and Gelatin. *Biophys. J.* **2007**, *92*, 17–19.
28. Davidenko, N.; Schuster, C.F.; Bax, D.V.; Farndale, R.W.; Hamaia, S.; Best, S.M.; Cameron, R.E. Evaluation of cell binding to collagen and gelatin: A study of the effect of 2D and 3D architecture and surface chemistry. *J. Mater. Sci. Mater. Med.* **2016**, *27*, 148. [CrossRef] [PubMed]

Disclaimer/Publisher's Note: The statements, opinions and data contained in all publications are solely those of the individual author(s) and contributor(s) and not of MDPI and/or the editor(s). MDPI and/or the editor(s) disclaim responsibility for any injury to people or property resulting from any ideas, methods, instructions or products referred to in the content.

Article

Chitosan Coatings Modified with Nanostructured ZnO for the Preservation of Strawberries

Dulce J. García-García [1], G. F. Pérez-Sánchez [1,*], H. Hernández-Cocoletzi [2,*], M. G. Sánchez-Arzubide [2], M. L. Luna-Guevara [2], E. Rubio-Rosas [3], Rambabu Krishnamoorthy [4] and C. Morán-Raya [1]

1. Ecocampus Valsequillo, ICUAP, Centro de Investigación en Fisicoquímica de Materiales, Benemérita Universidad Autónoma de Puebla, Edificio Val-3, San Pedro Zacachimapa, Puebla 72960, Mexico; dulce.garciagarcia@alumno.buap.mx (D.J.G.-G.); carolina.moran@correo.buap.mx (C.M.-R.)
2. Facultad de Ingeniería Química, Benemérita Universidad Autónoma de Puebla, Av. San Claudio y 18 sur S/N Edificio FIQ7 CU San Manuel, Puebla 72570, Mexico; madai.sancheza@correo.buap.mx (M.G.S.-A.); maria.luna@correo.buap.mx (M.L.L.-G.)
3. Centro Universitario de Vinculación y Transferencia de Tecnología, Benemérita Universidad Autónoma de Puebla, Prol. 24 sur S/N CU San Manuel, Puebla 72570, Mexico; efrain.rubio@correo.buap.mx
4. Department of Chemical Engineering, Khalifa University, Abu Dhabi P.O. Box 127788, United Arab Emirates
* Correspondence: francisco.perezsanchez@correo.buap.mx (G.F.P.-S.); heriberto.hernandez@correo.buap.mx (H.H.-C.)

Citation: García-García, D.J.; Pérez-Sánchez, G.F.; Hernández-Cocoletzi, H.; Sánchez-Arzubide, M.G.; Luna-Guevara, M.L.; Rubio-Rosas, E.; Krishnamoorthy, R.; Morán-Raya, C. Chitosan Coatings Modified with Nanostructured ZnO for the Preservation of Strawberries. *Polymers* **2023**, *15*, 3772. https://doi.org/10.3390/polym15183772

Academic Editor: Fengwei (David) Xie

Received: 25 July 2023
Revised: 30 August 2023
Accepted: 1 September 2023
Published: 15 September 2023

Copyright: © 2023 by the authors. Licensee MDPI, Basel, Switzerland. This article is an open access article distributed under the terms and conditions of the Creative Commons Attribution (CC BY) license (https://creativecommons.org/licenses/by/4.0/).

Abstract: Strawberries are highly consumed around the world; however, the post-harvest shelf life is a market challenge to mitigate. It is necessary to guarantee the taste, color, and nutritional value of the fruit for a prolonged period of time. In this work, a nanocoating based on chitosan and ZnO nanoparticles for the preservation of strawberries was developed and examined. The chitosan was obtained from residual shrimp skeletons using the chemical method, and the ZnO nanoparticles were synthesized by the close-spaced sublimation method. X-ray diffraction, scanning electron microscopy, electron dispersion analysis, transmission electron microscopy, and infrared spectroscopy were used to characterize the hybrid coating. The spaghetti-like ZnO nanoparticles presented the typical wurtzite structure, which was uniformly distributed into the chitosan matrix, as observed by the elemental mapping. Measurements of color, texture, pH, titratable acidity, humidity content, and microbiological tests were performed for the strawberries coated with the Chitosan/ZnO hybrid coating, which was uniformly impregnated on the strawberries' surface. After eight days of storage, the fruit maintained a fresh appearance. The microbial load was reduced because of the synergistic effect between chitosan and ZnO nanoparticles. Global results confirm that coated strawberries are suitable for human consumption.

Keywords: strawberry; food preservation; chitosan; ZnO nanoparticles; composites; edible coatings

1. Introduction

Strawberries are very attractive for their flavor and antioxidant content. They are perishable, with a preservation time dependent on the storage procedure. The damage to the strawberries is commonly due to incorrect handling and storage; during these steps, biological, chemical, and physical factors influence their quality. The production process involves the harvest, the transport, the load, and the download. In these steps, strawberries are exposed to dangerous biological agents, reducing their shelf life [1] and causing an economic loss of about 30% to growers and marketers.

Zinc oxide nanoparticles (ZnO-NPs) have attracted attention due to their antimicrobial properties, low cost of production, and low toxicity [2]. It is possible to obtain ZnO-NPs with a variety of chemical and physical methods in different shapes and sizes. In recent years, environmentally friendly methods (green synthesis) have been developed using different plant extracts for the preparation and characterization of chitosan–nano-ZnO composite films for the preservation of cherry tomatoes [3]. It has been demonstrated that nanostructured ZnO can inhibit the growth of bacteria and fungi, extending the shelf life of foods [4]. Unlike other metal oxides, it is recognized as a safe material by the Food and Drug Administration (FDA) in the US [5]. However, to date, there are no conclusive results; in an in vitro colon simulation, it was reported that nano-ZnO could alter the metabolism, microbiota, and resistome of the human gut [6]. Even though nanostructured ZnO has been used in food, medicine, and feed production and processing, it is generally inefficient to improve the antioxidant properties of edible films [7].

Chitosan is a linear polysaccharide scarcely found in nature. This biopolymer is obtained from chitin through a deacetylation reaction [8]. The main source of chitin is shrimp, lobster, and crab skeletons. Chitosan has been widely used in different areas, such as agriculture, medicine, food, cosmetics, textiles, pharmaceuticals, biotechnology, and wastewater treatment [9]. In the food industry, it is widely used as an edible semipermeable barrier; however, its mechanical properties reduce its performance in protecting fruits. The effectiveness of chitosan can be enhanced when combined with organic and inorganic compounds. Being loaded with metal oxides such as zinc oxide and titanium oxide, it is possible to improve firmness, maintain quality, extend post-harvest life, and reduce pesticide residues in fruits [10].

Nanobiocomposites open up an opportunity for the use of innovative, high-performance, lightweight, and ecological composite materials, which makes them ideal materials to replace traditional non-biodegradable plastics. Nanostructured ZnO is highly viable for developing composite materials because it has high antimicrobial activity and improved mechanical and barrier properties [11]. The antibacterial mechanism associated with ZnO nanoparticles depends both on the generation of reactive oxygen species and on the release of antimicrobial Zinc ions; likewise, it has been shown that its effectiveness increases as the size of the particles decreases due to the increase of its surface reactivity [11,12]. Chitosan/ZnO nanocomposites have attracted the attention of researchers due to their antibacterial activity [13]. According to the surveys, Chitosan/ZnO bionanocomposites are limited to films [3,14]. Al-naamani et al. [15] reported the method for preparing PE films coated with chitosan ZnO composites with effective antimicrobial effects, and Karkar et al. [16] developed a film with Nigella sativa extract for preventing grape spoilage. Chitosan/ZnO coatings have been recently applied to extend the shelf life of wild-simulated Korean ginseng root [17] as well as inhibit microbial growth on fresh-cut papaya [18]. More complex Chitosan/ZnO nanocoatings have been proposed, including compounds such as alginate, which were applied to the preservation of guavas [19].

Concerning strawberries, the studies developed to date are limited to investigating the antibacterial activity of the coatings [20]. However, studies are required in which the quality of the fruit during storage is evaluated and the physicochemical parameters related to the level of consumer acceptance are maintained. In addition, preserving fruits that contain antioxidant properties associated with human health is required. Environmentally friendly and easily applicable approaches for preserving strawberries during storage are required. Chitosan ZnO coatings are promising materials to be used in the food industry; however, more studies are required. Thus, this work aims to develop a bionanocoating containing inorganic nanostructured ZnO and the natural biopolymer chitosan with the potential to be used in the preservation of fruits. This is demonstrated by different microbiological (Aerobic Mesophilic Bacteria, molds, and yeasts) as well as physicochemical (moisture content, texture, pH, soluble solids, titratable acidity, and color) analyses.

2. Materials and Methods

2.1. Chitosan Synthesis

Chitosan (Ch) was prepared using a methodology described in our previous work [21]. Briefly, raw shrimp shells were obtained from a local seafood restaurant; heads and ends were not included. After washing with distilled water and drying at 90 °C for 3 h, the shells were pulverized to a particle size in the range of 44–53 μm. The as-obtained sample was demineralized using 0.6 M HCl with a ratio of dried shells to an acid solution of 1:11

(w/v) for 3 h at 30 °C and stirring at 300 rpm. Afterwards, to obtain chitin, the sample was sonicated with a high-frequency ultrasonic bath in deionized water for 40 min. Finally, NaOH (50%) was added to chitin in the proportion 1:4 (w/v) with a constant stirring speed of 700 rpm. It was initially heated and maintained at 70 °C for 2 h, followed by heating at 115 °C for another 2 h. The final product of each step was washed with distilled water until it became neutral and then dried at 90 °C for 3 h. The as-obtained chitosan has a molecular weight of 56 kDa and a deacetylation degree of 94%.

2.2. Infrared Spectroscopy Characterization

The identification of chitosan was made using infrared spectroscopy. A Perkin Elmer Spectrophotometer with a fast Fourier transform and ATR system was employed for this purpose. The scan was carried out in the range of 4000–500 cm^{-1}, employing eight scans with a 2 cm^{-1} resolution. Humidity (0%) and a temperature of 25 °C were maintained while obtaining the spectra to ensure proper determination of the –OH groups. Prior to the analysis, the samples were dried in a conventional heater at 105 °C for 4 h, and about 0.05 g of the sample was subjected to the analysis.

2.3. Synthesis of Nanostructured ZnO Using the Close-Spaced Sublimation Method

The zinc oxide nanostructures were synthesized using the close-space sublimation (CSS) method on glass substrates in two simple steps: sublimation of metallic zinc followed by thermal annealing at atmospheric pressure conditions. The experimental procedure is briefly described as follows: glass substrates of 2 cm^2 were cleaned sequentially with xylene, acetone, and propanol at 50 °C for 10 min in an ultrasonic cleaner; finally, they were rinsed with deionized water (18 MΩ-cm) and dried with high-purity nitrogen. For the synthesis of the zinc nanostructures by CSS, initially, circular pellets made from zinc powders (99%, mesh size −325 microns) were used as the source. Inside the CSS system, the zinc pellet was placed on a graphite heater, establishing a distance separation between the source and the substrate of 1 cm by means of a quartz ring (thickness 0.76 mm). The growth deposition parameters of the zinc nanostructured films were fixed at 350 °C, 10 min, and 5 × 10^{-3} torr for the reactor temperature, growth time, and vacuum, respectively. The as-grown films showed an opaque gray color (to the naked eye). Finally, the zinc nanostructures were oxidized in the air using a tubular furnace at a temperature of 500 °C for one hour at atmospheric pressure. At the end of the treatment, the films turned white, indicating a conversion of zinc-to-zinc oxide nanostructures. The procedure for the HR-TEM analysis consisted of the mechanical separation of ZnO nanostructured material from the substrates after thermal annealing in air. After that, 4 mg of the as-prepared ZnO powder were dispersed in 50 mL of ethanol solvent, and the solution was homogenized in a bath sonicator for 30 min. Then, a drop of the solution was cast on a copper grid (400 mesh size) and dried on a hotplate at 70 °C for less than a minute. Finally, the grid was introduced into the HR-TEM system, and measurements were performed using a voltage acceleration of 200 KV.

2.4. Preparation and Application of the Coating

A solution with 1 mL of acetic acid 1 M and 100 mL of water was prepared; after homogenization, 1 g of chitosan was dissolved into this solution at 80 °C for 30 min. A second solution was prepared under the same conditions; however, in this case, 0.15 g ZnO nanoparticles were incorporated with constant stirring for 10 min. The obtained solutions were applied by submerging the strawberries for 1 and 2 s. It is important to mention that the coating was firmly adhered to the strawberry surface with insignificant runoff. A strawberry without any coating was used as a control sample. All the experiments were performed at room temperature and in triplicate. The samples were stored at 5 °C and observed for eight days [22].

2.5. Characterization of the Coating

The solution used for coating strawberries was analyzed by FTIR spectroscopy. The coating, once applied to the strawberries, was characterized by scanning electron microscopy (SEM) (JEOL JSM-6610LV, Tokyo, Japan); an elemental and a mapping analysis were developed to identify the main components of the coating. The accelerating voltage was 20 kV in both morphological characterizations and EDX measurements. A 2 × 2 cm portion of the strawberry surface was cut and heated at 40 °C for 4 h; the as obtained sample was Au covered. The mapping analysis was developed in a 3 × 3 mm area in 3 different zones; all the image fields were considered.

2.6. Chitosan and Ch/ZnO-NPS Coating Functionality in the Preservation of Strawberries

2.6.1. Colorimetric Test

Strawberries with uniform size and commercial maturity were selected for the preservation study. They were purchased at a local market and used without any additional treatment. The sample luminosity was measured with a Hunter lab colorimeter every 24 h for eight days. With this parameter, the chroma (C_{ab}^*) for strawberries was obtained using Equation (1).

$$\text{the } C_{ab}^* = \sqrt{(a^{*2} + b^{*2})} \qquad (1)$$

where a^* represents the red and green colors; b^* the yellow and blue colors. The browning index (IP) was calculated with Equation (2) [23].

$$IP = \frac{100 \times (X - 0.31)}{0.172} \qquad (2)$$

where

$$X = \frac{a^* + 1.75L}{5.645L + a^* - 3.012b^*} \qquad (3)$$

2.6.2. Texture Test

The texture test was performed with a TA.XT2i (Stable Micro Systems) texture analyzer. For that, a plane cylindrical probe p/3 (3 mm diameter) was used with a speed of 1 mm/s and a resistance time of 5 s for the analysis.

2.6.3. pH, Titratable Acidity, Soluble Solids, and Humidity Content Determination

The pH, the soluble solids, and the titratable acidity were measured according to the AOAC 981.12, AOAC 932.12, and AOAC 942.15 methods [24], respectively. The pH of ground strawberries was measured with a WPA CD310 potentiometer. The titratable acidity was measured with titration equipment. The titration solution (8.9 mL) was prepared with NaOH at 0.1 N. For developing the titration, 10 mL of ground strawberries were incorporated into 10 mL of purified water; after mixing, two drops of phenolphthalein were added and mixed again. The obtained sample was titrated by adding two or three drops of the titration solution until observing a light pink color. The volumes of the NaOH solution were used for calculating the titratable acidity. For the determination of soluble solids content, strawberries were ground until they obtained a homogeneous mix. Each sample was placed in a digital refractometer (Atago, RX-100, Bellevue, DC, USA), and the measurements were expressed in °Bx. The humidity content was determined with the gravimetric method established in 934.06 of the AOAC standard. Each sample previously weighed was dried at 105 °C for 42 h in a drying oven; after that, the weight was determined. The humidity percentage was obtained with Equation (4).

$$\% \text{ humidity} = \frac{M_0 - M_f}{M_0} \times 100 \qquad (4)$$

where M_0 is the wet sample mass (g), and M_f is the dry sample mass (g).

2.6.4. Aerobic Mesophilic Bacteria, Fungi, and Yeast

For the microbiological tests, 10 g of each sample was introduced into sterile bags and homogenized for 1 min. The aerobic mesophilic bacteria (AMB) were measured with the plate pouring method on standard count agar; the plates were incubated at 35 °C for 48 h. The fungi and yeast quantification was realized by employing the plate pouring method on potato dextrose agar (PDA) acidified with 10% tartaric acid at a pH of 3.5; the plates were incubated at 25 °C for 72–120 h. The results were expressed as colony-forming units (CFU/g).

2.7. Statistical Analysis

All experiments and measurements were performed in triplicate, using a fully random design. An analysis of variance (ANOVA) of simple classification was applied, and the Tukey test was used to determine the difference between the means. The data analysis was carried out with Statistics Plus software (Statgraphics Centurion 19, Statistical Graphics Corp., Manugistics, Inc., Cambridge, MA, USA). A $p \leq 0.05$ significance level was established for all cases.

3. Results

3.1. X-ray Diffraction

Figure 1 shows the XRD pattern of the spaghetti-like ZnO nanoparticles (ZnO-NPs) synthesized by thermal annealing-assisted CSS. The diffraction peaks located at 2θ = 31.7, 34.4, 36.2, 47.5, 56.5, 62.8, 66.3, 67.9, and 68.9° were indexed to the planes (100), (002), (101), (102), (110), (103), (200), (112), and (201), which are characteristic of the wurtzite phase of ZnO in accordance with JCPDS file number 36-1451; no signs of secondary phases or impurities were observed, thus confirming the crystalline nature of the ZnO NPs powder [25]. The crystalline size of ZnO nanostructures was calculated using the Williamson-Hall (W-H) method using the software provided by the XRD system (High Score Plus for Crystallite Size Analysis). The W-H analysis is an integral breadth method where both size and strain-induced broadening are considered in the deconvolution of the peak versus 2θ [26]. It is important to mention that in this research only the crystallite size was considered because the W-H plot is more realistic than using the Debye-Scherrer equation; moreover, the result estimated from this analysis was in good agreement with the corresponding HR-TEM results; the size obtained was 40.1 nm. Similar values of crystal size (46 nm) were found in pure ZnO nanoparticles [27]; however, they were slightly higher than those obtained in ZnO nanoparticles with chitosan [24,27] and ZnO added with citrus extracts [28].

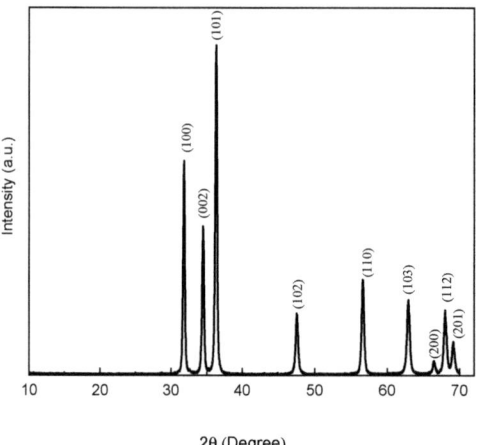

Figure 1. X-ray diffraction pattern of spaghetti-like ZnO nanoparticles.

3.2. Scanning Electron Microscopy

The SEM images of the as-synthesized nanostructured Zn by the CSS method are shown in Figure 2. The image at low magnification (Figure 2a) shows a homogeneous distribution of particle agglomerates with spaghetti-like morphology; their lengths reach approximately 5 µm, with a cross-section between 100 and 300 nm (Figure 2b,c). Figure 3 shows SEM images of the nanostructured ZnO layers (ZnO-NPs) obtained after the thermal treatment of Zn. In Figure 3a, the initial morphology of zinc nanostructures was not modified by the thermal stress. However, it could be observed after this process that the surface of the spaghetti-like nanostructures became rougher, along with the appearance of nanostructures of smaller dimensions on the surface (Figure 3b,c) and comparable results were reported elsewhere [29].

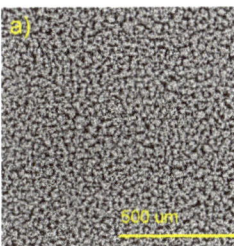 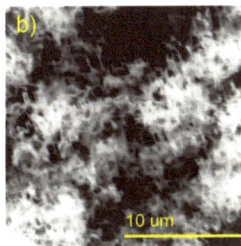 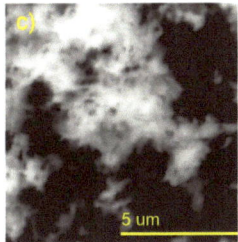

Figure 2. SEM images of the as-prepared Zn nanostructured films by the CSS technique: (**a**) morphology of the zinc spaghetti-like nanostructures at low magnification; (**b**,**c**) correspond to higher magnifications.

 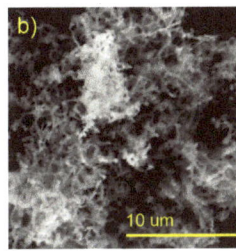 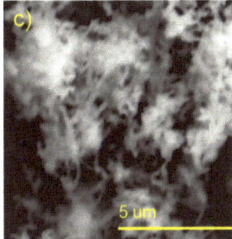

Figure 3. SEM images of the ZnO nanostructured films: (**a**) morphology of the spaghetti-like ZnO nanostructures after thermal annealing at low magnification; (**b**,**c**) are the micrographs at higher magnifications.

3.3. Transmission Electron Microscopy

TEM images of three different regions on the surface of ZnO nanostructures are shown in Figure 4. Figure 4a shows that agglomerates whose sizes are between 100 and 300 nm form ZnO nanostructures. On the other hand, Figure 4b reveals to us that the agglomerates are made up of nanoparticles whose average size corresponds to 50 nm, which is consistent with the previous XRD results. In addition, these nanoparticles tend to coalesce because of thermal annealing, forming nanorods that together give rise to the spaghetti-like morphology (Figure 4c), as observed in the SEM analysis. The agglomeration of nanoparticles is due to the densification process because of the narrow space between the metallic anions [30], giving rise to the formation of nanowires whose cross section is around 200 nm.

In general, the nanostructured ZnO prepared in this work showed higher values in nanoparticle size than those obtained in nanoparticles added with natural extracts (14–24 nm) [30,31] but like those obtained in commercial nanomaterials (50–60 nm) [14]. Finally, it was observed that there is a good correspondence between the results obtained by the X-ray diffraction and TEM techniques, given that in both cases, a crystallite size of ≈40 nm was obtained.

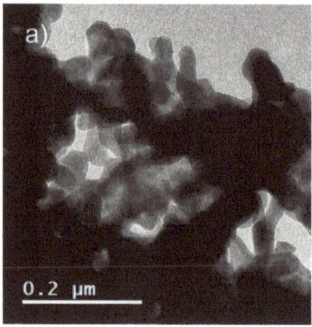 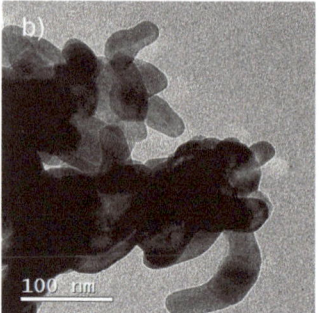 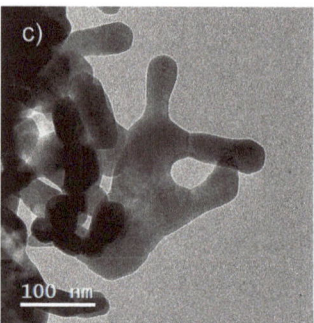

Figure 4. TEM images on the surface of the spaghetti-like ZnO nanostructures at different scale bars: (**a**) 0.2 µm, (**b**,**c**) 100 nm.

3.4. Fourier Transform Infrared Spectroscopy Analysis

Figure 5 shows the FTIR spectra of chitosan and Ch/ZnO-NPs bionanocomposite. The spectrum contains the characteristic absorption bands of chitosan at 3356 and 3398 cm^{-1}, which correspond to the stretching vibrations of the OH groups. At 2877 cm^{-1}, the vibrations attributed to the CH$_2$ groups are shown. At 1647 cm^{-1}, the bands corresponding to the bending vibrations from the N-H bonds and the stretching of the C-O bonds of the amine and amide groups are observed. The signals at ~1585 cm^{-1} are attributed to the vibration of the C=O bonds of the amide group [32]. At 1419 cm^{-1} the signals related to the CH$_2$ bonds are observed, and ~1377 cm^{-1}, the bands attributed to the C-O bond of the primary alcoholic group in the chitosan structure are observed. From 1060 to 1250 cm^{-1}, the bending vibrations of the C-O-C groups of glucose were identified; around 1026 cm^{-1}, the presence of the free amine groups (-NH$_2$) of glucosamine is observed. Finally, at 575 cm^{-1} the bending vibrations of the NH bonds are identified. According to these results, it is concluded that the chitosan obtained has the characteristic bands reported in the literature [33].

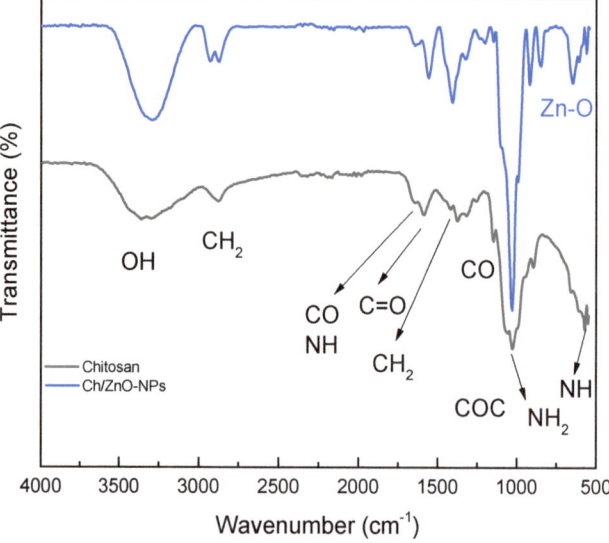

Figure 5. FTIR spectrum of chitosan and of the Ch/ZnO-NPs nanobiocomposite.

The FTIR spectrum of the Ch/ZnO-NPs bio-nanocomposite is also shown in Figure 5. A spectrum such as that of chitosan is observed; the vibrations located between 500 and 800 cm^{-1} correspond to the stretching of the O-Zn-O bonds, which are not present in the chitosan spectrum, confirming the presence of ZnO in the nanocomposite [27,34].

3.5. Energy Dispersive Spectroscopy Analysis

Elemental mapping of strawberries coated with chitosan during 1 min of immersion (FRCh-1) is shown in Figure 6. The spectrum revealed that the chitosan coating was composed of C (33.8%) and O (52.5%), whose distribution is seen uniformly on the strawberry surface [35]. The chemical composition of the Ch/ZnO-NPs coated strawberry immersed for 1 min immersion (FRBN-1) is presented in Figure 7. In addition to C (35.70%) and O (54.77%), a homogeneous distribution of Zn (0.56%) (Figure 7i) was found on the strawberry surface without severe aggregation [26,36]. Moreover, there is no evidence of the lump's formation, as has been observed in other cases where ZnO nanoparticles were used [16]. The homogeneous distribution of the Zn, as shown in Figure 7i, permits the inference that the ZnO nanostructures are uniformly distributed on the coating, which indicates a homogenous dispersion and integration into the chitosan matrix [26,36]. This also suggests high compatibility between the ZnO-NPs and the chitosan matrix due to the interactions between the free hydroxyl groups of chitosan and the available ZnO-NPs in the composite [37]. Thus, improved properties for the Ch/ZnO-NPs coating are expected for a longer shelf life of strawberries [38].

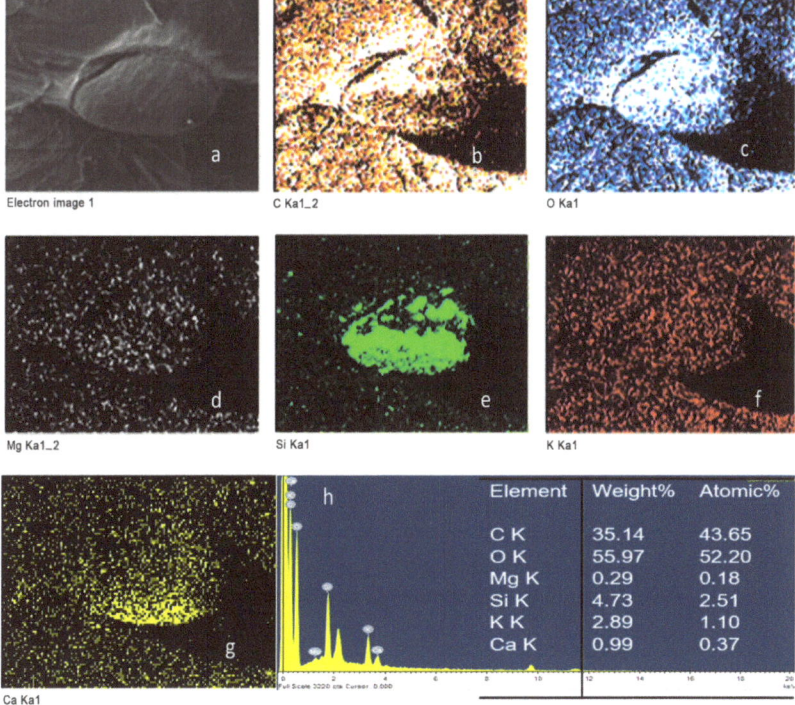

Figure 6. Elemental mapping of strawberries coated with chitosan after one minute of immersion (FRCh-1). (**a**) SEM micrograph, (**b**) carbon, (**c**) oxygen, (**d**) magnesium, (**e**) silicon, (**f**) potassium, (**g**) calcium, and (**h**) Quantitative elemental analysis of the coating.

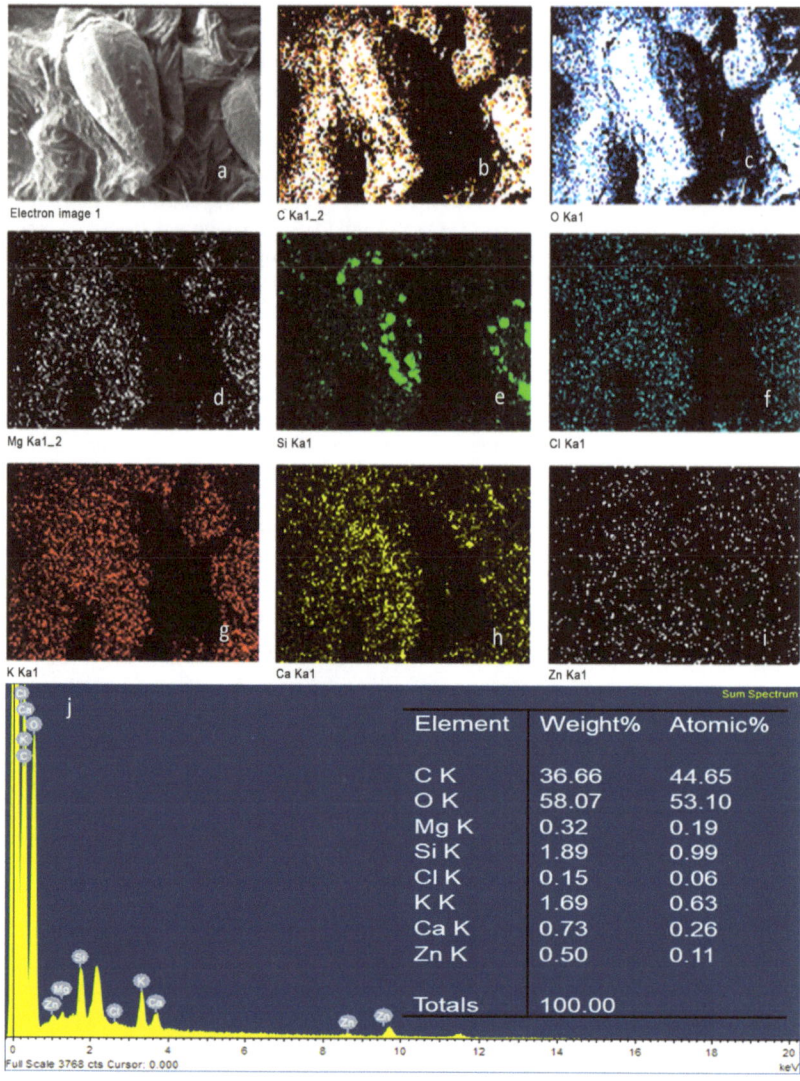

Figure 7. Elemental mapping of the bionanocomposite coated strawberry after one minute of immersion (FRBN-1). (**a**) SEM micrograph obtained, (**b**) carbon, (**c**) oxygen, (**d**) magnesium, (**e**) silicon, (**f**) chlorine, (**g**) potassium, (**h**) calcium, (**i**) zinc, and (**j**) quantitative elemental analysis of the bionanocomposite coating.

3.6. Coating Thickness

The thickness of the coating was measured from SEM images (Figure 8). For this, cross-sections of both pristine and ZnO nanocomposite coatings were prepared. The sample constituted by the chitosan coating immersed for 1 min (FRCh-1) presents an apparent longitudinal uniformity with an average thickness of 2.2 µm (Figure 8a). On the contrary, the thickness obtained after 2 min of immersion (FRCh-2) does not present uniformity in thickness, and their values are in the range of 2–3 µm (Figure 8b). In the case of Ch/ZnO NPs coating with an immersion time of 1 min (FRBN-1), SEM images at higher magnifications evidenced the appearance of randomly distributed particles on the surface of the coated strawberry, which is presumably an indication that the nanoparticles are incorporated throughout the chitosan matrix volume; the thickness was found in the range of 2–3 μm (Figure 8c). For the sample with 2 min of immersion (FRBN-2), it was observed that the cross section was rougher compared to the pristine coating, reinforcing the hypothesis that the ZnO nanoparticles lie on the surface as well as in the volume of the coating. The thickness does not differ significantly from the previous case, 2–3 μm (Figure 8d); therefore, we can conclude that the immersion times used in these experiments do not represent a notable change in the thickness of the coatings.

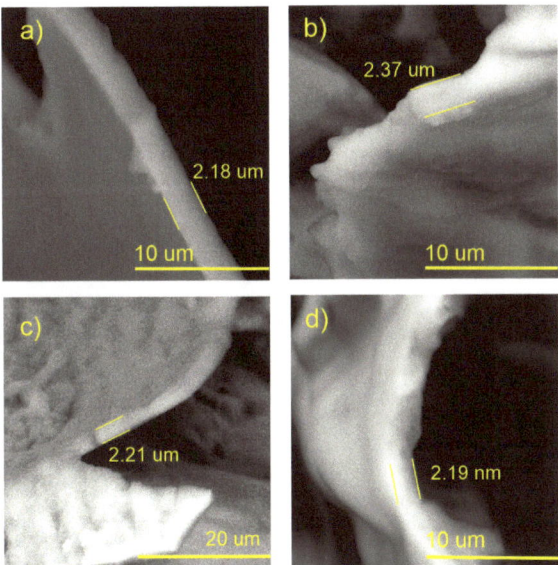

Figure 8. SEM images used for measuring the coating thickness. (**a**) FRCh-1; (**b**) FRCh-2; (**c**) FRBN-1; and (**d**) FRBN-2. 1 h.

3.7. Functionality of the Ch/ZnO-NSs Coating in the Preservation of Strawberries

Strawberries with and without coating were stored at 5 °C and 25 °C for 8 days (Tables 1 and 2). Table 1 shows that the microbial spoilage on the surface was lower in the strawberries coated with the Ch/ZnO-NPs nanobiocomposite than in the control strawberries; this is attributed to the antimicrobial properties of both chitosan and nanostructured ZnO. Cold storage reduces respiration rate and moisture loss, retarding microbial growth, allowing the shelf life to be extended and the quality of fruits to be preserved [39].

Table 2 shows that during storage of the coated strawberries (FRCh-1 and FRBN-1) at room temperature, they did not present microbial growth on the surface compared to the control strawberries. However, the coated strawberries have rough tissue on the surface, which suggests moisture loss during storage, which can give the appearance of deterioration. Tables 1 and 2 reveal that the coatings prepared in this work retarded weight loss and firmness for 8 days, attributed to their barrier properties. Films prepared with PVA and TP preserved strawberries for 5 days [40], while LDPE filled with LAE films extended the shelf life to 10 days [41].

The antifungal activity of FRBN-1 could be attributed to the previously identified action of zinc oxide. It has been reported that the antifungal activity of ZnO can be attributed to the generation of intracellular reactive oxygen species (ROS) by the nanobiocomposite in direct contact with the fungal cell wall [42,43]. The generated ROS results in elevated stress, leading to oxidative damage to the fungal cell wall and cellular components. Previous reports suggested that the antifungal activity of ZnO NPs was mediated by ROS production [44]. There exists a synergistic or complementary effect between the ZnO and the chitosan; the resulting synergetic interactions were effective against other pathogen microorganisms [45].

Table 1. Coated and uncoated strawberries stored at 5 °C for 8 days.

Table 2. Coated and uncoated strawberries stored at 25 °C for 8 days.

Visually, the presence of mold, the generation of fermentative odors, and the loss of turgidity and firmness were observed in strawberries subjected to both temperatures on day 1 and day 8 for the FRCh-1 treatment but not for the FRBN-1 treatment. The presence of fungi is more pronounced in the neat Ch coating (FRCh-1) compared to the Ch/ZnO coating (FRBN-1). This suggests that the incorporation of nanoZnO into the chitosan coating has a positive effect on inhibiting the presence of fungi in the strawberries. This observation indicates that the Ch/ZnO coating is more effective in preventing fungal growth compared to the neat chitosan coating.

3.8. Microbiological Tests

The results of the microbiological analysis (Aerobic Mesophilic Bacteria, molds, and yeasts) of the strawberries stored for 8 days at 5 and 25 °C are shown in Table 3. A progressive and significant increase in colonies of aerobic mesophilic bacteria (AMB) was observed in the control samples with storage time, mainly in the refrigerated strawberries. Coated strawberries under refrigeration (5 °C) also presented an increase in the AMB; this is expected because most of the AMB has psychotropic ability and may grow in low-temperature environments. The initial microbial load of the strawberry would also have contributed to the result obtained. As can be seen, the strawberries coated with chitosan follow the same trend. A better performance was obtained with the Ch/ZnO-NPs coating; despite the fact that the strawberries under refrigeration presented an increase in the AMB (in the order of 10^5), they are suitable for human consumption [31]. It is observed an increase in mold colonies and yeast in the control samples, mainly in those refrigerated. The FRCh-coated strawberries presented molds and yeast but in a reduced quantity; this confirmed the inhibition efficacy of chitosan against microorganisms on strawberries; the same effect has also been observed in tomatoes and ginseng [30,45].

Table 3. Effect of the different coatings on the microbiological changes of the treated strawberries stored at 5 and 25 °C.

Strawberry	t_s (Days)	T (°C)	AMB (CFU/g)	Molds and Yeasts (CFU/g)
Ctrl	1	5	$6.0 \times 10^2 \pm 112$ aC	$1.5 \times 10^2 \pm 10$ aB
		25	$2.0 \times 10^2 \pm 12$ bA	$3.0 \times 10^2 \pm 2.6$ aA
	8	5	INC	INC
		25	$4.0 \times 10^3 \pm 22$ bB	$1.04 \times 10^5 \pm 1.6$ aB
FRCh-1	1	5	50 ± 1.2 aA	0 ± 0.00 aA
		25	$2.0 \times 10^2 \pm 7.2$ bB	0 ± 0.00 aA
	8	5	$1.3 \times 10^6 \pm 15{,}612$ aB	$3.9 \times 10^6 \pm 13{,}212$ bB
		25	$4.3 \times 10^3 \pm 182$ bC	$2.9 \times 10^4 \pm 1122$ cC
FRCh-2	1	5	3.0×10^2 aB	0 ± 0.00 aA
		25	0 bC	0 ± 0.00 aA
	8	5	1.3×10^6 aA	$3.9 \times 10^6 \pm 10{,}100$ bB
		25	2.4×10^3 bA	$3.4 \times 10^3 \pm 123$ cC
* FRBN-1	1	5	0 ± 0.00 aA	0 ± 0.00 aA
		25	0 ± 0.00 aA	0 ± 0.00 aA
	8	5	$7.9 \times 10^5 \pm 600$ bC	$1.2 \times 10^6 \pm 1200$ bB
		25	0 ± 0.00 aA	0 ± 0.00 Aa
FRBN-2	1	5	0 ± 0.00 aA	0 ± 0.00 aA
		25	0 ± 0.00 aA	0 ± 0.00 aA
	8	5	$6.3 \times 10^5 \pm 0.00$ bC	$1.2 \times 10^6 \pm 1300$ bB
		25	0 ± 0.00 aA	0 ± 0.00 aA

Data are presented as mean ± standard deviation (SD). a, b, c superscripts in the same row indicate a significant difference between the means ($p \leq 0.05$) of temperature. A, B, C superscripts in the same row indicate a significant difference between the means ($p \leq 0.05$) of storage time. * Represents significant difference between the means ($p \leq 0.05$) of the treatments. INC: countless.

The incorporation of the nanostructured ZnO enhanced the inhibition capability of the composite; at room temperature (25 °C), no fungus or yeast was found after 8 days of storage. These results confirm the effectiveness of ZnO application as an antifungal agent incorporated into chitosan for preserving strawberries. This effect has also been observed in fresh foods, such as orange juice [46,47]. Likewise, it has been shown that the incorporation of nanomaterials in edible polymer coatings improves their physical properties, such as oxygen and moisture barrier properties, which inhibit the growth of microorganisms in coated fresh fruits [48].

3.9. Moisture Content

Table 4 shows the moisture content of the samples stored for 8 days (5 and 25 °C). The control sample presented a higher reduction in humidity. Coatings have the effect of maintaining the moisture of strawberries; better performance was reached under refrigeration, where this parameter decreased no significantly. At room temperature, a reduction of less than 10% in this parameter was shown in the coated strawberries. The coating solutions created a barrier on the surface of the strawberries that prevented moisture loss. This effect has also been reported in sodium alginate coatings, where the protective layer between fresh fruits and the surrounding atmosphere decreases moisture transfer and the exchange of O_2 and CO_2 [49].

Table 4. Moisture content of coated and uncoated strawberries stored at different temperatures for 8 days.

t_s (Days)	T (°C)			Moisture (%)		
		Ctrl	* FRCh-1	FRCh-2	FRBN-1	FRBN-2
1	5	91.9 ± 2.0 [aA]	92.6 ± 2.2 [aB]	92.2 ± 0.9 [aAB]	93.2 ± 1.1 [Aa]	91.6 ± 1.3 [aAB]
	25	89.1 ± 0.6 [aA]	89.0 ± 1.3 [aA]	82.0 ± 9.9 [aB*]	89.0 ± 2.2 [aA]	92.7 ± 4.0 [aAC]
8	5	89.3 ± 1.1 [aA]	92.3 ± 1.4 [aA]	91.4 ± 1.0 [aA]	92.9 ± 0.8 [Aa]	91.0 ± 1.4 [aA]
	25	45.3 ± 3.7 [bA*]	69.8 ± 1.5 [bB]	67.4 ± 2.7 [bB]	83.0 ± 1.6 [Bc]	84.5 ± 3.3 [bC]

Data are presented as mean ± standard deviation (SD). a, b, c superscripts in the same row indicate a significant difference between the means ($p \leq 0.05$) of the treatments. A, B, C Different superscripts in the same row indicate a significant difference between the means ($p \leq 0.05$) of storage time. * Represents a significant difference between the means ($p \leq 0.05$) of the treatments.

It should be noted that the moisture content was reduced when refrigeration conditions (5 °C) were used for storage. The application of low temperatures with complementary methods of food preservation, such as chemical treatments and edible coatings, generates lower moisture losses in the applied foods. It has been shown that the use of ZnO nanoparticles together with alginate significantly reduced ($p < 0.05$) the weight loss of burs [50].

3.10. Texture Test Analysis

Texture is one of the most important attributes for consumers when evaluating the quality of fruits. Firmness changes of strawberries with different coatings are shown in Figure 9; they were determined on Day 1 and Day 8 of storage. Firmness values obtained at the beginning of storage were 2.2 N (25 °C) and 0.6 N (5 °C). At refrigeration, the firmness value of most of the coated strawberries was lower than the control strawberry, except for those coated with the nanocomposite with an immersion time of 2 s (FRBN-2). The samples subjected to storage at room temperature (25 °C) presented a loss of firmness, impeding the measurement. Finally, the firmness of control strawberries showed significant differences ($p < 0.05$) compared to coated strawberry samples during cold storage.

In general, the control and coated strawberries showed a significant decrease in firmness. This decrease is due to the greater migration of water vapor on the surface of the fruit, which favors the growth of different fungi (*Botrytis cinera* and *Rhizopus stolonifer*). Both molds cause structural damage to the tissues and allow their softening [38]. Moreover, the decrease in firmness is related to the increase in moisture loss. In addition, the decrease in firmness which was observed during the first eight days of storage in the different evaluated coatings could be related to the degradation of the cortical parenchyma that forms the cell wall due to enzymatic degradation processes and to the same moisture loss during storage [39,44].

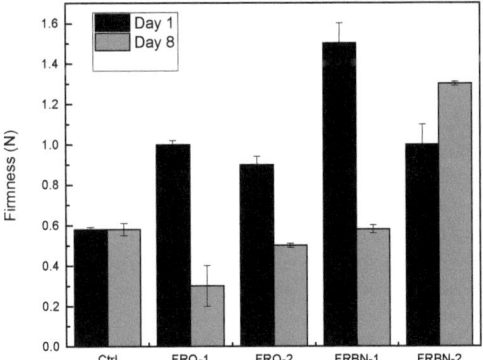

Figure 9. Firmness changes of coated and control strawberries stored at 5 °C for 1 and 8 days. Each data point are the mean of three replicates.

3.11. pH, Soluble Solids, and Titratable Acidity Analysis

Sweetness and acidity are among the essential indices for evaluating the flavor of fruits, which are evaluated by physicochemical parameters such as pH, total soluble solids (TSS), and titratable acidity (TA) [21]. Table 5 shows these values evaluated for coated and uncoated strawberries stored under refrigeration (5 °C) and at room temperature (25 °C). It is important to mention that these parameters can be influenced by factors such as crops, agricultural practices, agricultural region, season, etc. [51]. The pH of the uncoated samples increased slightly during storage, while no significant differences were observed in the coated samples (Table 5); similar results were reported by Nunes et al. (2006) [52].

Table 5. Physicochemical parameters of coated and uncoated strawberries stored at different temperatures for 8 days.

Strawberry	t_s (Days)	T (°C)	pH	TSS (°Bx)	TA(%)
Ctrl	1	5	3.59 ± 0.18 ^{aA}	8.5 ± 0.2 ^{aA}	0.76 ± 0.04 ^{aA}
		25	3.17 ± 0.03 ^{bB}	8.1 ± 0.1 ^{aA}	0.80 ± 0.02 ^{aA}
	8	5	3.31 ± 0.01 ^{aC}	8.0 ± 0.1 ^{aA}	0.76 ± 0.04 ^{aA}
		25	3.68 ± 0.03 ^{bA}	7.7 ± 0.2 ^{bA}	0.60 ± 0.01 ^{cB}
* FRCh-1	1	5	3.47 ± 0.02 ^{aA}	8.1 ± 0.1 ^{aA}	0.77 ± 0.05 ^{aA}
		25	3.38 ± 0.08 ^{aC}	8.2 ± 0.2 ^{aA}	0.79 ± 0.02 ^{aA}
	8	5	3.32 ± 0.01 ^{aC}	8.0 ± 0.1 ^{aA}	0.61 ± 0.01 ^{aA}
		25	3.76 ± 0.01 ^{bB}	6.9 ± 0.2 ^{bB}	0.69 ± 0.03 ^{bA}
FRCh-2	1	5	3.53 ± 0.06 ^{aA}	8.5 ± 0.5 ^{aA}	0.77 ± 0.03 ^{aA}
		25	3.31 ± 0.01 ^{bC}	7.8 ± 0.3 ^{bA}	0.80 ± 0.01 ^{aA}
	8	5	3.67 ± 0.06 ^{cA}	8.5 ± 0.3 ^{aA}	0.60 ± 0.01 ^{bB}
		25	3.80 ± 0.18 ^{cB}	6.7 ± 1.0 ^{bB}	0.68 ± 0.02 ^{cA}
* FRBN-1	1	5	3.48 ± 0.01 ^{aA}	9.1 ± 0.2 ^{cA}	0.80 ± 0.04 ^{aCA}
		25	3.37 ± 0.01 ^{aC}	9.5 ± 0.3 ^{aCA}	0.82 ± 0.02 ^{aCA}
	8	5	3.61 ± 0.01 ^{bA}	8.6 ± 0.2 ^{bA}	0.70 ± 0.01 ^{bA}
		25	3.85 ± 0.02 ^{aB}	8.5 ± 0.2 ^{Ba}	0.77 ± 0.02 ^{abA}
FRBN-2	1	5	3.22 ± 0.02 ^{aB}	8.9 ± 0.1 ^{aA}	0.80 ± 0.01 ^{aA}
		25	3.28 ± 0.02 ^{aB}	10.3 ± 0.5 ^{bCA}	0.80 ± 0.01 ^{aA}
	8	5	3.54 ± 0.05 ^{bA}	8.3 ± 0.3 ^{aA}	0.69 ± 0.01 ^{bA}
		25	3.98 ± 0.02 ^{aC}	9.2 ± 0.5 ^{bC}	0.75 ± 0.04 ^{abA}

Data are presented as mean ± standard deviation (SD). a, b, c superscripts in the same row indicate a significant difference between the means ($p \leq 0.05$) of temperature. A, B, and C superscripts in the same row indicate a significant difference between the means ($p \leq 0.05$) of storage time. * Represents significant difference between the means ($p \leq 0.05$) of the treatments.

Regarding TSS and TA, a decreasing trend was observed in the strawberries stored at room temperature. This means that the coatings control the strawberries' maturity, preventing an increase in the TSS [46,52]. The variations of TA and TSS values in coated and uncoated strawberries during storage were not statistically significant ($p > 0.05$). A tendency to decrease in acidity and TSS was observed with the increase in pH of strawberries, possibly because, in the case of edible coatings, they slow down the respiratory rate of strawberries and delay the utilization of organic acids in enzymatic reactions [53]. The obtained results coincide with those of different studies on the application of chitosan-based coatings [38,54]. The values obtained for the coated strawberries are comparable with the parameters of both quality and commercial acceptability of fresh strawberries (pH = 3.17–3.98; TA = 0.60–0.82%; TSS = 6.7–10.3°Bx) [55].

3.12. Color

Color is a critical parameter that affects the customer's selection and purchase decision for most fruits. Table 6 contains the color parameters L, a^*, b^*, and browning index (PI) of strawberries with and without coating during storage. As can be seen in Table 6, there is a significant difference in L in the different coatings, which is attributed to the process of oxidation and moisture loss that the strawberries suffered during storage [56–58].

Table 6. Color parameters of coated and uncoated strawberries stored at 5 and 25 °C after 8 days of storage.

Treatment	t_s (Days)	T (°C)	L	a^*	b^*	Browning Index
Ctrl	1	5	18.21 ± 0.87 aA	20.48 ± 2.43 aA	6.01 ± 1.08 aA	108.9 ± 11.1 aA
		25	22.42 ± 3.42 bC	22.28 ± 0.95 aC	6.57 ± 0.47 bA	97.7 ± 8.70 bB
	8	5	22.18 ± 3.70 bC	17.22 ± 1.10 aD	7.08 ± 1.17 aA	89.6 ± 8.30 aB
		25	20.05 ± 8.35 bB	08.01 ± 3.74 bE	3.95 ± 0.41 bB	56.0 ± 27.2 bC
FRCh-1	1	5	18.57 ± 4.00 aA	20.26 ± 5.10 aA	5.87 ± 2.69 aC	104.0 ± 13.5 aA
		25	18.83 ± 1.43 aA	17.71 ± 2.58 bD	6.40 ± 2.83 bA	101.2 ± 22.2 aA
	8	5	22.17 ± 1.63 bB	21.78 ± 3.42 aB	11.88 ± 10.29 aD	148.5 ± 99.1 aD
		25	18.42 ± 1.97 aA	13.36 ± 2.14 bE	4.34 ± 0.91 bC	73.6 ± 5.8 bEC
FRCh-2	1	5	19.59 ± 0.61 aAB	22.41 ± 0.88 aC	6.96 ± 1.12 aA	114.0 ± 7.5 aA
		25	21.65 ± 1.31 bBC	19.55 ± 2.20 aD	6.25 ± 1.29 aA	91.1 ± 8.8 bB
	8	5	20.32 ± 3.13 aB	49.03 ± 1.8 aF	11.97 ± 10.35 aD	152.1 ± 88.2 aD
		25	15.66 ± 1.29 bD	37.78 ± 0.86 bG	2.50 ± 0.67 bB	50.6 ± 3.2 Bc
FRBN-1	1	5	20.36 ± 3.10 aAB	22.06 ± 1.40 aC	6.84 ± 1.18 aA	108.4 ± 6.2 aA
		25	19.30 ± 2.38 aAB	17.81 ± 3.90 bD	5.17 ± 1.54 aAC	88.6 ± 12.5 bB
	8	5	21.71 ± 5.35 aBC	61.88 ± 1.9 aH	5.68 ± 2.41 aAC	93.9 ± 5.4 aB
		25	17.50 ± 2.50 bD	11.04 ± 3.5 bI	3.31 ± 0.88 Bb	61.5 ± 9.2 bC
FRBN-2	1	5	20.12 ± 1.32 aB	24.32 ± 1.17 aC	7.15 ± 0.40 Aa	117.4 ± 4.6 aA
		25	20.89 ± 1.65 bB	18.21 ± 0.77 bD	5.74 ± 1.15 bAC	87.6 ± 4.0 bB
	8	5	21.88 ± 2.07 aBC	21.88 ± 1.91 aC	7.31 ± 1.44 aA	103.1 ± 5.1 aA
		25	19.00 ± 2.03 Bab	12.07 ± 1.27 bE	8.82 ± 0.34 bA	124.6 ± 103.2 aA

Data are presented as mean ± standard deviation (SD). a, b, c superscripts in the same row indicate a significant difference between the means ($p \leq 0.05$) of temperature. A, B, C, D, E, F, G, H superscripts in the same row indicate a significant difference between the means ($p \leq 0.05$) of storage time.

Finally, it was found that the application of the coatings did not affect the luminosity of the strawberries with respect to the control sample. Regarding the chromatic coordinate a^* (the reddish hue of the strawberry epidermis), it did not show a statistical difference either at the beginning or at the end of storage; the measurements behaved as a homogeneous group. Looking at the data from Day-8 samples, a notable increase and subsequent decrease in the a^* values are observed, with a statistically significant difference. The increase in the shade of a^* is due to moisture loss during storage due to transpiration; the decrease in redness is probably due to an increase in respiratory and enzymatic activity that causes loss of quality due to oxidative browning [58].

Likewise, between the beginning and Day 8 of storage, no statistical difference was found ($p > 0.05$) between coated strawberries with respect to the chromatic coordinate b^* (the yellow hue of the strawberry's epidermis). On Day 8, a decrease in b^* was observed in the coated strawberries compared to the control; this is associated with enzymatic browning reactions [39,54,55]. Finally, the browning index values obtained were found to be in the range of 97–117 and 50–152 for Day-1 and Day-8 samples, respectively, which indicates an increasing trend with storage time. The results indicate that, despite the chitosan and zinc oxide coatings acting as a selective barrier that prevents the fruit from being exposed to ambient oxygen, they present possible oxidation reactions [53], as well as a decrease in ascorbic acid due to its degradation over time, which promotes enzymatic-browning reactions [57,58].

It should be noted that on Day 8, the browning index values for the Ch/ZnO-NPs-1 coating at temperatures of 5 and 25 °C were comparable to those obtained with the control treatment. In general, both the coated and the control strawberries presented statistical differences on day 8 with respect to the beginning (day 0), due to the effect of time, differences that are attributed to moisture loss throughout the storage period. This moisture loss is assumed to be caused by biological phenomena typical of plant tissues, such as transpiration [58]. In this regard, various authors mention that chitosan-based coatings can delay external color changes in strawberries, similar to other edible coatings based on natural biopolymers [59].

4. Conclusions

In this study, a nanocoating based on chitosan and nanostructured spaghetti-like ZnO nanostructures (Ch/ZnO-NPs) was developed. The coating showed important microbiological and physicochemical property enhancements for strawberry preservation. The wurtzite ZnO-NPs are spaghetti-like, with a crystal size of 40 nm. The Ch/ZnO-NPs coating, with a thickness of 2–3 μm, was uniformly impregnated on the strawberry surface; it is statistically independent of the immersion time. The nanocoating allowed the shelf life of strawberries to be increased at room temperature (25 °C) and at refrigeration temperature (5 °C). The homogeneous distribution of the nanostructured ZnO into the chitosan matrix favored the physicochemical and textural properties of strawberries, including for up to 8 days. The coating with nanostructured ZnO improved the antibacterial properties of chitosan, i.e., a synergistic effect between these two compounds was observed, reducing the microbial load. For strawberries stored at 5 °C, treatment FRBN-1 exhibited the highest moisture content on day 1 (93.2%) compared to the control (91.9%). Additionally, on day 8, treatment FRBN-1 (92.9%) also maintained a high moisture content compared to the control (89.3%). In contrast, for strawberries stored at 25 °C, on day 1, treatment FRCh-1 (82.0%) had the lowest moisture content, significantly different from the control (89.1%) and other treatments. However, on day 8, a significant improvement was observed in treatment FRCh-1 (69.8%), while the control (45.3%) experienced a significant decrease in moisture content compared to day 1. Moreover, results of pH, TSS, and TA are comparable with the parameters of both quality and commercial acceptability of fresh strawberries (pH = 3.17–3.98; TA = 0.60–0.82%; TSS = 6.7–10.3 °Bx) as well as acceptable sensory properties, which suggest the potential of the Ch/ZnO-NPs coating in improving the preservation of fresh strawberries.

Author Contributions: Conceptualization, H.H.-C., G.F.P.-S. and D.J.G.-G.; methodology, D.J.G.-G., G.F.P.-S., M.G.S.-A., M.L.L.-G., E.R.-R. and C.M.-R.; validation, M.G.S.-A. and M.L.L.-G.; formal analysis, D.J.G.-G., H.H.-C. and G.F.P.-S.; investigation, D.J.G.-G.; resources, H.H.-C. and G.F.P.-S.; data curation, E.R.-R.; writing—original draft preparation, H.H.-C. and K.R.; writing—review and editing, G.F.P.-S. and H.H.-C.; visualization, E.R.-R.; supervision, G.F.P.-S.; project administration, H.H.-C. and G.F.P.-S.; funding acquisition, G.F.P.-S., H.H.-C. and C.M.-R. All authors have read and agreed to the published version of the manuscript.

Funding: This work was supported by the Vicerrectoría de Investigación y Estudios de Posgrado, from Benemérita Universidad Autónoma de Puebla Project (100318500-VIEP2022).

Institutional Review Board Statement: Not applicable.

Data Availability Statement: The data presented in this study are available on request from the corresponding author.

Conflicts of Interest: The authors declare no conflict of interest.

References

1. Sridhar, A.; Ponnuchamy, M.; Kumar, P.S.; Kapoor, A. Food preservation techniques and nanotechnology for increased shelf life of fruits, vegetables, beverages and spices: A review. *Environ. Chem. Lett.* **2021**, *19*, 1715–1735. [CrossRef] [PubMed]
2. Li, X.; Ren, Z.; Wang, R.; Liu, L.; Zhang, J.; Ma, F.; Khan, Z.H.; Zhao, D.; Liu, X. Characterization and antibacterial activity of edible films based on carboxymethyl cellulose, Dioscorea opposita mucilage, glycerol and ZnO nanoparticles. *Food Chem.* **2021**, *349*, 129208. [CrossRef] [PubMed]
3. Li, Y.; Zhou, Y.; Wang, Z.; Cai, R.; Yue, T.; Cui, L. Preparation and characterization of chitosan–nano-zno composite films for preservation of cherry tomatoes. *Foods* **2021**, *10*, 3135. [CrossRef] [PubMed]
4. Ecuyer, J.L.; Audet, N.; Shink, D.; Triboulet, R.; Benz, K.W.; Fiederle, M.; Lincot, D.; Tomashik, V.N.; Tomashik, Z.F. Crystal Growth and Surfaces. In *CdTe and Related Compounds; Physics, Defects, Hetero- and Nano-Structures, Crystal Growth, Surfaces and Applications*; Elsevier Ltd.: Amsterdam, The Netherlands, 2010; pp. 1–144.
5. Espitia, P.J.P.; Batista, R.A.; Otoni, C.G.; Soares, N.F.F. Antimicrobial Food Packaging Incorporated with Triclosan: Potential Uses and Restrictions. In *Antimicrobial Food Packaging*; Elsevier Inc.: Amsterdam, The Netherlands, 2016; pp. 417–423. [CrossRef]
6. Zhang, T.; Zhu, X.; Guo, J.; Gu, A.Z.; Li, D.; Chen, J. Toxicity assessment of nano-ZnO exposure on the human intestinal microbiome, metabolic functions, and resistome using an in vitro colon simulator. *Environ. Sci. Technol.* **2021**, *55*, 6884–6896. [CrossRef]
7. Rajan, M.; Anthuvan, A.J.; Muniyandi, K.; Kalagatur, N.K.; Shanmugam, S.; Sathyanarayanan, S.; Chinnuswamy, V.; Thangaraj, P.; Narain, N. Comparative Study of Biological (*Phoenix loureiroi* Fruit) and Chemical Synthesis of Chitosan-Encapsulated Zinc Oxide Nanoparticles and their Biological Properties. *Arab. J. Sci. Eng.* **2020**, *45*, 15–28. [CrossRef]
8. Strand, S.P.; Issa, M.M.; Christensen, B.E.; Vårum, K.M.; Artursson, P. Tailoring of chitosans for gene delivery: Novel self-branched glycosylated chitosan oligomers with improved functional properties. *Biomacromolecules* **2008**, *9*, 3268–3276. [CrossRef]
9. Jiang, Y.; Li, J.; Jiang, W. Effects of chitosan coating on shelf life of cold-stored litchi fruit at ambient temperature. *LWT* **2005**, *38*, 757–761. [CrossRef]
10. Sharma, S.; Barman, K.; Siddiqui, M.W. Chitosan: Properties and roles in postharvest quality preservation of horticultural crops. In *Eco-Friendly Technology for Postharvest Produce Quality*; Elsevier Inc.: Amsterdam, The Netherlands, 2016; pp. 269–296. [CrossRef]
11. Folta, K.M.; Davis, T.M. Strawberry genes and genomics. *Crit. Rev. Plant Sci.* **2006**, *25*, 399–415. [CrossRef]
12. Sirelkhatim, A.; Mahmud, S.; Seeni, A.; Kaus, N.H.M.; Ann, L.C.; Bakhori, S.K.M.; Mohamad, D. Review on zinc oxide nanoparticles: Antibacterial activity and toxicity mechanism. *Nano-Micro Lett.* **2015**, *7*, 219–242. [CrossRef]
13. Indumathi, M.P.; Saral Sarojini, K.; Rajarajeswari, G.R. Antimicrobial and biodegradable chitosan/cellulose acetate phthalate/ZnO nano composite films with optimal oxygen permeability and hydrophobicity for extending the shelf life of black grape fruits. *Int. J. Biol. Macromol.* **2019**, *132*, 1112–1120. [CrossRef]
14. Thanh Huong, Q.T.; Hoai Nam, N.T.; Duy, B.T.; An, H.; Hai, N.D.; Kim Ngan, H.T.; Ngan, L.T.; Nhi, T.L.H.; Linh, D.T.Y.; Khanh, T.N.; et al. Structurally natural chitosan films decorated with *Andrographis paniculata* extract and selenium nanoparticles: Properties and strawberry preservation. *Food Biosci.* **2023**, *53*, 102647. [CrossRef]
15. Al-Naamani, L.; Dobretsov, S.; Dutta, J. Chitosan-zinc oxide nanoparticle composite coating for active food packaging applications. *Innov. Food Sci. Emerg. Technol.* **2016**, *38*, 231–237. [CrossRef]
16. Karkar, B.; Pat,r, İ.; Eyüboğlu, S.; Şahin, S. Development of an edible active chitosan film loaded with *Nigella sativa* L. extract to extend the shelf life of grapes. *Biocatal. Agric. Biotechnol.* **2023**, *50*, 102708. [CrossRef]
17. Kang, S.H.; Cha, H.J.; Jung, S.W.; Lee, S.J. Application of chitosan-ZnO nanoparticle edible coating to wild-simulated Korean ginseng root. *Food Sci. Biotechnol.* **2022**, *31*, 579–586. [CrossRef] [PubMed]
18. Lavinia, M.; Hibaturrahman, S.N.; Harinata, H.; Wardana, A.A. Antimicrobial activity and application of nanocomposite coating from chitosan and ZnO nanoparticle to inhibit microbial growth on fresh-cut papaya. *Food Res.* **2020**, *4*, 307–311. [CrossRef]
19. Arroyo, B.J.; Bezerra, A.C.; Oliveira, L.L.; Arroyo, S.J.; de Melo, E.A.; Santos, A.M.P. Antimicrobial active edible coating of alginate and chitosan add ZnO nanoparticles applied in guavas (*Psidium guajava* L.). *Food Chem.* **2020**, *309*, 125566. [CrossRef]
20. Khan, I.; Tango, C.N.; Chelliah, R.; Oh, D.H. Development of antimicrobial edible coating based on modified chitosan for the improvement of strawberries shelf life. *Food Sci. Biotechnol.* **2019**, *28*, 1257–1264. [CrossRef]
21. Vallejo-Domínguez, D.; Rubio-Rosas, E.; Aguila-Almanza, E.; Hernández-Cocoletzi, H.; Ramos-Cassellis, M.E.; Luna-Guevara, M.L.; Rambabu, K.; Manickam, S.; Munawaroh, H.S.H.; Show, P.L. Ultrasound in the deproteinization process for chitin and chitosan production. *Ultrason. Sonochem.* **2021**, *72*, 105417. [CrossRef]

22. Ochoa-Velasco, C.E.; Guerrero-Beltrán, J.Á. Postharvest quality of peeled prickly pear fruit treated with acetic acid and chitosan. *Postharvest Biol. Technol.* **2014**, *92*, 139–145. [CrossRef]
23. Huang, Z.; Li, J.; Zhang, J.; Gao, Y.; Hui, G. Physicochemical properties enhancement of Chinese kiwi fruit (*Actinidia chinensis* Planch) via chitosan coating enriched with salicylic acid treatment. *J. Food Meas. Charact.* **2017**, *11*, 184–191. [CrossRef]
24. Zhang, X.; Zhang, Z.; Wu, W.; Yang, J.; Yang, Q. Preparation and characterization of chitosan/Nano-ZnO composite film with antimicrobial activity. *Bioprocess Biosyst. Eng.* **2021**, *44*, 1193–1199. [CrossRef] [PubMed]
25. Camacho Feria, D.M.; Caviedes Rubio, D.I.; Delgado, D.R. Tratamientos para la remoción de antibacteriales y agentes antimicrobiales presentes en aguas residuales. *Logos Cienc. Y Tecnol.* **2017**, *9*, 43–62. [CrossRef]
26. Boura-Theodoridou, O.; Giannakas, A.; Katapodis, P.; Stamatis, H.; Ladavos, A.; Barkoula, N.M. Performance of ZnO/chitosan nanocomposite films for antimicrobial packaging applications as a function of NaOH treatment and glycerol/PVOH blending. *Food Packag. Shelf Life* **2020**, *23*, 100456. [CrossRef]
27. Mote, V.D.; Purushotham, Y.; Dole, B.N. Williamson-Hall Analysis in Estimation of Lattice Strain in Nanometer-Sized ZnO Particles. 2012. Available online: http://www.jtaphys.com/content/2251-7235/6/1/6 (accessed on 7 January 2023).
28. Priyadarshi, R.; Negi, Y.S. Effect of Varying Filler Concentration on Zinc Oxide Nanoparticle Embedded Chitosan Films as Potential Food Packaging Material. *J. Polym. Environ.* **2017**, *25*, 1087–1098. [CrossRef]
29. Zahiri Oghani, F.; Tahvildari, K.; Nozari, M. Novel Antibacterial Food Packaging Based on Chitosan Loaded ZnO Nano Particles Prepared by Green Synthesis from *Nettle* Leaf Extract. *J. Inorg. Organomet. Polym. Mater.* **2021**, *31*, 43–54. [CrossRef]
30. Gao, Y.; Xu, D.; Ren, D.; Zeng, K.; Wu, X. Green synthesis of zinc oxide nanoparticles using *Citrus sinensis* peel extract and application to strawberry preservation: A comparison study. *LWT* **2020**, *126*, 109297. [CrossRef]
31. Yuvaraj, D.; Narasimha Rao, K.; Nanda, K.K. Effect of oxygen partial pressure on the growth of zinc micro and nanostructures. *J. Cryst. Growth* **2009**, *311*, 4329–4333. [CrossRef]
32. Sogvar, O.B.; Koushesh Saba, M.; Emamifar, A.; Hallaj, R. Influence of nano-ZnO on microbial growth, bioactive content and postharvest quality of strawberries during storage. *Innov. Food Sci. Emerg. Technol.* **2016**, *35*, 168–176. [CrossRef]
33. Kumar, S.; Boro, J.C.; Ray, D.; Mukherjee, A.; Dutta, J. Bionanocomposite films of agar incorporated with ZnO nanoparticles as an active packaging material for shelf life extension of green grape. *Heliyon* **2019**, *5*, e01867. [CrossRef]
34. Barbosa-Cánovas, G.V. *Water Activity in Foods: Fundamentals and Applications*; John Wiley & Sons: Hoboken, NJ, USA, 2020; 616p.
35. Varma, R.; Vasudevan, S. Extraction, characterization, and antimicrobial activity of chitosan from horse mussel modiolus modiolus. *ACS Omega* **2020**, *5*, 20224–20230. [CrossRef]
36. Yusof, N.A.A.; Zain, N.M.; Pauzi, N. Synthesis of chitosan / zinc oxide nanoparticles stabilized by chitosan via microwave heating. *Bull. Chem. React. Eng. Catal.* **2019**, *14*, 450–458. [CrossRef]
37. Chagas, P.M.B.; Caetano, A.A.; Tireli, A.A.; Cesar, P.H.S.; Corrêa, A.D.; Guimarães, I.D.R. Use of an Environmental Pollutant from Hexavalent Chromium Removal as a Green Catalyst in The Fenton Process. *Sci. Rep.* **2019**, *9*, 12819. [CrossRef] [PubMed]
38. Youssef, A.M.; El-Sayed, S.M.; El-Sayed, H.S.; Salama, H.H.; Dufresne, A. Enhancement of Egyptian soft white cheese shelf life using a novel chitosan/carboxymethyl cellulose/zinc oxide bionanocomposite film. *Carbohydr. Polym.* **2016**, *151*, 9–19. [CrossRef]
39. Saral Sarojini, K.; Indumathi, M.P.; Rajarajeswari, G.R. Mahua oil-based polyurethane/chitosan/nano ZnO composite films for biodegradable food packaging applications. *Int. J. Biol. Macromol.* **2019**, *124*, 163–174. [CrossRef]
40. Restrepo, J.I.; Aristizábal, I.D. Conservación de fresa (*Fragaria x ananassa* Duch cv. *Camarosa*) mediante la aplicación de recubrimientos comestibles de gel mucilaginoso de penca sábila (*Aloe barbadensis* Miller) y cera de carnaúba. *Vitae* **2010**, *17*, 252–263. [CrossRef]
41. Lan, W.; Zhang, R.; Ahmed, S.; Qin, W.; Liu, Y. Effects of various antimicrobial polyvinyl alcohol/tea polyphenol composite films on the shelf life of packaged strawberries. *LWT* **2019**, *113*, 108297. [CrossRef]
42. Gaikwad, K.K.; Lee, S.M.; Lee, J.S.; Lee, Y.S. Development of antimicrobial polyolefin films containing lauroyl arginate and their use in the packaging of strawberries. *J. Food Meas. Charact.* **2017**, *11*, 1706–1716. [CrossRef]
43. Lakshmeesha, T.R.; Murali, M.; Ansari, M.A.; Udayashankar, A.C.; Alzohairy, M.A.; Almatroudi, A.; Alomary, M.N.; Asiri, S.M.M.; Ashwini, B.; Kalagatur, N.K.; et al. Biofabrication of zinc oxide nanoparticles from Melia azedarach and its potential in controlling soybean seed-borne phytopathogenic fungi. *Saudi J. Biol. Sci.* **2020**, *27*, 1923–1930. [CrossRef]
44. He, L.; Liu, Y.; Mustapha, A.; Lin, M. Antifungal activity of zinc oxide nanoparticles against *Botrytis cinerea* and *Penicillium expansum*. *Microbiol. Res.* **2011**, *166*, 207–215. [CrossRef]
45. Grande-Tovar, C.D.; Chaves-Lopez, C.; Serio, A.; Rossi, C.; Paparella, A. Chitosan coatings enriched with essential oils: Effects on fungi involve in fruit decay and mechanisms of action. *Trends Food Sci. Technol.* **2018**, *78*, 61–71. [CrossRef]
46. Sucharitha, K.V.; Beulah, A.M.; Ravikiran, K. Effect of chitosan coating on storage stability of tomatoes (*Lycopersicon esculentum* Mill). *Int. Food Res. J.* **2018**, *25*, 93–99.
47. Emamifar, A.; Bavaisi, S. Nanocomposite coating based on sodium alginate and nano-ZnO for extending the storage life of fresh strawberries (*Fragaria × ananassa* Duch.). *J. Food Meas. Charact.* **2020**, *14*, 1012–1024. [CrossRef]
48. Emamifar, A.; Mohammadizadeh, M. Preparation and Application of LDPE/ZnO Nanocomposites for Extending Shelf Life of Fresh Strawberries. *Food Technol. Biotechnol.* **2015**, *53*, 488–495. [CrossRef] [PubMed]
49. Pastor, C.; Sanchez-Gonzalez, L.; Marcilla, A.; Chiralt, A.; Cháfer, M.; Gonzalez-Martinez, C. Quality and safety of table grapes coated with hydroxypropylmethylcellulose edible coatings containing propolis extract. *Postharvest Biol. Technol.* **2011**, *60*, 64–70. [CrossRef]

50. Sanaa, S.; Entsar, N.; Entsar, S. Effect of Edible Coating on Extending the Shelf Life and Quality of Fresh Cut Taro. *Am. J. Food Technol.* **2017**, *12*, 124–131. [CrossRef]
51. Hernández Suárez, M.; Rodríguez Rodríguez, E.M.; Díaz Romero, C. Chemical composition of tomato (*Lycopersicon esculentum*) from Tenerife, the Canary Islands. *Food Chem.* **2008**, *106*, 1046–1056. [CrossRef]
52. Nunes, M.C.N.; Brecht, J.K.; Morais, A.M.M.B.; Sargent, S.A. Physicochemical changes during strawberry development in the field compared with those that occur in harvested fruit during storage. *J. Sci. Food Agric.* **2006**, *86*, 180–190. [CrossRef]
53. Varasteh, F.; Arzani, K.; Barzegar, M.; Zamani, Z. Pomegranate (*Punica granatum* L.) Fruit Storability Improvement Using Pre-storage Chitosan Coating Technique. *J. Agric. Sci. Tech.* **2017**, *389–400*, 54746420.
54. Tanada-Palmu, P.S.; Grosso, C.R.F. Effect of edible wheat gluten-based films and coatings on refrigerated strawberry (*Fragaria ananassa*) quality. *Postharvest Biol. Technol.* **2005**, *36*, 199–208. [CrossRef]
55. Jiang, Y.; Yu, L.; Hu, Y.; Zhu, Z.; Zhuang, C.; Zhao, Y.; Zhong, Y. The preservation performance of chitosan coating with different molecular weight on strawberry using electrostatic spraying technique. *Int. J. Biol. Macromol.* **2020**, *151*, 278–285. [CrossRef]
56. Huber, D.J. Strawberry Fruit Softening: The Potential Roles of Polyuronides and Hemicelluloses. *J. Food Sci.* **1984**, *49*, 1310–1315. [CrossRef]
57. Esmaeili, Y.; Zamindar, N.; Paidari, S.; Ibrahim, S.A.; Mohammadi Nafchi, A. The synergistic effects of aloe vera gel and modified atmosphere packaging on the quality of strawberry fruit. *J. Food Process. Preserv.* **2021**, *45*, e16003. [CrossRef]
58. Alvarado-Ambriz, S.; Lobato-Calleros, C.; Hernández-Rodríguez, L.; Vernon-Carter, E.J. Wet processing coffee waste as an alternative to produce extracts with antifungal activity: In vitro and in vivo valorization. *Rev. Mex. De Ing. Quim.* **2020**, *19*, 135–149. [CrossRef]
59. Hernández-Muñoz, P.; Almenar, E.; Ocio, M.J.; Gavara, R. Effect of calcium dips and chitosan coatings on postharvest life of strawberries (*Fragaria* x *ananassa*). *Postharvest Biol. Technol.* **2006**, *39*, 247–253. [CrossRef]

Disclaimer/Publisher's Note: The statements, opinions and data contained in all publications are solely those of the individual author(s) and contributor(s) and not of MDPI and/or the editor(s). MDPI and/or the editor(s) disclaim responsibility for any injury to people or property resulting from any ideas, methods, instructions or products referred to in the content.

Article

Chitosan Resin-Modified Glass Ionomer Cement Containing Epidermal Growth Factor Promotes Pulp Cell Proliferation with a Minimum Effect on Fluoride and Aluminum Release

Chanothai Hengtrakool [1], Supreya Wanichpakorn [2,3] and Ureporn Kedjarune-Leggat [2,3,*]

[1] Department of Conservative Dentistry, Faculty of Dentistry, Prince of Songkla University, Hat Yai, Songkhla 90112, Thailand; htkchino@gmail.com
[2] Department of Oral Biology and Occlusion, Faculty of Dentistry, Prince of Songkla University, Hat Yai, Songkhla 90112, Thailand; supreya.w@psu.ac.th
[3] Cell Biology and Biomaterials Research Unit, Faculty of Dentistry, Prince of Songkla University, Hat Yai, Songkhla 90112, Thailand
* Correspondence: ureporn.l@psu.ac.th

Citation: Hengtrakool, C.; Wanichpakorn, S.; Kedjarune-Leggat, U. Chitosan Resin-Modified Glass Ionomer Cement Containing Epidermal Growth Factor Promotes Pulp Cell Proliferation with a Minimum Effect on Fluoride and Aluminum Release. *Polymers* **2023**, *15*, 3511. https://doi.org/10.3390/polym15173511

Academic Editors: Sérgio Paulo Campana-Filho and William Facchinatto

Received: 21 July 2023
Revised: 17 August 2023
Accepted: 17 August 2023
Published: 23 August 2023

Copyright: © 2023 by the authors. Licensee MDPI, Basel, Switzerland. This article is an open access article distributed under the terms and conditions of the Creative Commons Attribution (CC BY) license (https://creativecommons.org/licenses/by/4.0/).

Abstract: The development of biomaterials that are able to control the release of bioactive molecules is a challenging task for regenerative dentistry. This study aimed to enhance resin-modified glass ionomer cement (RMGIC) for the release of epidermal growth factor (EGF). This RMGIC was formulated from RMGIC powder supplemented with 15% (w/w) chitosan at a molecular weight of either 62 or 545 kDa with 5% bovine serum albumin mixed with the same liquid component as the Vitrebond. EGF was added while mixing. ELISA was used to determine EGF release from the specimen immersed in phosphate-buffered saline at 1 h, 3 h, 24 h, 3 d, 1 wk, 2 wks, and 3 wks. Fluoride and aluminum release at 1, 3, 5, and 7 d was measured by electrode and inductively coupled plasma optical emission spectrometry. Pulp cell viability was examined through MTT assays and the counting of cell numbers using a Coulter counter. The RMGIC with 65 kDa chitosan is able to prolong the release of EGF for significantly longer than RMGIC for at least 3 wks due to its retained bioactivity in promoting pulp cell proliferation. This modified RMGIC can prolong the release of fluoride, with a small amount of aluminum also released for a limited time. This biomaterial could be useful in regenerating pulp–dentin complexes.

Keywords: chitosan resin-modified glass ionomer cement; EGF; fluoride; aluminum; chitosan

1. Introduction

Growth factors and active biological molecules play a crucial role in the successful regeneration and repair of dentin–pulp complexes [1,2]. The development of a possible new generation of biomaterials to control the release of bioactive molecules is, therefore, challenging work.

Glass ionomer cement (GIC) is an acid-base cement produced from the reaction of fluoro-aluminosilicate glass powder with poly(acrylic acid) and is widely used in dental and medical applications due to its biocompatibility, antibacterial properties, sealing ability, and capacity to prolong the release of fluoride. There have been some attempts to enhance the sustained release of substances from this cement [3], such as CPP-ACP [4], chlorhexidine [5], and surface pre-reacted glass ionomer (S-PRG) filler [6], especially in terms of protein [7]. Our group discovered that chitosan-fluoro-aluminosilicate glass ionomer cement could prolong the release of bovine serum albumin (BSA) without alteration of its molecular weight, and this cement does not increase toxicity toward pulp cells [7]. BSA was used as the released protein because of its good biocompatibility with cells and its function as a carrier protein, thus helping the release of other small proteins [8]. This material can prolong the release of protein, possibly due to the formation of a polyelectrolyte complex between the cationic group of chitosan and the anionic group of poly(acrylic acid) [9]. Chitosan is a biocompatible, biodegradable, natural biopolymer that is a copolymer of glucosamine and N-acetylglucosamine derived from chitin. Chitosan has been used widely in biomedical areas, such as bone scaffolds, tissue engineering, and controlled drug or biological molecule release [10]. This polymer can form an insoluble complex from the reaction of chitosan and polyacrylic acid, and this complex has been proven as a controlled drug delivery system [11,12]. Some studies have reported the antibacterial properties of chitosan-modified GIC cement [13,14]. Recently, chitosan in various forms has been studied to improve some properties of GIC, such as increasing osteogenic potential in osteosarcoma cells [15] and adding quaternized chitosan-coated mesoporous silica nanoparticles in order to improve mechanical properties, increase fluoride release, and enhance antibacterial effects [16].

Resin-modified glass-ionomer cement (RMGIC), or light-cured GIC, is the development of glass ionomer cement by adding resin, especially 2-hydroxyethyl methacrylate (HEMA), in order to control the setting of the cement by light-activated polymerization, which allows the clinician more time to work with the material. The main components of RMGIC are similar to conventional GIC, namely fluoro-aluminosilicate glass powder and poly(acrylic acid). Thus, it is feasible for modified RMGIC to control the release of growth factors through chitosan. A previous study revealed that chitosan-modified RMGIC supplemented with translationally controlled tumor protein, an anti-apoptotic protein, can reduce the cytotoxicity of residual HEMA from this RMGIC [17]. This study aimed to modify RMGIC with two molecular weights of chitosan and albumin in order to prolong the release of the growth factor. The growth factor that was used in this study was human epidermal growth factor (EGF). It is a low molecular weight protein (about 6.3 kDa) and has a mitogenic or cell proliferation property [18]. A recent study reported that EGF can also induce neural differentiation in dental pulp stem cells [19]. The important properties of the glass ionomer cement, especially fluoride and aluminum release, were also investigated.

2. Materials and Methods

2.1. Materials

The RMGIC used in this study was the Light Cure Glass Ionomer Liner/Base (Vitrebond TM) from 3M ESPE (3M, ESPE, St. Paul., MN, USA). The RMGIC powder was composed of fluoro-aluminum-zinc-silicate glass (batch no. 7LB2010-05 for the study of EGF release and batch no. 7 ME 2010-10 for Al and F measurements), and the liquid (batch no. 7HJ2010-02 and 7HP2010-04) was composed of 40% itaconacid-isocyanoethyl-methacrylate acrylic acid, 24% HEMA, and water. Chitosan was used at two molecular weights (Mw): 62 kDa, degree of deacetylation (DD) = 89% (Ta Ming Enterprises, Samutsakorn, Thailand), and 545 kDa, DD = 79% (Fluka, Steinheim, Switzerland, batch no. 50494). Cell culture medium and supplements were products of Gibco (Invitrogen Corporation, Grand Island, NY, USA). EGF was sourced from Promega (Promega, Madison, WI, USA).

2.2. Specimen Preparation

Two different diameters of specimens (but with the same 1 mm thickness) were used here. One was a diameter of 10 mm, which was used for the study of EGF, fluoride, and aluminum release, and another was 5 mm, which was used for cell culture experiments.

The powder of the innovative RMGIC was composed of the RMGIC powder mixed with 5% by weight bovine albumin and 15% by weight chitosan. There were two groups of the innovative RMGIC according to the type of chitosan: the first group using chitosan at an Mw of 545 kDa, referred to as the GI+C(F) group, and the second group using chitosan at a lower Mw of 62 kDa, referred to as GI+C(K). The liquid part consisted of the same components as the commercial resin-modified glass ionomer cement, representing a control group (GI), as described above. The composition of the powder in the groups of different RMGICs is shown in Table 1.

Table 1. Powder compositions for various groups of RMGICs specimens.

Group of Specimens	Powder Compositions
GI	Fluoroaluminosilicate glass.
GI+EGF	Fluoroaluminosilicate glass with added EGF during mixing the cement.
GI+C(F)	Fluoroaluminosilicate glass, 15% chitosan Mw 545 kDa, and 5% BSA.
GI+C(F)+EGF	Fluoroaluminosilicate glass, 15% chitosan Mw 545 kDa, and 5% BSA with added EGF during cement mixing.
GI+C(K)	Fluoroaluminosilicate glass, 15% chitosan Mw 62 kDa, and 5% BSA.
GI+C(K)+EGF	Fluoroaluminosilicate glass, 15% chitosan Mw 62 kDa, and 5% BSA with added EGF during cement mixing.

The types of cement were dispensed according to the manufacturer's instruction of RMGIC, with a 1.4:1 powder:liquid weight ratio. The EGF release groups were supplemented with 4 or 8 µg/mL of EGF in 1% Albumin in PBS, which was added during mixing the powder and the liquid part in order to have about 40 ng of total EGF for each specimen. The mixed cement was hand spatulated to form a uniform mix and transferred into ring Teflon molds. Polythene sheets and glass slides were then placed over the filled mold, after which light hand pressure was applied. Then, the cement was photocured for 40 s on both sides of the mold with curing light (sds Kerr, LE Dementron, Kerr Corporation, Danbury, CT, USA) at a wavelength of 450–470 nm. Specimens were retained in the molds for 1 h during storage in an incubator at 37 °C. This procedure was to complete the maturation of the material prior to further investigations.

2.3. Determination of EGF Releasing

There were three groups of the RMGICs, GI+EGF, GI+C (F) +EGF, and GI+C(K)+EGF. Each group had 6 specimens. After specimen storage in an incubator, they were removed from their molds, and then each specimen was weighed to an accuracy of 0.0001 g using a digital balance (Sartorius MC210, Goettingen, Germany). After weighing, each specimen was stored at 37 °C in individual pots containing 1 mL of phosphate-buffered saline (PBS) of pH 7.4 with 1 mM of phenylmethanesulfonyl-fluoride, which is a proteinase inhibitor for preventing EGF degradation. The storage pot was continually shaken at low speed (50 rpm) in an incubator at 37 °C. All storage solutions were replaced with a similar volume of fresh PBS at 1 h, 3 h, 24 h, 3 days, 1 week, 2 weeks, and 3 weeks, respectively. The amount of the released EGF was determined using the sandwich ELISA technique. High-binding ELISA plates (Nunc, Roskilde, DK) were coated with 5 µg/mL of monoclonal anti-human EGF antibody in 100 µL of phosphate-buffered saline (PBS) per well (R&D Systems, Minneapolis, MN, USA) for 24 h. The liquid was removed and rinsed twice with 300 µL/well of wash buffer (0.05% Tween 20 in PBS pH 7.4) using a Titertek Microplate Washer (Flow Laboratories, Lugano, Switzerland) before each new addition. The plates were blocked with 300 µL/well of PBS, which contained 1% BSA, 5% sucrose, and 0.05% NaN3, and were incubated at room temperature for 1 h before washing. Then, 100 µL of each sample, including standard EGF (Promega, Madison, WI, USA), at a concentration between 5 and 200 pg/mL, was diluted with diluent and incubated for 2 h at room temperature. After washing, the plates were left with 100 µL/well of 50 µg/mL of biotinylated anti-human EGF antibody and left for 20 min before washing. Then, 100 µL/well of streptavidin HRP (R&D Systems) diluted at 1:200 with PBS containing 1% BSA was added and incubated at room temperature for 20 min. Following this, 100 µL/well of substrate solution (1:1 of Color reagent A, H2O2, and Color reagent B, Tetramethylbenzidine, R&D Systems) was added and left in a dark place for 20 min before stopping the reaction with 1 M H2SO4 and reading the optical density immediately by a microplate reader (Titertek Multiskan, Flow Laboratories, Lugano, Switzerland), using dual-wavelength at 450 and 570 nm. The optical density was calculated by subtracting the readings at 570 nm from the readings at

450 nm. The data was linearized using log transformation of absorbance and concentration of standards, and then the EGF of samples was calculated using regression analysis.

2.4. Fluoride and Aluminum Release Measurement

Specimens were allocated to one of 6 groups, as shown in Table 1. Each group was composed of 6 specimens. After the removal of the specimens from the molds, the excess material was expelled. The specimens were weighed using a digital balance (± 0.0001 g), and the dimensions were measured in order to confirm the same size of each specimen prior to immersion in each plastic container containing 10 mL of deionized water. The sealed plastic container was continually shaken at low speed (50 rpm) in an incubator at 37 °C. All storage solution was replaced with a similar volume of fresh deionized water commencing at 1, 3, 5, and 7 days.

2.4.1. Fluoride Analysis

The fluoride concentration of the storage solution was measured using an ion-selective electrode system consisting of an ion analyzer (Expandable ion analyzer EA 940, Orion Research, Cambridge, MA, USA) and a combination fluoride electrode (Orion Research, Cambridge, MA, USA). The measurement solution was performed by mixing 0.5 mL of each sample solution with 1.5 mL of 0.1 M hydrochloric acid in a plastic container [20]. Standard solutions with fluoride ranging from 1 to 100 ppm were used to calibrate the system prior to sample measurement and recalibrated every 1 h to compensate for local temperature and humidity changes. Three readings of fluoride concentrations from each solution were recorded in parts per million. The amount of fluoride was calculated as milligrams of fluoride released per gram of glass-ionomer cement (mg F/g cement).

2.4.2. Aluminum Analysis

The aluminum concentration of the storage solution was measured using an inductively coupled plasma-optical emission spectrometer, ICP-OES (Perkin Elmer Optima 4300 DV, Valencia, CA, USA). Three readings of aluminum concentration from each solution were recorded in parts per million. The amount of aluminum released per gram of cement (mg Al-/g cement) was calculated afterward.

2.5. Ethical Statement

Human dental pulp tissue collection performed in this study was approved by the Human Research Ethics Committee of the Faculty of Dentistry, Prince of Songkla University (No. of Approval: MOE 0521.1.03/998). For the use of pulp tissue samples, written informed consent was obtained from a human subject who participated.

2.6. Cell Culture

Pulp cells were cultured from normal human third molar from one adult patient aged about 18 years seen at the Dental Hospital, Faculty of Dentistry, Prince of Songkla University, with the approval of the Research Ethics Committee, Faculty of Dentistry, Prince of Songkla University (No. of Approval: MOE 0521.1.03/998). Primary culture of pulp cells was performed using an enzymatic method. Briefly, the pulp tissue was minced into pieces and digested in a solution of 3 mg/mL of collagenase Type I (Gibco, Invitrogen Corporation, Grand Island, NY, USA) and 4 mg/mL of dispase (Gibco, Invitrogen Corporation, Grand Island, NY, USA) for 1 h at 37 °C. After centrifugation, cells were cultured in alpha-modified Eagle's medium (αMEM), supplemented with 20% FCS, 100 µM L-ascorbic acid 2-phosphate, 2 mM L-glutamate, 100 units/mL penicillin, and 100 µg/mL streptomycin, and incubated at 37 °C with 5% CO_2. Pulp cells from passages 3 to 8 were used to test cytotoxicity in this study.

2.7. Cytotoxicity Assay

MTT assay was used to investigate the cytotoxicity of these RMGICs to pulp cells. There were 6 groups of specimens composed of different components, as shown in Table 1. Disc specimens with 5 mm diameter and 1 mm height were prepared using the same method described above, but the total amount of EGF was 40 ng per specimen for the groups with added EGF. There were also 2 groups, control and positive control, which were cells cultured in normal medium, and another group was cells cultured in medium with 10 ng/mL of EGF, respectively. Human dental pulp (HDP) cells were seeded at 1×10^5 cells/well on the bottom compartment of a 12-well Transwell cluster plate (Costar; Corning Inc., Corning, NY, USA). Cells were fed with 1.5 mL of α-MEM supplemented with 10% FCS, 100 μM L-ascorbic acid 2-phosphate, 2 mM L-glutamate, 100 units/mL penicillin, and 100 μg/mL streptomycin, and incubated under 5% CO_2 at 37 °C. After incubation for 24 h, the culture media was refreshed, and each specimen was placed on the upper compartment (Transwell insert, with 12 mm membrane diameter and 0.4 μm pore size), and 0.5 mL of the culture medium was added. Transwell plates and the culture medium allowed the released substances from the upper compartment to diffuse through the cells at the bottom compartment. Two MTT assays [21] were performed. The first assay was cells exposed to the specimens for 3 days. Another assay was performed after cells were exposed to specimens for 6 days. In the second assay, the media was refreshed after cells were exposed to the specimens for 3 days, and the experiment was continued until the sixth day of exposure. The experiments were repeated three times in each group.

2.8. Proliferative Assay

The effect of the released substances from the specimens on cell proliferation was investigated by counting cell numbers using a Coulter counter (Coulter® Z™, New York, NY, USA) after cells were exposed to the specimens for three periods each over the last two days. There were 6 groups of the specimens, the same groups as in the cytotoxicity test, and the control group was cells culture with normal media. HDP cells were seeded at 1.5×10^5 cells/well on the bottom compartment of a 6-well Transwell cluster plate 24 h prior to the placement of the specimen with the same method as described above. The specimens were placed on the inserted part for two days before the specimen was moved to place on the new Transwell plate, which had already seeded the same number of cells 24 h before; the experiment was repeated for another two periods. After each period, cells were washed with PBS 2 times before trypsinization, and the cell number of each well was counted using the Coulter counter. The result was reported as percentages of cell numbers compared to the control of each time period, which was set as 100%.

2.9. Statistical Analysis

The logarithmic transformation of the EGF release plus one was applied to overcome the large variation of the EGF release and some zero results. ANOVA with repeated measures and the Tukey HSD hoc test [22] was used to analyze the transformed variable, Lg10 (EGF release+1). The releases of fluoride and aluminum from different formulas of modified RMGICs were compared using ANOVA with repeated measures and the Tukey HSD post hoc test. The results of MTT and proliferative assays were analyzed using a two-way analysis of variance (two-way ANOVA) and the Tukey HSD post hoc test with statistical significance set at $p < 0.05$.

3. Results

3.1. EGF Release

The result of the EGF release rate has high variation; the median and the value ranging from minimum to maximum of each group have been summarized in Table 2. The cumulative release of EGF has also been shown in Figure 1. By using logarithmic transformation, it was demonstrated that RMGIC modified by adding 15% of chitosan K and 5% of albumin supplemented with EGF, GI+C(K)+EGF, was the best group that gave a significantly higher release rate than the commercial RMGIC or GI group ($p < 0.05$, ANOVA with repeated measure and the Tukey HSD post hoc test). The specimens group GI+C(F) +EGF did not have any statistically significant difference in the EGF release rate compared to the other two groups. It was noted that the variation of all groups was high; however, the average released rate of EGF in the GI group was less than 5 pg/g of specimen per h in the first 2 and 4 h and reduced to 0 after 24 h, while the GI+C(K) group had a much higher release rate than 100 pg/g/h at the first 2 and 4 h like a burst release. After 24 h, the released rate dramatically reduced, but was still higher than 5 pg/g/h, and still released EGF in small, but detectable amounts for at least 3 weeks.

Table 2. Median and (minimum–maximum) rate of EGF release (pg/g cement/h) from a different formula of RMGICs with added EGF at seven time periods.

Time (h)	Type of Material		
	GI+EGF [a]	GI-C(F)+EGF [a,b]	GI-C(K)+EGF [b]
2	3.362	131	253.31
	(0.00–6.38)	(50.06–277.62)	(31.44–529.67)
4	3.865	2.788	70.296
	(0.00–13.56)	(0.91–74.44)	(5.75–238.50)
24	0	0.773	1.71
	(0.00–0.01)	(0.16–2.42)	(0.27–23.82)
72	0	0.03	0
	(0.00–0.38)	(0.00–0.31)	(0.00–1.68)
168	0	0.004	0
	(0.00–0.00)	(0.00–0.06)	(0.00–0.61)
336	0	0.001	0.001
	(0.00–0.00)	(0–0.01)1	(0.00–0.02)
504	0.008	0.01	0.001
	(0.00–0.22)	(0.00–0.05)	(0.00–0.33)

[a,b] Different letters are statistically significantly different at $p < 0.05$.

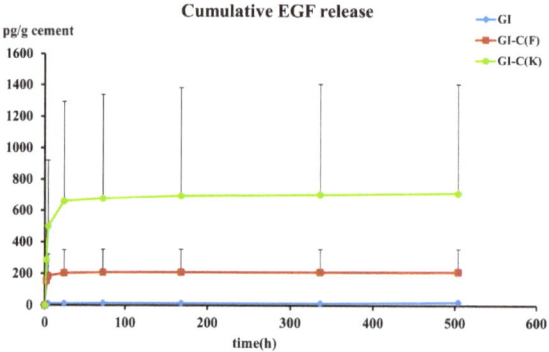

Figure 1. Cumulative release of EGF of different RMGICs with added EGF.

3.2. Fluoride and Aluminum Release

The results of fluoride and aluminum release have been presented as daily release rates, which are summarized in Table 3. The cumulative fluoride and aluminum release have been shown in Figure 2A and B, respectively. It was noticed that the release patterns of fluoride from all groups were not much different. Two phases of fluoride release were observed in all groups, an initial rapid fluoride washout phase (Day 1 to Day 3) followed by a slower steady elution of fluoride. The GI group gave the lowest average release rate, while the GI+EGF had the highest average release rate ($p < 0.05$). The average fluoride release rates of GI+C(F)+EGF, GI+C(K)+EGF, and GI+C(K) groups were not significantly different, but GI+C(F) had a significantly lower release rate than GI+C(F)+EGF.

Table 3. Fluoride and aluminum release rates (mg/g cement/day) from various formulations of glass-ionomer cement over a period of 7 days presented as means (SD).

	Day1	Day3	Day5	Day7
	Rate of fluoride release (mg/g cement/day)			
GI [a]	0.71 (0.08)	0.25 (0.03)	0.15 (0.00)	0.14 (0.01)
GI+EGF [d]	1.09 (0.08)	0.53 (0.02)	0.29 (0.01)	0.22 (0.01)
GI+C(F) [c]	0.99 (0.07)	0.41 (0.02)	0.25 (0.01)	0.24 (0.01)
GI+C(F)+EGF [b]	0.79 (0.03)	0.42 (0.02)	0.25 (0.01)	0.21 (0.01)
GI+C(K) [b,c]	0.95 (0.08)	0.37 (0.02)	0.24 (0.02)	0.23 (0.01)
GI+C(K)+EGF [b,c]	0.86 (0.11)	0.46 (0.06)	0.25 (0.01)	0.21 (0.01)
	Rate of aluminum release (mg/g cement/day)			
GI [a]	0.15 (0.01)	0.01 (0.00)	0.00 (0.00)	0.00 (0.00)
GI+EGF [b]	0.30 (0.01)	0.02 (0.00)	0.01 (0.00)	0.00 (0.00)
GI+C(F) [a]	0.16 (0.02)	0.02 (0.00)	0.01 (0.00)	0.01 (0.00)
GI+C(F)+EGF [a]	0.13 (0.01)	0.02 (0.00)	0.01 (0.00)	0.00 (0.00)
GI+C(K) [a]	0.17 (0.02)	0.02 (0.00)	0.01 (0.00)	0.00 (0.00)
GI+C(K)+EGF [a]	0.17 (0.04)	0.02 (0.00)	0.00 (0.00)	0.00 (0.00)

[a,b,c] Different letters are statistically significantly different at $p < 0.05$.

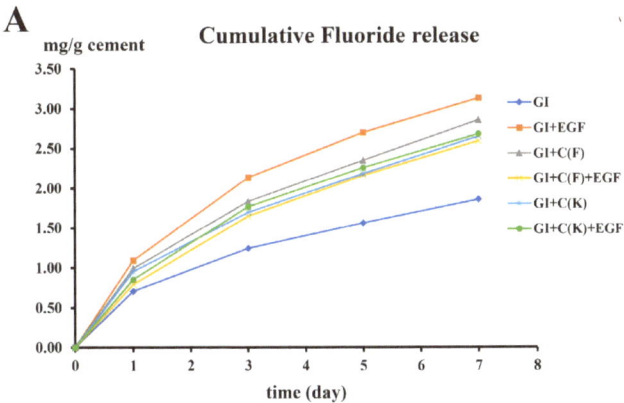

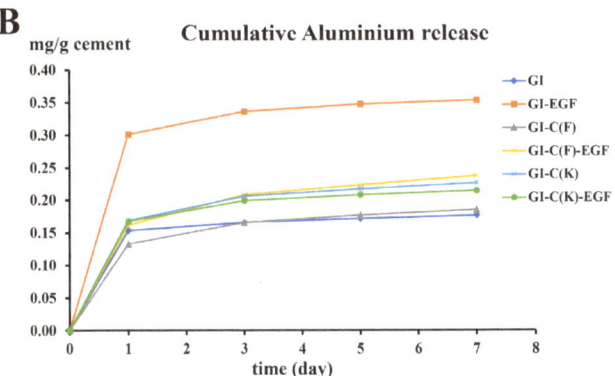

Figure 2. Cumulative release of fluoride in (**A**) and cumulative release of aluminum in (**B**). The GI+EGF group had the highest cumulative release of both fluoride and aluminum.

The daily aluminum release from all RMGICs had the same pattern, which had the burst of release on the first day then reduced rapidly after 3 days and cannot be detected on day 5 in GI and GI+C(K) groups, while other groups also released at only about 0.01 mg/g/day. It was noticed that the GI+EGF had the significantly highest aluminum release rate ($p < 0.05$), while other groups had no statistically significant difference ($p > 0.05$).

3.3. Cytotoxicity Assay

The cytotoxicity of the specimens was investigated at two time intervals, 3 and 6 days, with an MTT assay, as shown in Figure 3. The result was analyzed with 2-way ANOVA, and the Tukey post hoc test demonstrated that the GI+C(K)+EGF group had the highest percentages of cell viability ($p < 0.05$). Cells exposed to the specimens for 6 days had significantly higher percentages of survival cells than 3 days in all groups, which means that the toxicity of the specimens reduced significantly when exposing cells for longer than 3 days. The positive control groups that were cells cultured in media supplemented with 10 ng/mL of EGF for 3 and 6 days had percentages of cell viability at 109.5 ± 14.8 and 101.3 ± 2.2 mg/g cement, respectively.

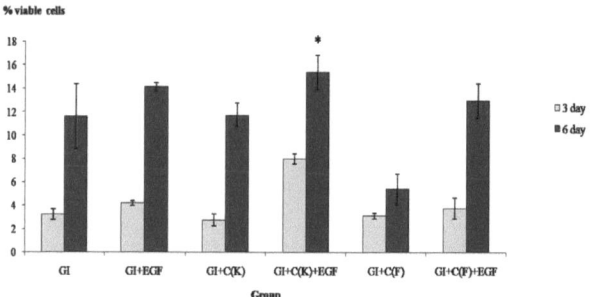

Figure 3. MTT assay of HDP cells exposed to the specimens for 3 and 6 days. * indicated significant ($p < 0.05$).

3.4. Proliferative Assay

Figure 4 revealed the results of percentages of cell number after cells were cultured with different RMGICs at three time periods. The specimens in group GI+C(K)+EGF had the highest average percentages of cell numbers ($p < 0.05$), and the third incubation (6 d) of the specimens gave a significantly higher ($p < 0.05$) percentage of cell numbers than the first (2 d) and the second (4 d), which may demonstrate the promotion of prolonged cell proliferation of the GI+C(K)+EGF group for at least up to 6 days.

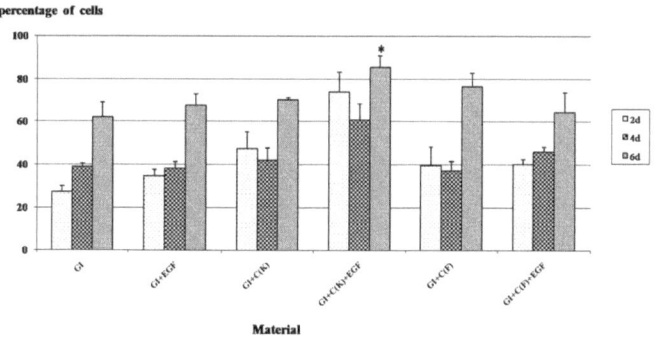

Figure 4. Percentages of cell number using Coulter counter after cells were cultured with different RMGICs for three periods. * indicated significant ($p < 0.05$).

4. Discussion

This study attempted to develop a glass-ionomer cement, which is able to control the release of bioactive molecules and also maintain the important properties of the glass-ionomer cement. These are particular to fluoride and aluminum release. Aluminum should be released only in a short period of time with a non-toxic level. The percentages of chitosan and albumin used in this study were the result of several prior studies, which had been performed by adjusting the percentages of chitosan and albumin in order to obtain a satisfactory prolonged release rate.

The pattern of the EGF release could be observed in the modified RMGICs. The GI-C(K)+EGF group appeared clearly composed of two phases. The first phase is the high release rate of the burst effect, whereas EGF can be detected in high concentrations at 2 and 4 h after immersion of the specimens. In the second phase (after 24 h), the amount of EGF was dramatically reduced, but could still be detected until 3 weeks. This result was similar to our prior study that reported that chitosan-modified glass-ionomer cement was able to prolong the release of protein [7]. Moreover, this investigation here also showed that the molecular weight of chitosan influenced releasing property of the cement. The results suggested that RMGIC modified by adding chitosan K at 15% and albumin at 5% by weight released a high amount of EGF and can prolong the release for at least 3 weeks. Chitosan can prolong protein release, possibly due to the formation of a polyelectrolyte complex between the cationic group of chitosan and the anionic group of poly (acrylic acid) [23,24]. Albumin was added to prolong the release of growth factor [25]. Bovine serum albumin was added to RMGIC due to its action as a carrier protein [8]. It can help the release of small proteins like EGF without any cytotoxic to HDP cells. It was concerned that EGF release from this chitosan-modified RMGIC still showed some variation amount of EGF release. The degradation of this small protein (EGF) or the poor distribution of this growth factor in the specimens should be the focus of further investigation.

The results of cytotoxicity (MTT assay) from Figure 3 revealed that the GI+C(K)+EGF group had the least cytotoxicity. Moreover, this group released substances that can promote the growth of pulp cells (Figure 4) compared to the control groups for at least 6 days, which may be referred to as the prolonged release of EGF, as demonstrated in Table 2. This group still released a detectable amount of EGF after 7 days and had its bioactivity in promoting cell growth.

The result clearly showed that all the chitosan-modified RMGICs have the same fluoride release pattern as general RMGIC. They all released fluoride at the highest rate on the first day as a burst effect and reduced rapidly after 2, 3, and 5 days and started to be slowly released, but continuously thereafter. Two mechanisms have been proposed for fluoride release from glass ionomers into an aqueous environment. One mechanism is a short-term reaction, which involves rapid dissolution from the outer surface into solution, whereas the second step is more gradual and results in the sustained diffusion of species through the bulk material [26,27]. It has been found that after the initial burst effect, resin-modified glass-ionomer cement continues to release small amounts of fluoride in vitro for 1-2.7 years [26,28].

Aluminum release of these different RMGICs had the highest rate as a burst effect on the first day, then rapidly reduced. It could not be detected at the end period of seven days. It was noticed that the GI+EGF group released the highest amount of average fluoride and aluminum, especially on the first day. This may be because the EGF protein might influence the setting reaction or polymerization of RMGIC. Considering the safety of the released aluminum [29], it was reported that the current recommended level of the US Department of Agriculture for aluminum's maximum permissible dose per day is 0.10–0.12 mg Al/kg/day [30–32]. The recommended amount is much higher than the released aluminum from the modified RMGIC used here in this study. It was found that a single glass-ionomer filling would provide roughly 0.5% of the maximum permissible dose per day, and it was believed that there should be no adverse health effects from glass-ionomer fillings [33]. In addition, a study showed that the release of aluminum from glass-ionomer material is able to assist the biological effect of fluoride against the acidogenicity of *S. mutans* biofilms [34]. In this study, chitosan-modified RMGICs groups release greater rates of fluoride than control RMGIC, the same result as a recent study, which reported that the inclusion of chitosan in RMGIC increased fluoride release [35]. Chitosan also enhanced fluoride release in conventional GIC [16,36].

It was interesting that the GI+C(K)+EGF group, where the C(K) had lower molecular weight with a higher degree of deacetylation, seems to give better results in EGF release and, in particular, had significantly lower cytotoxicity and could promote cell proliferation. This corresponded with the studies reporting that chitosan with low molecular weight and a high degree of deacetylation was suitable for drug delivery [37,38]. Chitosan-based polyelectrolyte complexes have been broadly considered for drug delivery systems, which include the polyelectrolyte complexes formed by chitosan and poly (acrylic) acid [12]. This modified glass ionomer cement revealed the potential property of prolonged release of bioactive molecules. However, this study is a preliminary in vitro experiment, which has limitations in interpretation. Further investigations will be required to examine such aspects as its chemical, physical and mechanical properties, as well as the proper type and amount of bioactive molecules that should be added for specific therapeutic purposes.

5. Conclusions

This study demonstrated that chitosan resin-modified glass ionomer cement added with EGF could prolong the release of EGF; even though the amount released was very small after 72 h, it still retained its bioactivity. This modified RMGIC also had the property of prolonging fluoride release and the release of a limited amount of aluminum. The capability to control the release of bioactive molecules in this biomaterial may assist in regenerating pulp–dentin complexes.

Author Contributions: Conceptualization, C.H. and U.K.-L.; methodology, S.W. and U.K.-L.; software, C.H. and U.K.-L.; validation, C.H., S.W. and U.K.-L.; formal analysis, U.K.-L.; investigation, C.H.; writing—original draft preparation, U.K.-L. and C.H.; writing—review and editing, C.H., S.W. and U.K.-L.; project administration, U.K.-L.; funding acquisition, U.K.-L. All authors have read and agreed to the published version of the manuscript.

Funding: This research was funded by Prince of Songkla University grant number. DEN 511990007S and grant number DEN 540020S-1-2. It was also funded by the Cell Biology and Biomaterial Research Unit.

Institutional Review Board Statement: Not applicable.

Data Availability Statement: Data will be available on request.

Acknowledgments: We thank Peter Leggat, James Cook University, Australia, for his kind assistance with proofreading the manuscript.

Conflicts of Interest: The authors declare no conflict of interest.

References

1. Kumar, N.; Maher, N.; Amin, F.; Ghabbani, H.; Zafar, M.S.; Rodríguez-Lozano, F.J.; Oñate-Sánchez, R.E. Biomimetic Approaches in Clinical Endodontics. *Biomimetics* **2022**, *7*, 229. [CrossRef]
2. Yu, S.; Zheng, Y.; Guo, Q.; Li, W.; Ye, L.; Gao, B. Mechanism of Pulp Regeneration Based on Concentrated Growth Factors Regulating Cell Differentiation. *Bioengineering* **2023**, *10*, 513. [CrossRef] [PubMed]
3. Hafshejani, T.M.; Zamanian, A.; Venugopal, J.R.; Rezvani, Z.; Sefat, F.; Saeb, M.R.; Vahabi, H.; Zarrintaj, P.; Mozafari, M. Antibacterial glass-ionomer cement restorative materials: A critical review on the current status of extended release formulations. *J. Control. Release* **2017**, *262*, 317–328. [CrossRef] [PubMed]
4. Sobh, E.G.; Hamama, H.H.; Palamara, J.; Mahmoud, S.H.; Burrow, M.F. Effect of CPP-ACP modified-GIC on prevention of demineralization in comparison to other fluoride-containing restorative materials. *Aust. Dent. J.* **2022**, *67*, 220–229. [CrossRef]
5. de Castilho, A.R.; Duque, C.; Negrini Tde, C.; Sacono, N.T.; de Paula, A.B.; de Souza Costa, C.A.; Spolidorio, D.M.; Puppin-Rontani, R.M. In vitro and in vivo investigation of the biological and mechanical behaviour of resin-modified glass-ionomer cement containing chlorhexidine. *J. Dent.* **2013**, *41*, 155–163. [CrossRef]

6. Tomiyama, K.; Ishizawa, M.; Watanabe, K.; Kawata, A.; Hamada, N.; Mukai, Y. Antibacterial effects of surface pre-reacted glass-ionomer (S-PRG) filler eluate on polymicrobial biofilms. *Am. J. Dent.* **2023**, *36*, 91–94.
7. Limapornvanich, A.; Jitpukdeebodintra, S.; Hengtrakool, C.; Kedjarune-Leggat, U. Bovine serum albumin release from novel chitosan-fluoro-aluminosilicate glass ionomer cement: Stability and cytotoxicity studies. *J. Dent.* **2009**, *37*, 686–690. [CrossRef] [PubMed]
8. Zhou, L.; Lu, N.; Pi, X.; Jin, Z.; Tian, R. Bovine Serum Albumin as a Potential Carrier for the Protection of Bioactive Quercetin and Inhibition of Cu(II) Toxicity. *Chem. Res. Toxicol.* **2022**, *35*, 529–537. [CrossRef]
9. Hoang, H.T.; Jo, S.H.; Phan, Q.T.; Park, H.; Park, S.H.; Oh, C.W.; Lim, K.T. Dual pH-/thermo-responsive chitosan-based hydrogels prepared using "click" chemistry for colon-targeted drug delivery applications. *Carbohydr. Polym.* **2021**, *260*, 117812. [CrossRef]
10. Budiarso, I.J.; Rini, N.D.W.; Tsalsabila, A.; Birowosuto, M.D.; Wibowo, A. Chitosan-Based Smart Biomaterials for Biomedical Applications: Progress and Perspectives. *ACS Biomater. Sci. Eng.* **2023**, *9*, 3084–3115. [CrossRef]
11. Paker, E.S.; Senel, M. Polyelectrolyte Multilayers Composed of Polyethyleneimine-Grafted Chitosan and Polyacrylic Acid for Controlled-Drug-Delivery Applications. *J. Funct. Biomater.* **2022**, *13*, 131. [CrossRef] [PubMed]
12. Hamman, J.H. Chitosan based polyelectrolyte complexes as potential carrier materials in drug delivery systems. *Mar. Drugs* **2010**, *8*, 1305–1322. [CrossRef] [PubMed]
13. Ibrahim, M.A.; Neo, J.; Esguerra, R.J.; Fawzy, A.S. Characterization of antibacterial and adhesion properties of chitosan-modified glass ionomer cement. *J. Biomater. Appl.* **2015**, *30*, 409–419. [CrossRef] [PubMed]
14. Debnath, A.; Kesavappa, S.B.; Singh, G.P.; Eshwar, S.; Jain, V.; Swamy, M.; Shetty, P. Comparative Evaluation of Antibacterial and Adhesive Properties of Chitosan Modified Glass Ionomer Cement and Conventional Glass Ionomer Cement: An in Vitro Study. *J. Clin. Diagn. Res.* **2017**, *11*, ZC75–ZC78. [CrossRef] [PubMed]
15. Ranjani, M.S.; Kavitha, M.; Venkatesh, S. Comparative Evaluation of Osteogenic Potential of Conventional Glass-ionomer Cement with Chitosan-modified Glass-ionomer and Bioactive Glass-modified Glass-ionomer Cement An in vitro Study. *Contemp. Clin. Dent.* **2021**, *12*, 32–36. [CrossRef] [PubMed]
16. Elshenawy, E.A.; El-Ebiary, M.A.; Kenawy, E.R.; El-Olimy, G.A. Modification of glass-ionomer cement properties by quaternized chitosan-coated nanoparticles. *Odontology* **2023**, *111*, 328–341. [CrossRef]
17. Wanachottrakul, N.; Chotigeat, W.; Kedjarune-Leggat, U. Effect of novel chitosan-fluoroaluminosilicate resin modified glass ionomer cement supplemented with translationally controlled tumor protein on pulp cells. *J. Mater. Sci. Mater. Med.* **2014**, *25*, 1077–1085. [CrossRef] [PubMed]
18. Kobayashi, R.; Hoshikawa, E.; Saito, T.; Suebsamarn, O.; Naito, E.; Suzuki, A.; Ishihara, S.; Haga, H.; Tomihara, K.; Izumi, K. The EGF/EGFR axis and its downstream signaling pathways regulate the motility and proliferation of cultured oral keratinocytes. *FEBS Open Bio* **2023**, *13*, 1469–1484. [CrossRef]
19. Lott, K.; Collier, P.; Ringor, M.; Howard, K.M.; Kingsley, K. Administration of Epidermal Growth Factor (EGF) and Basic Fibroblast Growth Factor (bFGF) to Induce Neural Differentiation of Dental Pulp Stem Cells (DPSC) Isolates. *Biomedicines* **2023**, *11*, 255. [CrossRef]
20. Tyler, J.E.; Poole, D.F. The rapid measurement of fluoride concentrations in stored human saliva by means of a differential electrode cell. *Arch. Oral Biol.* **1989**, *34*, 995–998. [CrossRef]
21. Mosmann, T. Rapid colorimetric assay for cellular growth and survival: Application to proliferation and cytotoxicity assays. *J. Immunol. Methods* **1983**, *65*, 55–63. [CrossRef]
22. Tabachnick, B.G.; Fidell, L.S. *Using Multivariate Analysis*; Pearson: Boston, MA, USA, 2007; pp. 98–99.
23. Cao, J.; Wang, Y.; He, C.; Kang, Y.; Zhou, J. Ionically crosslinked chitosan/poly(acrylic acid) hydrogels with high strength, toughness and antifreezing capability. *Carbohydr. Polym.* **2020**, *242*, 116420. [CrossRef] [PubMed]
24. de Oliveira, H.C.; Fonseca, J.L.; Pereira, M.R. Chitosan-poly(acrylic acid) polyelectrolyte complex membranes: Preparation, characterization and permeability studies. *J. Biomater. Sci. Polym. Ed.* **2008**, *19*, 143–160. [CrossRef]
25. Murray, J.B.; Brown, L.; Langer, R.; Klagsburn, M. A micro sustained release system for epidermal growth factor. *In Vitro* **1983**, *19*, 743–748. [CrossRef] [PubMed]
26. Williams, J.A.; Billington, R.W.; Pearson, G.J. A long term study of fluoride release from metal-containing conventional and resin-modified glass-ionomer cements. *J. Oral Rehabil.* **2001**, *28*, 41–47. [CrossRef] [PubMed]
27. Xu, X.; Burgess, J.O. Compressive strength, fluoride release and recharge of fluoride-releasing materials. *Biomaterials* **2003**, *24*, 2451–2461. [CrossRef]
28. Rolim, F.G.; de Araújo Lima, A.D.; Lima Campos, I.C.; de Sousa Ferreira, R.; da Cunha Oliveira-Júnior, C.; Gomes Prado, V.L.; Vale, G.C. Fluoride Release of Fresh and Aged Glass Ionomer Cements after Recharging with High-Fluoride Dentifrice. *Int. J. Dent.* **2019**, *2019*, 9785364. [CrossRef]
29. Savarino, L.; Cervellati, M.; Stea, S.; Cavedagna, D.; Donati, M.E.; Pizzoferrato, A.; Visentin, M. In vitro investigation of aluminum and fluoride release from compomers, conventional and resin-modified glass-ionomer cements: A standardized approach. *J. Biomater. Sci. Polym. Ed.* **2000**, *11*, 289–300. [CrossRef]
30. Renke, G.; Almeida, V.B.P.; Souza, E.A.; Lessa, S.; Teixeira, R.L.; Rocha, L.; Sousa, P.L.; Starling-Soares, B. Clinical Outcomes of the Deleterious Effects of Aluminum on Neuro-Cognition, Inflammation, and Health: A Review. *Nutrients* **2023**, *15*, 2221. [CrossRef]

31. Hunt, C.D.; Meacham, S.L. Aluminum, boron, calcium, copper, iron, magnesium, manganese, molybdenum, phosphorus, potassium, sodium, and zinc: Concentrations in common western foods and estimated daily intakes by infants; toddlers; and male and female adolescents, adults, and seniors in the United States. *J. Am. Diet. Assoc.* **2001**, *101*, 1058–1060. [CrossRef]
32. Malik, J.; Frankova, A.; Drabek, O.; Szakova, J.; Ash, C.; Kokoska, L. Aluminium and other elements in selected herbal tea plant species and their infusions. *Food Chem.* **2013**, *139*, 728–734. [CrossRef]
33. Nicholson, J.W.; Czarnecka, B. Review paper: Role of aluminum in glass-ionomer dental cements and its biological effects. *J. Biomater. Appl.* **2009**, *24*, 293–308. [CrossRef]
34. Hayacibara, M.F.; Rosa, O.P.; Koo, H.; Torres, S.A.; Costa, B.; Cury, J.A. Effects of fluoride and aluminum from ionomeric materials on S. mutans biofilm. *J. Dent. Res.* **2003**, *82*, 267–271. [CrossRef] [PubMed]
35. Patel, A.; Dhupar, J.K.; Jajoo, S.S.; Shah, P.; Chaudhary, S. Evaluation of Adhesive Bond Strength, and the Sustained Release of Fluoride by Chitosan-infused Resin-modified Glass Ionomer Cement: An in Vitro Study. *Int. J. Clin. Pediatr. Dent.* **2021**, *14*, 254–257. [CrossRef]
36. Nishanthine, C.; Miglani, R.; Indira, R.; Poorni, S.; Srinivasan, M.R.; Robaian, A.; Albar, N.H.M.; Alhaidary, S.F.R.; Binalrimal, S.; Almalki, A.; et al. Evaluation of Fluoride Release in Chitosan-Modified Glass Ionomer Cements. *Int. Dent. J.* **2022**, *72*, 785–791. [CrossRef] [PubMed]
37. Ahn, S.; Lee, I.H.; Lee, E.; Kim, H.; Kim, Y.C.; Jon, S. Oral delivery of an anti-diabetic peptide drug via conjugation and complexation with low molecular weight chitosan. *J. Control. Release* **2013**, *170*, 226–232. [CrossRef] [PubMed]
38. Affes, S.; Aranaz, I.; Acosta, N.; Heras, Á.; Nasri, M.; Maalej, H. Chitosan derivatives-based films as pH-sensitive drug delivery systems with enhanced antioxidant and antibacterial properties. *Int. J. Biol. Macromol.* **2021**, *182*, 730–742. [CrossRef]

Disclaimer/Publisher's Note: The statements, opinions and data contained in all publications are solely those of the individual author(s) and contributor(s) and not of MDPI and/or the editor(s). MDPI and/or the editor(s) disclaim responsibility for any injury to people or property resulting from any ideas, methods, instructions or products referred to in the content.

Article

Injectable Lyophilized Chitosan-Thrombin-Platelet-Rich Plasma (CS-FIIa-PRP) Implant to Promote Tissue Regeneration: In Vitro and Ex Vivo Solidification Properties

Fiona Milano [1], Anik Chevrier [2], Gregory De Crescenzo [1,2] and Marc Lavertu [1,2,*]

1. Biomedical Engineering Institute, Polytechnique Montreal, Montréal, QC H3T 1J4, Canada
2. Chemical Engineering Department, Polytechnique Montreal, Montréal, QC H3T 1J4, Canada
* Correspondence: marc.lavertu@polymtl.ca

Abstract: Freeze-dried chitosan formulations solubilized in platelet-rich plasma (PRP) are currently evaluated as injectable implants with the potential for augmenting the standard of care for tissue repair in different orthopedic conditions. The present study aimed to shorten the solidification time of such implants, leading to an easier application and a facilitated solidification in a wet environment, which were direct demands from orthopedic surgeons. The addition of thrombin to the formulation before lyophilization was explored. The challenge was to find a formulation that coagulated fast enough to be applied in a wet environment but not too fast, which would make handling/injection difficult. Four thrombin concentrations were analyzed (0.0, 0.25, 0.5, and 1.0 NIH/mL) in vitro (using thromboelastography, rheology, indentation, syringe injectability, and thrombin activity tests) as well as ex vivo (by assessing the implant's adherence to tendon tissue in a wet environment). The biomaterial containing 0.5 NIH/mL of thrombin significantly increased the coagulation speed while being easy to handle up to 6 min after solubilization. Furthermore, the adherence of the biomaterial to tendon tissues was impacted by the biomaterial-tendon contact duration and increased faster when thrombin was present. These results suggest that our biomaterial has great potential for use in regenerative medicine applications.

Keywords: chitosan; thrombin; platelet-rich plasma; injectable implant; tissue regeneration; solidification properties

Citation: Milano, F.; Chevrier, A.; De Crescenzo, G.; Lavertu, M. Injectable Lyophilized Chitosan-Thrombin-Platelet-Rich Plasma (CS-FIIa-PRP) Implant to Promote Tissue Regeneration: In Vitro and Ex Vivo Solidification Properties. *Polymers* **2023**, *15*, 2919. https://doi.org/10.3390/polym15132919

Academic Editors: William Facchinatto and Sérgio Paulo Campana-Filho

Received: 6 June 2023
Revised: 26 June 2023
Accepted: 28 June 2023
Published: 30 June 2023

Copyright: © 2023 by the authors. Licensee MDPI, Basel, Switzerland. This article is an open access article distributed under the terms and conditions of the Creative Commons Attribution (CC BY) license (https:// creativecommons.org/licenses/by/ 4.0/).

1. Introduction

Our laboratory developed a freeze-dried (FD) formulation of chitosan (CS) intended to be solubilized in platelet-rich plasma (PRP) to form injectable implants that coagulate in situ [1]. This biomaterial is intended to be used as a surgical adjuvant in diverse applications in orthopedic regenerative medicine.

Chitosan (CS) is a cationic linear polymer consisting of monomers of N-acetyl glucosamine and glucosamine monomers, which is usually obtained through alkaline deacetylation of chitin [2]. CS has been found to exhibit a greater affinity for cells and growth factors than other natural polymers, most likely because of its ability to mediate attractive electrostatic interactions [3]. CS is biodegradable and biocompatible [2,4], and it has numerous applications in the biomedical field [5]. These applications include drug delivery, gene therapy, vaccines, tissue engineering [6], and wound healing [7]. CS's interesting properties in wound healing include its antimicrobial, hemostatic, and analgesic character [8]. Moreover, CS influences the coagulation cascade by mediating hemagglutination and promoting platelet activity [6]. CS molecular weight and degree of deacetylation (DDA, the fraction of glucosamine monomers) impact the innate immune responses [6], and chitosan is considered as an excellent candidate as a mucosal adjuvant [9]. Overall, the unique properties of CS make it an attractive polymer for a variety of biomedical applications.

PRP is a plasma fraction containing a platelet concentration several times higher than that of whole blood (WB) [10]. Upon activation, these platelets release several growth factors [11], which are believed to promote tissue repair by inducing cell proliferation, cell differentiation, and angiogenesis at the injury site [12]. However, the literature on the efficacy of PRP in improving tissue repair has yielded inconsistent results, and there is currently insufficient clinical evidence to support its use [13]. These inconsistencies may arise from the lack of standardization of platelet separation techniques, the variability in formulations of platelet separation techniques, the variability in formulations of PRP used, as well as the poor stability and residency of PRP in vivo [13,14].

Using PRP instead of WB to rehydrate FD-CS formulations is expected to enhance the bioactivity of the resulting hybrid implants and improve tissue repair outcomes [1]. Our previous research has demonstrated that CS increases the release of platelet-derived growth factor, prolongs the release of platelet-derived growth factor, prolongs the residence time, and enhances the bioactivity of PRP in vivo while inhibiting the platelet-mediated clot retraction observed with the use of PRP alone [15]. Animal studies have confirmed that CS-PRP implants are biodegradable, biocompatible, and effective in augmenting rotator cuff, cartilage, and meniscus repair (increase cell recruitment, vascularization, remodeling, and tissue repair) in both small and large animal models [14,16,17]. Moreover, CS-PRP implants possess important characteristics for soft tissue repair, such as injectability, stickiness, paste-like handling properties, and the ability to solidify in situ.

However, mixing FD-CS with PRP results in a hybrid clot implant that solidifies slowly, especially for human PRP. When applied directly after mixing, the wound's or defect's wet environment can hinder the biomaterial from adhering to the site of injection. In an ongoing rotator cuff multicenter, prospective, randomized clinical trial (ClinicalTrials.gov, NCT number NCT05333211, Sponsor: ChitogenX Inc., Kirkland, QC, Canada), the biomaterial is required to be incubated in a syringe for 30–45 min before reaching a viscosity suitable for injection at the repaired site. A more rapid and controlled implant solidification would facilitate its handling, its administration, and ultimately its clinical implementation.

The objective of this study was to develop a novel formulation of FD-CS with accelerated solidification. We hypothesized that the solidification of CS-PRP implants can be accelerated by incorporating thrombin (FIIa), as previously reported for CS-glycerol phosphate/blood implants [18]. Our selection of this coagulation factor was based on its well-known ability to promote the repair process as well as its established use in other clinical contexts [19,20]. A second hypothesis was that thrombin could be added prior to FD to create a stable CS-FIIa-lyophilized formulation. Adding thrombin prior to FD (1) ensures that thrombin is evenly dispersed in the biomaterial upon rehydration in PRP, and (2) simplifies the procedure compared to in situ mixing of the CS-PRP implant with a thrombin formulation/solution. However, this approach presents a challenge: the formulation should coagulate fast enough to be applied in a wet environment upon mixing, but not so quickly that it becomes difficult, if not impossible, to handle/inject. To achieve this balance, CS-PRP biomaterials containing thrombin concentrations of up to 1.0 NIH/mL were studied in vitro (using thromboelastography, rheology, indentation, syringe injectability, and thrombin activity tests) as well as ex vivo (by assessing the implant's adherence to tendon tissue in a wet environment).

2. Materials and Methods

2.1. Preparation of Freeze-Dried Chitosan Formulations

The FD-CS formulations preparation were based on the method developed by Chevrier et al. (2018) [1]. Raw chitosan (Primex, Siglufjordur, Iceland, Product N°43030) underwent heterogeneous alkaline deacetylation to reach a DDA of 82 ± 2 % (mean value ± SD for 3 batches, measured by nuclear magnetic resonance spectroscopy [21]). It was then depolymerized with nitrous acid to reach a molar mass (M_n) of 39.3 ± 0.4 kDa (mean value ± SD for 3 batches, measured by size-exclusion chromatography/multi-angle laser light scattering [22]). CS was dissolved in HCl overnight at room temperature to obtain a solution containing 1% (w/v) CS. The concentration of this HCl solution was adjusted to reach 60% protonation of the CS amino groups. Calcium chloride (Sigma-Aldrich, St. Louis, MO, USA, Product N°C-5670) was added as a solidifying agent to reach a final concentration of 42.2 mM, and then trehalose (Sigma-Aldrich, St. Louis, MO, USA, Product N°T-0167) was added as a lyoprotectant to reach a final 1% (w/v) content. The resulting solution was passed through 0.45 μm filters and dispensed in glass vials. The pH of the chitosan solutions was 6.35 ± 0.16 and the osmolality was 199 ± 21 (mean value ± SD for 9 batches).

For formulations containing thrombin, thrombin diluted in a 0.1% (w/v) BSA solution was added to each glass vial to reach a thrombin activity of 0.25 NIH/mL, 0.50 NIH/mL, or 1.0 NIH/mL and a final BSA concentration of 0.01% (w/v). All vials were then freeze-dried through the following cycle using a Millrock freeze-dryer (Laboratory Series LD85S3, Millrock Technology, Kingston, NY, USA): (1) step freezing (to 20 °C then isothermal for 15 min, to 5 °C then isothermal for 30 min, to −5 °C then isothermal for 30 min) then ramp freezing to −40 °C at −1 °C/min, then isothermal for 2 h; (2) ramp heating at 0.5 °C/min up to 8 °C at 76 mTorr, then isothermal for 650 min; (3) ramp heating at 0.15 °C/min until 40 °C at 76 mTorr, then isothermal for 6 h. A less aggressive cycle was initially used, heating up to only 25 °C during step 3 (secondary drying), but for 10 h. However, no differences were observed in the resulting FD-CS formulations, and the cycle heating up to 40 °C was preferred for the rest of the study. Of note, some of the data used in Section 3.1 (thromboelastography results) were obtained from the less aggressive FD cycle.

2.2. Isolation of Platelet-Rich Plasma

Commercially sterile citrated sheep blood (Cedarlane, Burlington, ON, Canada, Product N°CL2581-500C) was used to isolate the leucocyte-rich PRP. Briefly, whole blood was distributed in 10 mL aliquots in 15 mL Falcon tubes, then centrifuged at $800 \times g$ for 20 min (unforced deceleration). The supernatant, buffy coat, and first 1-2 mm of erythrocytes were collected in new 15-mL Falcon tubes, and a second centrifugation was then performed at $600 \times g$ for 10 min (unforced deceleration).

2.3. Rehydration of the CS-FD Formulation

Freeze-dried CS cakes were solubilized in PRP by vigorous manual shaking for 30 s. The cakes were made from various volumes of CS solutions, depending on the final amount of biomaterial needed (1 mL, 3 mL, 5 mL, 6 mL, or 8 mL), and rehydrated in the same volume of PRP.

2.4. Assessment of Clotting Properties of CS-PRP Formulations by Thromboelastography

After dissolution of the CS cakes in PRP, 360 μL of each formulation was immediately loaded in a thromboelastograph (TEG) cup using a 1 mL syringe and an 18 G needle (allowing the viscous formulation to be easily retrieved from the vial). TEG tracings were recorded for 1 h, and three parameters were used to compare the samples: (1) The clot reaction time, R, is the time in minutes to reach a 2 mm divergence between the two branches of the TEG curve and is indicative of the time needed for the coagulation process to start. (2) The maximal amplitude, MA, is the maximal distance reached between the two branches of the TEG curve (in mm) and is indicative of the clot strength. (3) The K-value, which is the duration between R and the time when a 20 mm divergence between the two branches of the TEG curve is reached and is indicative of the coagulation speed. Two PRP solutions were used as controls. The first consisted of PRP recalcified in 42.2 mM of calcium chloride. The second consisted of PRP recalcified in 42.2 mM of calcium chloride and 0.5 NIH/mL of thrombin. A TEG Model 5000 (Haemoscope Corp., Niles, IL, USA) was used for all samples.

2.5. Assessment of Clotting Properties of CS-PRP Formulations through Rheology Measurements

Clotting properties were measured using a Physica MCR 501 rheometer (Anton Paar, Graz, Austria). Concentric cylinders with rough geometry were used for measurements (CC17/T200/SS/p, cup diameter of 18.08 mm, and bob diameter of 16.66 mm). After mixing the CS-FD formulation (8 mL) with PRP, 5 mL of the formulation was placed in the cup. Rheological properties were measured during a time sweep over the course of either 1 h or 15 min at a fixed 1% strain and a fixed 5 rad/s frequency. The choice of these parameters is explained and justified in Appendix A. All measurements were performed at 37 °C.

2.6. Assessment of the Force Required to Eject the Biomaterial

The protocol used was implemented on the multi-axis mechanical tester Mach-1, Model V500css (Biomomentum Inc., Laval, QC, Canada), with a 25 kg uniaxial load cell and a flat circular indenter with a diameter of 30 mm. 6 mL of each sample were placed in a 10 mL syringe with an 18 G × 3.50 IN quincke spinal needle (BD Product N°405184) right after rehydration of the FD-CS formulation in PRP. The syringe was then fixed vertically below the indenter through a custom system. Starting right after the syringe's fixation, 0.5 mL of the biomaterial was pushed every 2 min, 10 times, at a speed of 1 mm/s (resulting in 3 s of movement). Force data were saved at a frequency of 100 Hz for the whole duration of the test.

2.7. Assessment of Hybrid-Clot Mechanical Properties through Indentation

The indentation protocol used was implemented on the multi-axis mechanical tester Mach-1, Model V500css (Biomomentum Inc., Laval, QC, Canada) with a 150 g uniaxial load cell and a spherical indenter with a radius of 1.5 mm (<25% of the sample height—~8 mm in our case—as previously recommended in the literature [23,24]). 0.3 mL of each sample was placed in a 96-well microplate and allowed to coagulate for 24 h at room temperature. The bottom of the microplate was first probed in three empty wells to fit a plane and compute the position of the bottom in each well. For each sample, the surface of the biomaterial was then found by lowering the probe at a speed of 0.05 mm/s until a force of 0.1 gf was detected. A single indentation of 0.7 mm (<10% of the total sample height, as previously recommended in the literature [24,25] to prevent the influence of the sample height) was then performed at a velocity of 0.07 mm/s. Force data were saved at a frequency of 100 Hz during the indentation and the relaxation that followed, until the relaxation rate reached 1 gf/min (slope measured for 10 s). Sample stiffness and force at equilibrium were computed for each sample.

2.8. Assessment of Hybrid-Clot Homogeneity through Histology and MASQH Algorithm

After rehydration of CS-FD cakes in PRP, the formulations were placed in glass test tubes (250 µL) at 37 °C and solidified for 1 h. The resulting clots were fixed in 10% neutral buffered formalin (NBF) before being dehydrated, cleared, and paraffin embedded. They were then sectioned into 5 µm-thick slices, dewaxed, and stained. Cibacron Brilliant Red (Glentham Life Sciences, Product N°GT9393) and Iron-Hematoxylin (Sigma, Products N°HT107, and N°HT109) were used to color chitosan, red, and white blood cells, as per the protocol described in Rossomacha et al. [26]. The colored slices were digitally scanned (Hamamatsu Nanozoomer RS) at 10× and 40×. The homogeneity of the chitosan distribution in the hybrid clot was evaluated using the MASQH algorithm [27].

2.9. Assessment of CS-PRP Formulations Adhesion to Tendon Tissues Using an Ex Vivo Test

The objective of this test was to determine whether the biomaterial could adhere to wet tissues. Ex vivo evaluation of the CS-PRP formulations adhesion to tendon tissues was performed using bovine tendons obtained from a local butcher shop. Prior to testing, tendons were defrosted and heated to 37 °C in a 0.9% NaCl bath, a frequently used fluid during shoulder arthroscopic surgeries [28]. Right after mixing a CS-FD cake with PRP, the formulation was retrieved from its vial with a 10 mL syringe and an 18 G needle (Sigma, Product N°305180), and a waiting time varying between 0 and 7 min was observed to mimic the time taken by the surgeon between mixing the biomaterial and its use. Next, 3 mL of the biomaterial was delivered to the tendon, and a second waiting time varying between 0 and 7 min was observed. During the surgery, after suturing the tendon to the bone, the irrigation fluid is stopped, and then the surgical site is dried with a combination of fluid aspiration and sterile cotton swabs. The biomaterial is then delivered at the tendon-bone interface and on the tendon before closing the articulation. The second waiting time, therefore, mimics the conditions after the biomaterial application and the end of the surgery, as the physiological fluid does not return instantly to the articulation. After this second waiting time, the tendon and adhered biomaterial were submerged in a 1 L 0.9% NaCl bath and twisted in the fluid for 1 min to simulate the presence of physiological fluids and movement of the shoulder. The quantity of biomaterial remaining on the tendon was then visually observed (qualitative evaluation). The color of the 0.9% NaCl bath was quantitatively evaluated through image analysis: a photograph of the bath was taken, and the weight of its red channel was computed using the ImageJ software [29].

2.10. Assessment of Thrombin Activity in Freeze-Dried Chitosan Formulations

The activity of thrombin in the CS formulations was quantified before FD and in CS-FD cakes using a fluorometric thrombin activity assay kit (Abcam, ab197006). FD samples were rehydrated in water instead of PRP. All samples were diluted at 1/2 and 1/5 in assay buffer. As per Abcam protocol, six standards were used to construct a standard curve, with concentrations ranging from 0.26 NIH/mL to 1.3 NIH/mL. Fluorescence was measured at Ex/Em = 350/450 at 37 °C in a kinetic mode every 2 min for 60 min using an Infinite M200 reader (Tecan Trading AG, Männedorf, Switzerland).

2.11. Statistical Analysis

To investigate whether thrombin accelerates coagulation significantly, the TEG results were subjected to statistical analysis using the Mann-Whitney U test. Similarly, the same test was performed on the TEG results to determine if there was any alteration in the biomaterial activity due to the storage time. Differences were considered significant for p-values < 0.05. The figure's caption accompanying each result contains information on the number of independent experiments conducted (N) as well as the number of replicates for each sample in these experiments (n).

3. Results and Discussion

3.1. Chitosan Freeze-Dried with Thrombin Easily Solubilizes in PRP, Exhibits Thrombin Activity, and Maintains Stability for at Least 2 Months at Room Temperature

Freeze-dried chitosan formulations (Figure 1a) were slightly retracted from the vial walls for all thrombin concentrations (this phenomenon was expected and previously observed by our research group [1]). They rapidly solubilized in PRP without any apparent influence on the thrombin concentration.

The pH of whole blood was 7.69 ± 0.13, and the pH of PRP was 7.97 ± 0.18 (mean value $\pm$ SD for 12 batches). The osmolality of whole blood was 316 ± 3 and the osmolality of PRP was 317 ± 2 (mean value $\pm$ SD for 8 batches). The pH of the CS-PRP formulation, with or without thrombin, was 7.20 ± 0.16 (mean value $\pm$ SD for 15 batches).

An activity loss of ~20–25% was observed after lyophilization for samples containing 0.5 NIH/mL of thrombin, which was the only formulation assessed using the thrombin activity kit. In the remainder of the manuscript, the thrombin activity mentioned will refer to the activity before lyophilization (i.e., not taking into account the ~25% activity loss). The 0.5 NIH/mL formulation was also tested for thrombin activity with and without CS before freeze-drying, and no difference was observed. This suggests that the interaction between thrombin and CS is weak or nonexistent, as previously reported in [30,31]. This expected result is likely due to the cationic character of chitosan and thrombin at the pH of the solution. The histology of CS-FIIa-PRP implants confirmed that CS was evenly distributed among blood components in the presence of thrombin. A MASQH homogeneity score of 0.92 ± 0.03 was computed for CS-PRP implants and of 0.93 ± 0.02 for CS-FIIa-PRP implants (mean value $\pm$ SD for 5 samples in each case).

A stability study was conducted for a period of two months, and thrombin activity was assessed at three specific time points (t = 0, 28, and 56 days) using TEG. All the samples were stored at room temperature (RT). Some FDA-approved biomaterials containing thrombin were stored at temperatures below $-18\ °C$ (they contain solubilized human thrombin for topical use, e.g., Evithrom®, Tisseel®, and Evicel® pre-filled syringes), and others were stored at RT (they contain lyophilized human thrombin powder, e.g., Recothrom®, and Tisseel® kit).

This study investigated the stability of biomaterials without thrombin (controls) and of biomaterials containing 0.5 NIH/mL of thrombin. Additionally, freshly extracted PRP controls without thrombin or with 0.5 NIH/mL of thrombin were also included. Examples of typical TEG curves are presented in Figure 1b for biomaterial samples and in Figure 1c for PRP samples. No other thrombin concentration was tested, as 0.5 NIH/mL was previously identified as the preferred concentration (this choice is justified in Section 3.2 below). The results showed that thrombin had a statistically significant impact on the biomaterial's clot reaction time compared to the biomaterial without thrombin (control) over a period of up to 56 days (p-value < 0.01 for t = 0 and t = 2 months and p-value < 0.05 for t = 1 month, as shown in Figure 1d). The average clot reaction time (R) for the control biomaterial was 4.04 min, while biomaterial with 0.5 NIH/mL of thrombin had an average R of 3.02 min over the three time points. No significant differences in clot reaction time were observed between control samples at each time point, which was expected as the biomaterial is already known to be stable for at least three years (unpublished data). Furthermore, no significant differences in clot reaction time were observed between samples with thrombin at each time-point, suggesting that no activity loss occurred during this 2-month period. No statistically significant differences were observed for the K-value between samples with and without thrombin for all three-time points (Figure 1e). Significant differences were, however, observed between samples with thrombin after two months and both samples observed at time 0 (p-value < 0.05). This could be explained by the small sample size (N = 2 and n = 3) for the two-month samples. Further studies are needed to confirm the observed trend. K had a mean duration of 3.45 min for control samples and 3.11 min for samples with 0.5 NIH/mL of thrombin across the three time points. Notably, MA had a consistent value of approximately 67 mm at all time points, irrespective of the thrombin concentration. This indicates that the addition of thrombin did not affect the strength of the final clot (no significant differences were observed, as shown in Figure 1f), as expected.

Compared to PRP controls, biomaterial samples (with or without thrombin) had a smaller R, a smaller K, and a larger MA (p-value < 0.01). These findings are consistent with previous research conducted by Chevrier et al. [1], who also demonstrated that the addition of CS to blood samples enhanced clot strength and led to faster clot initiation. Furthermore, the addition of thrombin to PRP resulted in a significant decrease in clot reaction time but did not have a significant impact on K and MA.

Long-term stability studies at $-20\ °C$, $4\ °C$, and RT using thrombin activity quantification kits are ongoing.

Visual observation of the samples coagulation in plastic weighing boats confirmed that the coagulation process is faster when thrombin is present. Although the difference in coagulation time may appear moderate when examining the TEG results, it is significant when the handling properties of the samples are assessed. Supplementary Materials Video S1 includes videos demonstrating this phenomenon.

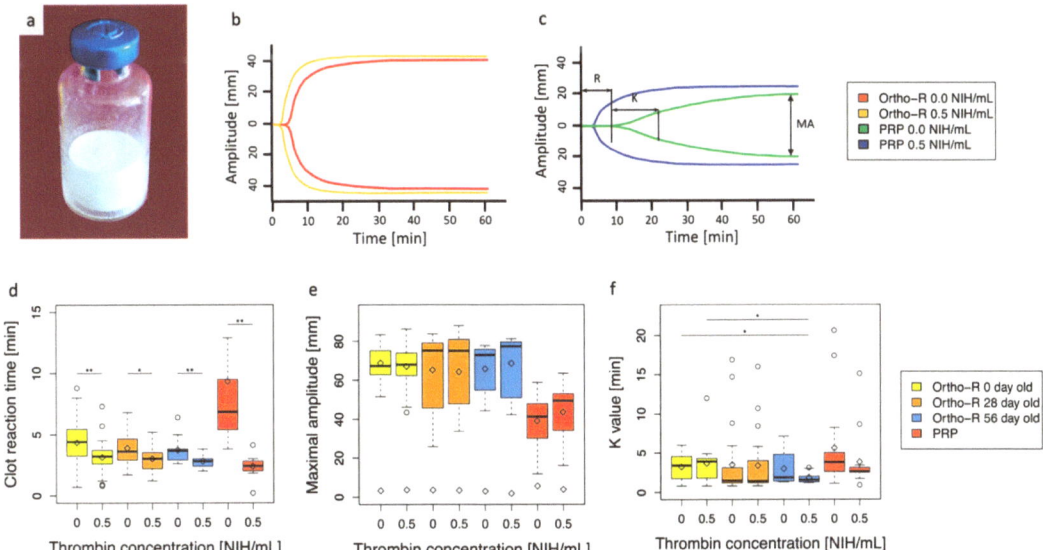

Figure 1. (**a**) The biomaterial, which is a lyophilized formulation of chitosan, is designed to be mixed with platelet-rich plasma (PRP) to create a hybrid biomaterial that coagulates in situ. To ensure rapid coagulation in situ during arthroscopic surgery, thrombin was added to the existing formulation. (**b**) Typical thromboelastograph curves obtained for biomaterials with and without thrombin. (**c**) Typical thromboelastograph curves obtained for PRP with and without thrombin. (**d**) clot reaction time (R) (time needed to reach an amplitude of 2 mm), (**e**) clot maximal amplitude (MA) (indicative of clot strength), and (**f**) K value (time needed after R to reach an amplitude of 20 mm). * indicates a p-value < 0.05, ** indicates a p-value < 0.01, and ◊ indicates the mean value. Note that statistically significant differences were also observed for the R values between samples with and without thrombin issued from different timepoints but are not indicated in the figure for clarity's sake. For the biomaterial with and without thrombin at $t = 0$, $N = 7$ and $n = 3$, at $t = 1$ month, $N = 6$ and $n = 3$, and at $t = 2$ months, $N = 2$ and $n = 3$. For PRP controls with and without thrombin, $N = 6$ and $n = 2$.

3.2. Biomaterial Containing 0.5 NIH/mL of Thrombin Is Easily Dispensed through an 18-Gauge Needle and 10-cc Syringe up to Six Min after Rehydration of the Lyophilized Formulation in PRP

For every thrombin concentration, the force needed to eject the biomaterial out of a 10-cc syringe equipped with an 18-gauge needle increases over time, as expected since the coagulation process is ongoing in the syringe. After a few minutes (sooner for higher thrombin concentrations), the force required to eject the biomaterial reaches a maximum before dropping and/or stabilizing around 25 N.

We believe that this drop occurs when the applied force breaks the clot's structure being formed (around 30–40 N on average, but reached values as high as 70–95 N, as shown in Figure 2). Once the structure is broken, the biomaterial becomes easier to eject, but the coagulation process is likely impaired, resulting in altered mechanical properties (as previously observed for PRP after stress [32] or for fibrin after a stretch or biaxial confinement [33,34]).

Even before reaching this maximal force, the high force required to eject the biomaterial might not allow the surgeon to perform precise movements, preventing him from precisely deliver/inject the biomaterial as expected. Indeed, a study by R. Watt et al. recommends targeting no more than 20 N of injection force to guarantee easy expulsion of syringe contents [35]. A force exceeding 20 N was observed as early as the 8th minute for controls and samples with 0.25 NIH/mL of thrombin, on the 6th minute for samples with 0.5 NIH/mL of thrombin, and between the 4th and 5th minutes for samples with 1.0 NIH/mL of thrombin, as shown in Figure 2. Based on this analysis, the 0.5 NIH/mL sample was identified as the most promising as it allows a significant increase in coagulation rate while preserving adequate handling properties for at least 6 min, providing enough time for the surgeon to use/deliver the biomaterial after its reconstitution.

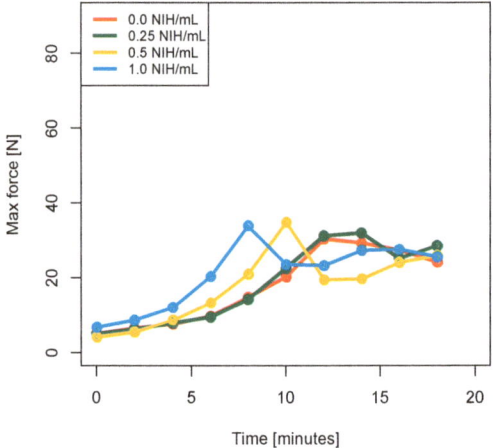

Figure 2. Force (lines indicate mean values and envelopes indicate min and max values) necessary to push 500 µL of the biomaterial out of a 10-cc syringe equipped with a spinal 18-gauge needle every two minutes, 20 times ($N = 6$ and $n = 2$). The 0.5 NIH/mL solution allows for easy ejection of the biomaterial as the force required for biomaterial ejection remains below 20 N for at least 6 min. This makes it easier to manipulate than the 1.0 NIH/mL solution, for which a force exceeding 20 N is needed for biomaterial ejection after 4–5 min.

3.3. Rheometry Reveals That Thrombin Concentration Has a Significant Impact on Storage and Loss Modulus, as Well as on the Time to Gelation Point

Biomaterial samples containing 0, 0.25, 0.5, and 1.0 NIH/mL of thrombin were studied through rheometry. The raw results for storage (G') and loss (G'') moduli are presented in Figure 3a. It was observed that after 60 min, values for G' and G'' were not yet stabilized (Figure 3a has a semi-logarithmic scale), indicating that the coagulation process has not been completed for any of the samples.

The gelation point (cross-over between G' and G'') occurred on average at 4.15 min after rehydration of FD samples in PRP for the controls and after 2.40, 1.03, and 0.35 min for the formulation containing, respectively, 0.25, 0.5, and 1.0 NIH/mL of thrombin. Of note, for 0.5 NIH/mL and 1.0 NIH/mL, this gelation point occurred before the start of the test, as it took about 75 s to start the rheology experiment after rehydration of the FD samples in PRP. The mixture was therefore already in a viscous solid state ($G' > G''$) when the test was started. The time to gelation for these formulations was determined by extrapolating the raw data (G' and G'' as a function of time), as shown in Figure 3b. For the fit, an exponential function of the form $y = A + Be^{Rt}$ was used over the first 3 min of the test, where $R \geq 0$, y is either the loss or the storage modulus and t is the time.

The time to gelation decreased significantly with thrombin concentration and followed a negative exponential, as shown in Figure 3c. For example, a thrombin concentration of 0.5 NIH/mL allowed for a reduction of the time to gelation point by a factor of 4 compared to samples without thrombin, with a time of 1.03 min versus 4.14 min). Furthermore, the negative exponential extrapolation of the data suggests that increasing the thrombin concentration beyond 1.0 NIH/mL would bring only a limited decrease in the time to the gelation point.

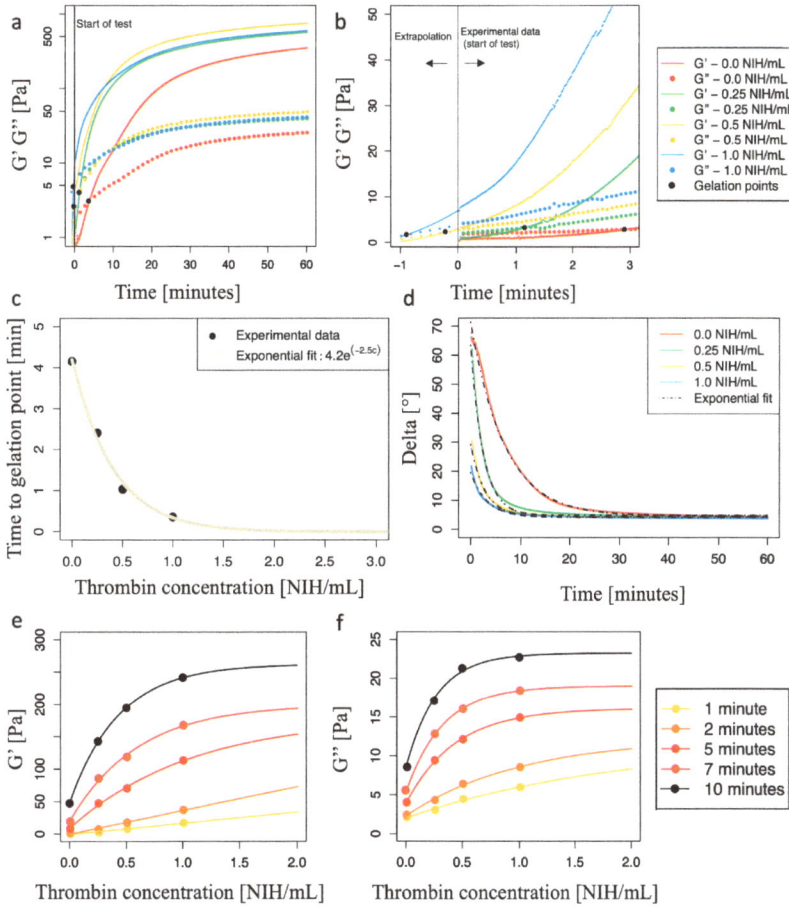

Figure 3. Rheological results. (**a**) Vertical black line indicates the beginning of the test (t = 0 min). On the right of this vertical line: Time sweeps mean raw data for storage (G′, full lines) and loss (G″, dotted lines) modulus of biomaterial with thrombin (0.0, 0.25, 0.5, 1.0 NIH/mL) for a displacement of 1% and a frequency of 5 rad/s. On the left of the vertical black line, data were extrapolated based on the first 3 min of the test for thrombin concentrations of 0.5 and 1.0 NIH/mL to determine their gelation point. The equation $y = A + Be^{Rt}$ was used for the fit, where $R \geq 0$, y is either the loss or the storage modulus, and t is the time. Black dots indicate the gelation point (G′ = G″) for each thrombin concentration. (**b**) Zoom for t = −1 to 3 min. (**c**) Black dots show the time needed to reach the gelation point for biomaterials with each thrombin concentration (0.0, 0.25, 0.5, and 1.0 NIH/mL). In order to account for the time elapsed between the rehydration of the biomaterial and the start of the test, 1.25 min were added to the experimental time observed in panel b. The experimental data were fitted using the equation $y = Ae^{Rx}$, where $R \leq 0$ (solid curves), y is the time to gelation and x is the thrombin concentration. (**d**) Results of phase angle (δ°) for biomaterials (0.0, 0.25, 0.5, or 1.0 NIH/mL of thrombin) over 60 min. $\delta = \tan^{-1}(G''/G')$. The experimental data was fitted using the equation $\delta = A + Be^{Rx}$, where $R \leq 0$ (solid curves), and x is the thrombin concentration. (**e**) Values of G′ (Pa) and (**f**) values of G″ (Pa) for biomaterials with or without thrombin (0.0, 0.25, 0.5, 1.0 NIH/mL) at 1, 2, 5, 7, and 10 min (● are experimental data). The values of G′ and G″ (Pa) at those times were empirically fitted with the equation $y = A - Be^{Rx}$, where $R \leq 0$ (solid curves), and x is the thrombin concentration. For results presented in (**a**,**d**), N = 2 and n = 2. For results presented in (**b**,**c**,**e**,**f**), N = 5 and n = 2.

Due to the visco-elastic nature of the samples, there is a time lag between the movement of the rheometer and the response signal. This time lag, δ, was calculated using the formula $\tan^{-1}(G''/G')$. A δ value closer to 90° indicates a mainly viscous behavior, as seen in the formulation containing 0 and 0.25 NIH/mL of thrombin at the beginning of the test. Conversely, a δ value closer to 0° indicates mainly elastic behavior. When modeled as a negative exponential, δ reached a constant value of approximately 4.5° for all thrombin concentrations used, as shown in Figure 3d. This indicates that the coagulated material was in an elastic solid state, as expected. Thrombin-containing biomaterials reached this near equilibrium faster than controls.

To model the relationship between thrombin concentration and each modulus at a given time, a model of the form $y = A - Be^{Rx}$, where $R \leq 0$ and x is the thrombin concentration, was empirically chosen. A represents the maximal modulus value at a given time (the asymptotic value obtained with an infinite thrombin concentration), $A - B$ is the modulus value at a given time when no thrombin is added to the biomaterial (as thrombin is naturally present in the blood, the biomaterial is still able to coagulate, so $(A - B) > 0$ and increases with time).

Here, the values of G' and G'' characterize the coagulation "state". Within the first 5 min of coagulation, increasing thrombin concentration had an almost linear impact on the coagulation state at a given time, as observed in Figure 3e for G' and in Figure 3f for G''. However, after 5 min, the effect of increasing thrombin concentration on the coagulation state was saturated for high thrombin concentrations. Based on the chosen exponential model, after 5 min, the 0.5 NIH/mL sample had already reached 40% of the maximal G' value and 76% of the maximal G'' value, where the maximal values were extrapolated using an infinite thrombin concentration. In comparison, after 5 min, the control only reached 5% of the maximal G' value and 25% of the maximal G'' value, while the 1.0 NIH/mL samples reached 63% of the maximal G' value and 93% of the maximal G'' value. All these percentages increased with time, as expected. It should be noted here that the 0.5 NIH/mL samples allowed a significant increase in G' and G'' compared to controls and 0.25 NIH/mL samples while still being easy to manipulate, as discussed in Section 3.2. In contrast, the 1.0 NIH/mL samples allowed an even more significant increase in storage and loss modulus but were less easy to manipulate. Using the exponential model, the percentages of maximal G' or G'' values reached after 5, 7, and 10 min for all thrombin concentrations tested are presented in Table 1.

Table 1. Percentage of maximal G' or G'' value reached after a given time for a given thrombin concentration.

Time [min]	0.0 NIH/mL		0.25 NIH/mL		0.5 NIH/mL		1.0 NIH/mL	
	G'	G''	G'	G''	G'	G''	G'	G''
5	5%	25%	25%	58%	40%	76%	63%	93%
7	10%	30%	41%	67%	61%	85%	83%	97%
10	18%	37%	54%	75%	74%	90%	92%	98%

3.4. Indentation Experiments Indicate That There Is No Significant Difference in Stiffness and Equilibrium Force between the Coagulated Biomaterials Containing Thrombin and the Control Samples

Despite its ability to increase the speed of the coagulation process, additional thrombin does not seem to have a significant impact on the mechanical properties of the fully coagulated clot, as the force at equilibrium and stiffness were stable regardless of the thrombin concentration tested, and no correlation was observed between either of them and thrombin concentration (Figure 4). The average stiffness varied only slightly from 3.7 gf/mm (material with 0.5 NIH/mL of thrombin) to 4.1 gf/mm (material without thrombin), as shown in Figure 4a, and the averaged equilibrium force varied only slightly from 1.00 gf (material with 0.25 NIH/mL of thrombin) to 1.18 gf (material with 1.0 NIH/mL of thrombin), as shown in Figure 4b. A difference is, however, noticeable between the stiffness of the biomaterial (mean of 3.9 gf/mm across all 4 concentrations) and the stiffness of PRP (mean of 3.1 gf/mm), indicating that the presence of CS increases the resistance of the biomaterial to compression. CS does not appear to have an impact on the equilibrium force.

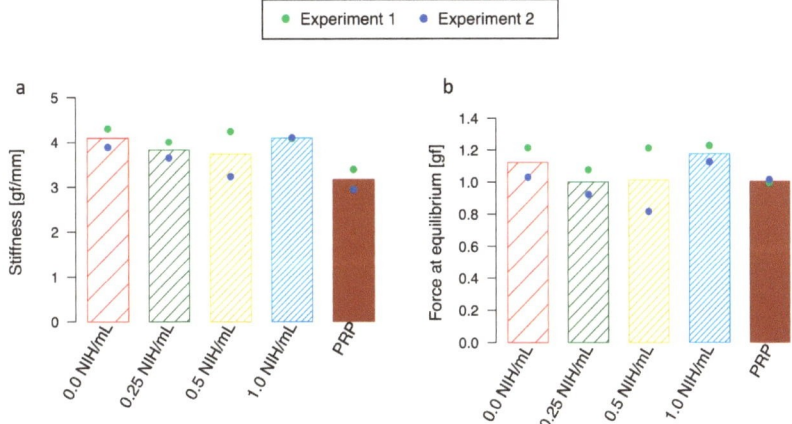

Figure 4. (**a**) Average stiffness and (**b**) force at equilibrium in response to an indentation compressive strain (amplitude of 0.7 mm—10% of sample's height—speed of 0.07 mm/s) for biomaterials with different concentrations of thrombin (0.0, 0.25, 0.5, and 0.1 NIH/mL) as well as for PRP. The presence of additional thrombin appears to leave the mechanical properties of the completely coagulated clot unchanged. Both the stiffness and the force at equilibrium remained stable across every thrombin concentration tested. A difference is, however, noticeable between the stiffness of the biomaterial and the stiffness of PRP, indicating that the presence of CS increases the resistance of the biomaterial to compression. CS does not appear to have an impact on the equilibrium force. $N = 2$, and $n = 3$.

3.5. The Adherence of the Biomaterial to Tendon Tissues Is Impacted by the Biomaterial-Tendon Contact Duration and Increases Faster When Thrombin Is Present

Ex vivo studies were performed to investigate the adherence between the biomaterial and tendon tissues, examining the impact of the biomaterial-tendon contact duration and the biomaterial incubation time in the syringe before being delivered to the tendon. We wanted to study (1) if the adherence increased with the contact time and (2) if the adherence decreased if the coagulation process had already evolved for a few minutes in the syringe. Results indicated that the adherence of the biomaterial to tendon tissues was significantly affected by the contact duration, with a longer waiting time resulting in less biomaterial dissolution in the fluid, as shown in Figure 5a. The adherence increased faster with the presence of thrombin, which accelerated the coagulation process. In contrast, our results indicate that the time the biomaterial spends in the syringe before application has less impact on adhesion than the biomaterial-tendon contact duration. Moreover, increasing the incubation time in the syringe did not reduce adhesion but instead enhanced it, as shown in Figure 5b. As a result, surgeons can safely wait a few minutes after rehydrating and placing the biomaterial in the syringe before injecting it.

Figure 5c–e show the biomaterial on the tendon immediately after injection, after a 1-min plunge in fluid following its application, and after a 1-min plunge in fluid following a 5-min wait after its application, respectively. Most of the biomaterial dissolved in the fluid after 1 min in the absence of a waiting period, while most of it remained on the tendon after a 5-min waiting period.

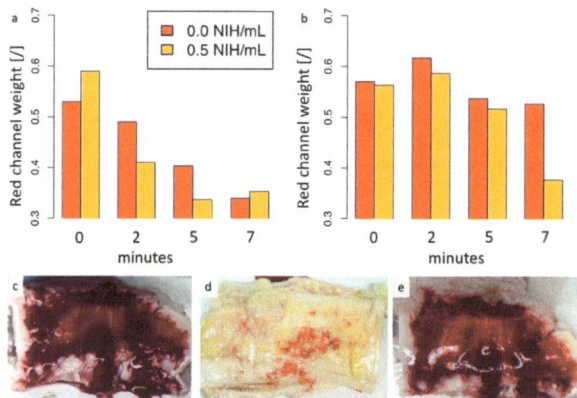

Figure 5. Quantification of the adherence of the biomaterial to tendon tissues when soaked in 0.9% NaCl for biomaterial with and without 0.5 NIH/mL of thrombin. After soaking, a photo of the NaCl solution was taken, and the amount of biomaterial detached from the tendon was quantified by its red channel weight. (**a**) The biomaterial was injected into the tendon immediately after reconstitution with PRP and left for 0, 2, 5, or 7 min. The tendon was then soaked in 0.9% NaCl. (**b**) The biomaterial was left for 0, 2, 5, or 7 min in a 10-cc syringe before being placed on the tendon. A 2 min wait was then observed before soaking the tendon in 0.9% NaCl. (**c**) Aspect of the biomaterial just after its injection on the tendon, before soaking. (**d**) When no waiting time is observed after injection of the biomaterial, the biomaterial almost completely dissolves in 0.9% NaCl when soaked. (**e**) When a 5-min waiting time is observed after injection of the biomaterial containing 0.5 NIH/mL of thrombin, the biomaterial does not dissolve when soaked in 0.9% NaCl, and its aspect on the tendon seems unchanged. For results presented in (**a**,**b**), $N = 2$ and $n = 2$.

3.6. Limitations

One limitation of this study is the use of sheep blood to investigate a biomaterial that will eventually be prepared with human PRP and used in humans. Although sheep's blood was found to be more similar to human blood than pig, rabbit, or dog blood, significant differences in coagulation exist between these blood types [36]. In humans, the onset of coagulation occurs later than in sheep, and the coagulation process is slower [36]. Furthermore, human thrombin was used with sheep blood, and it is important to note that thrombin from one species may not produce the same results in another species as interferences may occur in the clotting reaction [37–39]. Given the above, we anticipate that the incorporation of human thrombin into our biomaterial will have a more pronounced effect when used in conjunction with human PRP. Currently, in the clinical trials, the biomaterial must rest in a syringe for 30–45 min after rehydration to obtain the desired consistency, which can be challenging and time-consuming during surgical procedures. Nevertheless, our results strongly suggest that the inclusion of thrombin will resolve this issue and facilitate the use of our biomaterial in clinical settings.

4. Conclusions

This proof-of-concept study successfully validated our two initial hypotheses. Firstly, it demonstrated that thrombin has the potential to accelerate the solidification of CS-PRP FD implants. Secondly, thrombin can be added to the formulation before FD, and a balance can be achieved to obtain a biomaterial that is both fast to coagulate and easy to handle/inject. Moving forward, it is important to conduct animal studies to provide further evidence that CS-FIIa-PRP implants are safe and possess the same regenerative potential as CS-PRP implants. Additionally, future in vitro studies using the methodology developed here will allow us to investigate the solidification of CS-FIIa biomaterial rehydrated in human PRP and determine the optimal thrombin concentration. We are also conducting ongoing stability studies at $-20\ °C$, $4\ °C$, and RT to further evaluate the performance of our biomaterial. With continued research and evaluation, we are confident that our CS-FIIa-PRP biomaterial has great potential for use in regenerative medicine applications.

Supplementary Materials: The following supporting information can be downloaded at: https://zenodo.org/record/7992063. Video S1 shows the coagulation of the biomaterial for the analyzed thrombin concentrations.

Author Contributions: Conceptualization, F.M., A.C. and M.L.; data curation, F.M. and A.C.; formal analysis, F.M.; funding acquisition, G.D.C., and M.L.; investigation, F.M.; methodology, F.M., A.C. and M.L.; project administration, M.L.; resources, M.L.; software, F.M.; supervision, G.D.C. and M.L.; validation, F.M., A.C. and M.L.; visualization, F.M.; writing—original draft, F.M.; writing—review and editing, A.C., G.D.C. and M.L. All authors have read and agreed to the published version of the manuscript.

Funding: This research was funded by ChitogenX (research contract), the FRQNT (stipend allocated to F.M. via the PBEEE program, funding number 304445), and the Natural Sciences and Engineering Research Council of Canada (stipend allocated to F.M. via the NSERC-CREATE PrEEmiuM program, funding number 511957-2018).

Data Availability Statement: The data presented in this study are available on request from M.L. (the corresponding author). The data are not publicly available due to privacy restrictions.

Acknowledgments: The authors would like to thank Laure Siebierski for TEG data acquisition.

Conflicts of Interest: Anik Chevrier and Marc Lavertu hold shares of ChitogenX.

Appendix A. Assessment of Clotting Properties of CS-PRP Formulations through Rheology Measurements

The linear viscoelastic region of the samples was observed to be from 0 to 2% strain, which is in accordance with previous studies on blood coagulation [40,41]. A fixed 1% strain was therefore chosen for time sweeps tests. To determine the frequency to use in time sweeps, a frequency sweep was performed. The sample appeared as a viscous solid at high frequencies (>40 rad/s) before appearing as a viscoelastic liquid at lower frequencies. This can be indicative of the sample's behavior at high frequencies: it was previously reported for non-coagulated blood and other macromolecular solutions that G' (storage modulus) > G'' (loss modulus) for high frequencies (>10 rad/s) [42,43]. A hypothesis given for this behavior is that for low frequencies, the system predominantly loses energy since the translational movement of molecules through the solvent predominates over contortion. For higher frequencies, the movements of chain segments become more important, and G' increases. This increase in G' can be attributed to the mechanism for storing elastic energy, which is probably the contortion of plasma proteins and the deformation of individual RBCs. This hypothesis is given by Alves et al. [42]. For safety measures, a fixed frequency of 5 rad/s was therefore used in time sweeps tests. In light of those results, rheological properties were measured during a time sweep over the course of either 1 h or 15 min at a fixed 1% strain and a fixed 5 rad/s frequency. All measurements were performed at 37 °C.

References

1. Chevrier, A.; Darras, V.; Picard, G.; Nelea, M.; Veilleux, D.; Lavertu, M.; Hoemann, C.D.; Buschman, M.D. Injectable chitosan-platelet-rich plasma implants to promote tissue regeneration: In vitro properties, in vivo residence, degradation, cell recruitment and vascularization. *J. Tissue Eng. Regen. Med.* **2018**, *12*, 217–228. [CrossRef] [PubMed]
2. Cheung, R.C.; Ng, T.B.; Wong, J.H.; Chan, W.Y. Chitosan: An Update on Potential Biomedical and Pharmaceutical Applications. *Mar Drugs* **2015**, *13*, 5156–5186. [CrossRef] [PubMed]
3. Wang, K.; Wang, H.; Pan, S.; Fu, C.; Chang, Y.; Li, H.; Yang, X.; Qi, Z. Evaluation of New Film Based on Chitosan/Gold Nanocomposites on Antibacterial Property and Wound-Healing Efficacy. *Adv. Mater. Sci. Eng.* **2020**, *2020*, 6212540. [CrossRef]
4. Duceac, I.A.; Vereștiuc, L.; Coroaba, A.; Arotăriței, D.; Coseri, S. All-polysaccharide hydrogels for drug delivery applications: Tunable chitosan beads surfaces via physical or chemical interactions, using oxidized pullulan. *Int. J. Biol. Macromol.* **2021**, *181*, 1047–1062. [CrossRef] [PubMed]

5. Desai, N.; Rana, D.; Salave, S.; Gupta, R.; Patel, P.; Karunakaran, B.; Sharma, A.; Giri, J.; Benival, D.; Kommineni, N. Chitosan: A Potential Biopolymer in Drug Delivery and Biomedical Applications. *Pharmaceutics* **2023**, *15*, 1313. [CrossRef] [PubMed]
6. Hoemann, C.; Fong, D. Immunological responses to chitosan for biomedical applications. In *Chitosan Based Biomaterials Volume 1*; Elsevier: Amsterdam, The Netherlands, 2017; pp. 45–79.
7. Patrulea, V.; Ostafe, V.; Borchard, G.; Jordan, O. Chitosan as a starting material for wound healing applications. *Eur. J. Pharm. Biopharm.* **2015**, *97*, 417–426. [CrossRef]
8. Bano, I.; Arshad, M.; Yasin, T.; Ghauri, M.A.; Younus, M. Chitosan: A potential biopolymer for wound management. *Int. J. Biol. Macromol.* **2017**, *102*, 380–383. [CrossRef]
9. Lavelle, E.C.; Ward, R.W. Mucosal vaccines—Fortifying the frontiers. *Nat. Rev. Immunol.* **2022**, *22*, 236–250. [CrossRef]
10. Lai, H.; Chen, G.; Zhang, W.; Wu, G.; Xia, Z. Research trends on platelet-rich plasma in the treatment of wounds during 2002–2021: A 20-year bibliometric analysis. *Int. Wound J.* 2022, *Online ahead of print*. [CrossRef]
11. Pavlovic, V.; Ciric, M.; Jovanovic, V.; Stojanovic, P. Platelet rich plasma: A short overview of certain bioactive components. *Open Med.* **2016**, *11*, 242–247. [CrossRef]
12. Thu, A.C. The use of platelet-rich plasma in management of musculoskeletal pain: A narrative review. *J. Yeungnam Med. Sci.* **2022**, *39*, 206–215. [CrossRef]
13. Popescu, M.N.; Iliescu, M.G.; Beiu, C.; Popa, L.G.; Mihai, M.M.; Berteanu, M.; Ionescu, A.M. Autologous platelet-rich plasma efficacy in the field of regenerative medicine: Product and quality control. *BioMed Res. Int.* **2021**, *2021*, 4672959. [CrossRef]
14. Depres-Tremblay, G.; Chevrier, A.; Snow, M.; Rodeo, S.; Buschmann, M.D. Freeze-dried chitosan-platelet-rich plasma implants improve supraspinatus tendon attachment in a transosseous rotator cuff repair model in the rabbit. *J. Biomater. Appl.* **2019**, *33*, 792–807. [CrossRef]
15. Dépres-Tremblay, G.; Chevrier, A.; Tran-Khanh, N.; Nelea, M.; Buschmann, M.D. Chitosan inhibits platelet-mediated clot retraction, increases platelet-derived growth factor release, and increases residence time and bioactivity of platelet-rich plasma in vivo. *Biomed. Mater.* **2017**, *13*, 015005. [CrossRef]
16. Chevrier, A.; Hurtig, M.; Lacasse, F.; Lavertu, M.; Potter, H.; Pownder, S.; Rodeo, S.; Buschmann, M. Freeze-dried chitosan solubilized in platelet-rich plasma in a sheep model of rotator cuff repair. *Orthop. Proc.* **2020**, *102-B*, 57.
17. Dwivedi, G.; Chevrier, A.; Hoemann, C.D.; Buschmann, M.D. Injectable freeze-dried chitosan-platelet-rich-plasma implants improve marrow-stimulated cartilage repair in a chronic-defect rabbit model. *J. Tissue Eng. Regen. Med.* **2019**, *13*, 599–611. [CrossRef]
18. Marchand, C.; Rivard, G.E.; Sun, J.; Hoemann, C.D. Solidification mechanisms of chitosan–glycerol phosphate/blood implant for articular cartilage repair. *Osteoarthr. Cartil.* **2009**, *17*, 953–960. [CrossRef]
19. Sung, Y.K.; Lee, D.R.; Chung, D.J. Advances in the development of hemostatic biomaterials for medical application. *Biomater. Res.* **2021**, *25*, 37. [CrossRef]
20. Zamora, M.; Robles, J.P.; Aguilar, M.B.; Romero-Gómez, S.d.J.; Bertsch, T.; Martinez de la Escalera, G.; Triebel, J.; Clapp, C. Thrombin cleaves prolactin into a potent 5.6-kDa vasoinhibin: Implication for tissue repair. *Endocrinology* **2021**, *162*, bqab177. [CrossRef]
21. Lavertu, M.; Xia, Z.; Serreqi, A.; Berrada, M.; Rodrigues, A.; Wang, D.; Buschmann, M.; Gupta, A. A validated 1H NMR method for the determination of the degree of deacetylation of chitosan. *J. Pharm. Biomed. Anal.* **2003**, *32*, 1149–1158. [CrossRef]
22. Nguyen, S.; Winnik, F.M.; Buschmann, M.D. Improved reproducibility in the determination of the molecular weight of chitosan by analytical size exclusion chromatography. *Carbohydr. Polym.* **2009**, *75*, 528–533. [CrossRef]
23. Zheng, Y.; Mak, A.F.; Lue, B. Objective assessment of limb tissue elasticity: Development of a manual indentation procedure. *J. Rehabil. Res. Dev.* **1999**, *36*, 71–85. [PubMed]
24. Forte, A.E.; Galvan, S.; Manieri, F.; y Baena, F.R.; Dini, D. A composite hydrogel for brain tissue phantoms. *Mater. Des.* **2016**, *112*, 227–238. [CrossRef]
25. Van Dommelen, J.; Van der Sande, T.; Hrapko, M.; Peters, G. Mechanical properties of brain tissue by indentation: Interregional variation. *J. Mech. Behav. Biomed. Mater.* **2010**, *3*, 158–166. [CrossRef] [PubMed]
26. Rossomacha, E.; Hoemann, C.D.; Shive, M.S. Simple methods for staining chitosan in biotechnological applications. *J. Histotechnol.* **2004**, *27*, 31–36. [CrossRef]
27. Milano, F.; Chevrier, A.; Crescenzo, G.D.; Lavertu, M. Robust Segmentation-Free Algorithm for Homogeneity Quantification in Images. *IEEE Trans. Image Process.* **2021**, *30*, 5533–5544. [CrossRef]
28. Memon, M.; Kay, J.; Gholami, A.; Simunovic, N.; Ayeni, O.R. Fluid Extravasation in Shoulder Arthroscopic Surgery: A Systematic Review. *Orthop. J. Sport. Med.* **2018**, *6*, 2325967118771616. [CrossRef]
29. Abràmoff, M.D.; Magalhães, P.J.; Ram, S.J. Image processing with ImageJ. *Biophotonics Int.* **2004**, *11*, 36–42.
30. Liu, W.; Zhang, J.; Cao, Z.; Xu, F.; Yao, K. A chitosan-arginine conjugate as a novel anticoagulation biomaterial. *J. Mater. Sci. Mater. Med.* **2004**, *15*, 1199–1203. [CrossRef]
31. Liu, W.; Zhang, J.; Cheng, N.; Cao, Z.; Yao, K. Anticoagulation activity of crosslinked N-sulfofurfuryl chitosan membranes. *J. Appl. Polym. Sci.* **2004**, *94*, 53–56. [CrossRef]
32. Aleksakhina, E.; Parfenov, A.; Priyatkin, D.; Fomina, N.; Tomilova, I. Spectral and structural properties of clotting factor proteins under mechanical stress. *J. Phys. Conf. Ser.* **2021**, *2094*, 022044. [CrossRef]

33. Cone, S.J.; Fuquay, A.T.; Litofsky, J.M.; Dement, T.C.; Carolan, C.A.; Hudson, N.E. Inherent fibrin fiber tension propels mechanisms of network clearance during fibrinolysis. *Acta Biomater.* **2020**, *107*, 164–177. [CrossRef]
34. Li, Y.; Li, Y.; Prince, E.; Weitz, J.I.; Panyukov, S.; Ramachandran, A.; Rubinstein, M.; Kumacheva, E. Fibrous hydrogels under biaxial confinement. *Nat. Commun.* **2022**, *13*, 3264. [CrossRef]
35. Watt, R.P.; Khatri, H.; Dibble, A.R.G. Injectability as a function of viscosity and dosing materials for subcutaneous administration. *Int. J. Pharm.* **2019**, *554*, 376–386. [CrossRef]
36. Lechner, R.; Helm, M.; Mueller, M.; Wille, T.; Riesner, H.-J.; Friemert, B. In-vitro study of species-specific coagulation differences in animals and humans using rotational thromboelastometry (ROTEM). *BMJ Mil. Health* **2019**, *165*, 356–359. [CrossRef]
37. Seegers, W.H.; Smith, H. Factors which influence the activity of purified thrombin. *Am. J. Physiol.-Leg. Content* **1942**, *137*, 348–354. [CrossRef]
38. Siller-Matula, J.M.; Plasenzotti, R.; Spiel, A.; Quehenberger, P.; Jilma, B. Interspecies differences in coagulation profile. *Thromb. Haemost.* **2008**, *100*, 397–404. [CrossRef]
39. Doolittle, R.F.; Oncley, J.L.; Surgenor, D.M. Species differences in the interaction of thrombin and fibrinogen. *J. Biol. Chem.* **1962**, *237*, 3123–3127. [CrossRef]
40. Evans, P.A.; Hawkins, K.; Williams, P.R. Rheometry for blood coagulation studies. *Rheol. Rev.* **2006**, *2006*, 255–291.
41. Tomaiuolo, G.; Carciati, A.; Caserta, S.; Guido, S. Blood linear viscoelasticity by small amplitude oscillatory flow. *Rheol. Acta* **2016**, *55*, 485–495. [CrossRef]
42. Alves, M.M.; Rocha, C.; Gonçalves, M.P. Study of the rheological behaviour of human blood using a controlled stress rheometer. *Clin. Hemorheol. Microcirc.* **2013**, *53*, 369–386. [CrossRef] [PubMed]
43. Kar, S.; Kar, A.; Chaudhury, K.; Maiti, T.K.; Chakraborty, S. Formation of blood droplets: Influence of the plasma proteins. *ACS Omega* **2018**, *3*, 10967–10973. [CrossRef] [PubMed]

Disclaimer/Publisher's Note: The statements, opinions and data contained in all publications are solely those of the individual author(s) and contributor(s) and not of MDPI and/or the editor(s). MDPI and/or the editor(s) disclaim responsibility for any injury to people or property resulting from any ideas, methods, instructions or products referred to in the content.

MDPI AG
Grosspeteranlage 5
4052 Basel
Switzerland
Tel.: +41 61 683 77 34

Polymers Editorial Office
E-mail: polymers@mdpi.com
www.mdpi.com/journal/polymers

Disclaimer/Publisher's Note: The title and front matter of this reprint are at the discretion of the Guest Editors. The publisher is not responsible for their content or any associated concerns. The statements, opinions and data contained in all individual articles are solely those of the individual Editors and contributors and not of MDPI. MDPI disclaims responsibility for any injury to people or property resulting from any ideas, methods, instructions or products referred to in the content.